RELIABILITY OF LARGE SYSTEMS

Elsevier Titles of Related Interest (Books and CD-ROMs):

Probabilistic Safety Assessment and Management
Edited by: **E.J. Bonano, A.L. Camp, M.J. Majors, R.A. Thompson**
2180 pages
Hardbound *(ISBN: 0-08-044122-X)*
CD-ROM *(ISBN: 0-08-044120-3)*
Set *(ISBN: 0-08-044121-1)*

Advances in Steel Structures (ICASS '99)
Edited by: **Siu-Lai Chan, J.G. Teng**
1246 pages, 2-Volume Set
Hardbound *(ISBN: 0-08-043015-5)*

Response of Structures to Extreme Loading
Edited by: **A. Ghobarah**
1141 pages
CD-ROM *(ISBN: 0-08-044322-2)*

An Elementary Guide To Reliability
Fifth Edition
G.W.A. Dummer, R.C. Winton, M. Tooley
112 pages
Paperback *(ISBN: 0750635533)*

Risk-Based Management
R.B. Jones
282 pages
Hardbound *(ISBN: 0884157857)*

Practical Industrial Safety, Risk Assessment and Shutdown Systems
D. Macdonald
384 pages
Paperback *(ISBN: 0750658045)*

Reliability, Maintainability and Risk
D. Smith
370 pages
Paperback *(ISBN: 0750651687)*

Handbook of Statistics 20: Advances in Reliability
Edited by: **N. Balakrishnan, C.R. Rao**
886 pages
Hardbound *(ISBN: 0-444-50078-2)*

Related Journals:

Free specimen copy gladly sent on request. Elsevier Ltd, The Boulevard, Langford Lane, Kidlington, Oxford, OX5 1GB, UK. Pay per view available for all journals - www.elsevier.com

Reliability Engineering & System Safety
Probabilistic Engineering Mechanics
Structural Safety
Engineering Structures
Computers and Structures
Computer Methods in Applied Mechanics and Engineering

To Contact the Publisher
Elsevier Science welcomes enquiries concerning publishing proposals: books, journal special issues, conference proceedings, etc. All formats and media can be considered. Should you have a publishing proposal you wish to discuss, please contact, without obligation, the publisher responsible for Elsevier's mechanics and structural integrity publishing programme:

Isabelle Kandler
Publishing Editor, Materials Science & Engineering
Elsevier Ltd
The Boulevard, Langford Lane   Phone: +44 1865 843378
Kidlington, Oxford   Fax: +44 1865 843987
OX5 1GB, UK   E.mail: i.kandler@elsevier.com

General enquiries, including placing orders, should be directed to Elsevier's Regional Sales Offices – please access the Elsevier homepage for full contact details (homepage details at the top of this page).

RELIABILITY OF LARGE SYSTEMS

Krzysztof Kołowrocki
Gdynia Maritime University, Gdynia, Poland

ELSEVIER

2004
Amsterdam – Boston – Heidelberg – London – New York – Oxford
Paris – San Diego – San Francisco – Singapore – Sydney – Tokyo

ELSEVIER B.V.
Sara Burgerhartstraat 25
P.O. Box 211,
1000 AE Amsterdam
The Netherlands

ELSEVIER Inc.
525 B Street, Suite 1900
San Diego, CA 92101-4495
USA

ELSEVIER Ltd
The Boulevard, Langford Lane
Kidlington, Oxford OX5 1GB
UK

ELSEVIER Ltd
84 Theobalds Road
London WC1X 8RR
UK

1st. edition 2004

Library of Congress Cataloging in Publication Data
A catalog record is available from the Library of Congress.

British Library Cataloguing in Publication Data
A catalogue record is available from the British Library.

ISBN: 0-08-044429-6

⊗ The paper used in this publication meets the requirements of ANSI/NISO Z39.48-1992 (Permanence of Paper).
Printed in The Netherlands.

To my Wife

Barbara

CONTENTS

PREFACE

The book is concerned with the application of limit reliability functions to the reliability evaluation of large systems. Two-state and multi-state large systems composed of independent components are considered. The main emphasis is on multi-state systems with degrading (ageing) components because of the importance of such an approach in safety analysis, assessment and prediction, and analysing the effectiveness of operation processes of real technical systems.

Many technical systems belong to the class of complex systems as a result of the large number of components they are built of and their complicated operating processes. This complexity very often causes evaluation of system reliability and safety to become difficult. As a rule these are series systems composed of large number of components. Sometimes the series systems have either components or subsystems reserved and then they become parallel-series or series-parallel reliability structures. We meet large series systems, for instance, in piping transportation of water, gas, oil and various chemical substances. Large systems of these kinds are also used in electrical energy distribution. A city bus transportation system composed of a number of communication lines each serviced by one bus may be a model series system, if we treat it as not failed, when all its lines are able to transport passengers. If the communication lines have at their disposal several buses we may consider it as either a parallel-series system or an "*m* out of *n*" system. The simplest example of a parallel system or an "*m* out of *n*" system may be an electrical cable composed of a number of wires, which are its basic components, whereas the transmitting electrical network may be either a parallel-series system or an "*m* out of *n*"-series system. Large systems of these types are also used in telecommunication, in rope transportation and in transport using belt conveyers and elevators. Rope transportation systems like port elevators and ship-rope elevators used in shipyards during ship docking and undocking are model examples of series-parallel and parallel-series systems.

Taking into account the importance of the safety and operating process effectiveness of such systems it seems reasonable to expand the two-state approach to multi-state approach in their reliability analysis. The assumption that the systems are composed of multi-state components with reliability states degrading in time without repair gives the possibility for more precise analysis of their reliability, safety and operational processes' effectiveness. This assumption allows us to distinguish a system reliability critical state to exceed which is either dangerous for the environment or does not assure the necessary level of its operational process effectiveness. Then, an important system reliability characteristic is the time to the moment of exceeding the system reliability critical state and its distribution, which is called the system risk function. This distribution is strictly related to the system multi-state reliability function that is a basic characteristic of the multi-state system.

In the case of large systems, the determination of the exact reliability functions of the systems and the system risk functions leads us to very complicated formulae that are often useless for reliability practitioners. One of the important techniques in this situation is the asymptotic approach to system reliability evaluation. In this approach, instead of the preliminary complex formula for the system reliability function, after assuming that the number of system components tends to infinity and finding the limit reliability of the system, we obtain its simplified form.

The mathematical methods used in the asymptotic approach to the system reliability analysis of large systems are based on limit theorems on order statistics distributions considered in very wide literature ([3], [9]–[11], [13], [20]–[21], [26]–[29], [33]–[36], [39]–[40], [84], [100], [106], [108], [112], [114]). These theorems have generated the investigation concerned with limit reliability functions of the systems composed of two-state components ([5], [7], [23]–[27], [41]–[43], [50]–[65], [71]–[72], [79]–[84], [94]–[95], [105], [109]–[111], [115]). The main and fundamental results on this subject that determine the three-element classes of limit reliability functions for homogeneous series systems and for homogeneous parallel systems have been established by Gniedenko in [36]. These results are also presented, sometimes with different proofs, for instance in subsequent works [7], [13], [21], [28], [56] and [71]. The generalisations of these results for homogeneous "*m* out of *n*" systems have been formulated and proved by Smirnow in [108], where the seven-element class of possible limit reliability functions for these systems has been fixed. Some partial results obtained by Smirnow may be found in [71] and additionally with the solution of the speed of convergence problem in [29]. As it has been done for homogeneous series and parallel systems classes of limit reliability functions have been fixed by Chernoff and Teicher in [21] for homogeneous series-parallel and parallel-series systems. Their results were concerned with so-called "quadratic" systems only. They have fixed limit reliability functions for the homogeneous series-parallel systems with the number of series subsystems equal to the number of components in these subsystems, and for the homogeneous parallel-series systems with the number of parallel subsystems equal to the number of components in these subsystems. These results may also be found for instance in later works [7] and [56].

All the results so far described have been obtained under the linear normalisation of the system lifetimes. Of course, there is a possibility to look for limit reliability functions of large systems under other than linear standardisation of their lifetimes. In this context, the results obtained by Pantcheva ([100]) and Cichocki ([25]) are exemplary. Pantcheva in [100] has fixed the seven-element classes of limit reliability functions of homogeneous series and parallel systems under power standardisation for their lifetimes. Cichocki in [25] has generalised Pantcheva's results to hierarchical series-parallel and parallel-series systems of any order.

The book contains the results described above and their newest generalisations for large two-state systems and their developments for multi-state systems' asymptotic reliability analysis under the linear standardisation of the system lifetimes and the system sojourn times in the state subsets, respectively.

Generalisations presented here of the results on limit reliability functions of two-state homogeneous series, and parallel systems for these systems in case they are non-homogeneous, are mostly taken from [71] and [74]. A more general problem is concerned with fixing the classes of possible limit reliability functions for so-called "rectangular" series-parallel and parallel-series systems. This problem for homogeneous series-parallel and parallel-series systems of any shapes, with different number of subsystems and numbers of components in these subsystems, has been progressively solved in [53]–[56], [59] and [61]. The main and new result of these works was the determination of seven new limit reliability functions for homogeneous series-parallel systems as well as for parallel-series systems. This way, new ten-element classes of all possible limit reliability functions for these systems have been fixed. Moreover, in these works it has been pointed out that the type of the system limit reliability function strongly depends on the system shape. These results allow us to evaluate reliability characteristics of homogeneous series-parallel and parallel-series systems with regular reliability structures, i.e. systems composed of subsystems having the same numbers of components. The extensions of these results for non-homogeneous series-parallel and parallel-series systems have been formulated and proved successively in [56], [60]–[63] and [74]. These generalisations additionally allow us to evaluate reliability characteristics of the series-parallel and parallel-series systems with non-regular structures, i.e. systems with subsystems having different numbers of components. In some of the cited works, as well as the theoretical considerations and solutions, numerous practical applications of the asymptotic approach to real technical system reliability evaluation may also be found ([27], [41]–[43], [52], [57], [62], [64], [71]–[72], [109], [111], [115]).

More general and practically important complex systems composed of multi-state and degrading in time components are considered among others in [1]–[2], [4]–[6], [8], [12], [14]–[19], [30]–[32], [38], [45]–[49], [65]–[71], [73]–[79], [83], [85]–[91], [93], [96]–[99], [101], [104], [107] and [116]–[119]. An especially important role they play in the evaluation of technical systems reliability and safety and their operating process effectiveness is defined in the book for large multi-state systems with degrading components. The most important results regarding generalisations of the results on limit reliability functions of two-state systems dependent on transferring them to series, parallel, "*m* out of *n*", series-parallel and parallel-series multi-state systems with degrading components are given in [65]–[77]. Some of these publications also contain practical applications of the asymptotic approach to the reliability evaluation of various technical systems ([65]–[71], [73]–[74], [76]–[77]).

The results concerned with the asymptotic approach to system reliability analysis have become the basis for the investigation concerned with domains of attraction for the limit reliability functions of the considered systems ([23], [71]–[72], [81]–[82]). In a natural way they have led to investigation of the speed of convergence of the system reliability function sequences to their limit reliability functions ([71]). These results have also initiated the investigation of limit reliability functions of "*m* out of *n*"-series, series-"*m* out of *n*" systems ([23], [94]–[95]), and systems with hierarchical reliability structures ([23]–[25]), as well as investigations on the problems of the system reliability improvement and optimisation ([82]–[83]).

The aim of the book is to deliver a complete elaboration of the state of art on the method of asymptotic approach to reliability evaluation for as wide as possible a range of large systems. Pointing out the possibility of this method's extensive practical application in the operating processes of these systems is also an important reason for this book. The book contains complete current theoretical results of the asymptotic approach to reliability evaluation of large two-state and multi-state series, parallel, "*m* out of *n*", series-parallel, and parallel-series systems together with their practical applications to the reliability evaluation of a wide range of technical systems. Additionally some recent partial results on the asymptotic approach to reliability evaluation of "*m* out of *n*"-series, series-"*m* out of *n*" and hierarchical systems, and their application to large systems reliability improvement and to large systems reliability analysis in their operation processes are presented in the book.

The following construction of the book has been assumed. In chapters concerned with two-state systems the results and theorems are presented without the proofs but with exact reference to the literature where their proofs may be found. Moreover, the procedures of the results' practical applications are described and applied to the model two-state systems reliability evaluation. In chapters concerned with multi-state systems the recent theorems about their multi-state limit reliability functions are formulated and briefly justified. Next, the procedures of the result applications are presented and applied to real technical systems reliability and risk evaluation. Moreover, the possibility of the computer aided reliability evaluation of these systems is suggested and its use is presented.

The book contains complete actual solutions of the formulated problems for the considered large systems reliability evaluation in the case of any reliability functions of the system components.

The book consists of this Preface, eight chapters, Summary and Bibliography.

In Chapter 1, some basic notions necessary for further considerations are introduced. The asymptotic approach to the system reliability investigation and the system limit reliability function are defined.

In Chapter 2 two-state homogeneous and non-homogeneous series, parallel, "*m* out of *n*", series-parallel and parallel-series systems are defined. Their exact reliability functions are also determined.

Basic notions of the system multi-state reliability analysis are introduced in Chapter 3. Further the multi-state homogeneous and non-homogeneous series, parallel, "*m* out of *n*", series-parallel and parallel-series systems with degrading components are defined and their exact reliability functions are determined. Moreover, the notions of the multi-state limit reliability function of the system, its risk function and other multi-state system reliability characteristics are introduced.

Chapter 4 is concerned with limit reliability functions of two-state systems. Three-element classes of limit reliability functions for homogeneous and non-homogeneous series systems are fixed. Some auxiliary theorems that allow us to justify facts on the methods of those systems' reliability evaluation are formulated and proved. The chapter also contains the application of one of the proven facts to the reliability evaluation of a non-homogeneous gas pipeline that is composed of components with Weibull reliability functions. The accuracy of this evaluation is also illustrated. Three-element classes of possible limit reliability functions for homogeneous and non-homogeneous parallel systems are fixed as well. Some auxiliary theorems that allow us to justify facts on the methods of these systems' reliability evaluation are formulated and proved. The chapter also contains the application of one proved fact to the reliability evaluation of a homogeneous energetic cable used in the overhead electrical energy distribution that is composed of components with Weibull reliability functions. The accuracy of this evaluation is illustrated in a table and a figure. The class of limit reliability functions for a homogeneous "*m* out of *n*" system is fixed and the "16 out of 35" lighting reliability is evaluated in this chapter. Chapter 4 contains also the results of investigations on limit reliability functions of two-state homogeneous and non-homogeneous series-parallel systems. Apart from formulated and proved auxiliary theorems that allow us to justify facts on the methods of these systems' reliability evaluation, their ten-element classes of possible limit reliability functions are fixed. In this chapter, in the part concerned with applications there are two formulated and proved facts that determine limit reliability functions of series-parallel systems in the cases where they are composed of components having the same and different Weibull reliability functions. On the basis of those facts the reliability characteristics of a homogeneous gas pipeline composed of two lines of pipe segments and a non-homogeneous water supply system composed of three lines of pipe segments are evaluated. The results of investigations on limit reliability functions of two-state homogeneous and non-homogeneous parallel-series systems are given in this chapter as well. Theorems, which determine ten-element classes of possible limit reliability functions for these systems in the cases where they are composed of identical and different components, are formulated and justified. Moreover, some auxiliary theorems that are necessary in practical reliability evaluation of real technical systems are formulated and proved. In the part concerned with applications one fact is formulated and proved and then applied to evaluation of the reliability of a model homogeneous parallel-series system.

Generalisations of the results of Chapter 4 on limit reliability functions of two-state systems consisting in their transferring to multi-state series, parallel, "*m* out of *n*", series-parallel and parallel-series systems are done in Chapter 5. The classes of all possible limit reliability functions for these systems in the cases when they are composed of identical and different (in the reliability sense) components are fixed. The newest theorems that allow us to evaluate the reliability of large technical systems of those kinds are formulated and proved in Chapter 5 as well. Apart from the main theorems fixing the classes of multi-state limit reliability functions of the considered system, some auxiliary theorems and corollaries allowing their direct applications to reliability evaluation of real technical objects are also formulated and proved. Moreover, in this chapter there are wide applications depending on the results applying to the evaluation of reliability characteristics and risk functions of different multi-state transportation systems. The results concerned with multi-state series systems are applied to the reliability evaluation and risk function determination of homogeneous and non-homogeneous pipeline transportation systems, a homogeneous model telecommunication network and a homogeneous bus transportation system. The results concerned with multi-state parallel systems are applied to reliability evaluation and risk function determination of an energetic cable used in an overhead electrical energy distribution network and to reliability and durability evaluation of a three-level steel rope used in rope transport. Results on limit reliability functions of a homogeneous multi-state "*m* out of *n*" system are applied to durability evaluation of a steel rope. A model homogeneous series-parallel system and homogeneous and non-homogeneous series-parallel pipeline systems composed of several lines of pipe segments are estimated as well. Moreover, the reliability evaluation of the model homogeneous parallel-series electrical energy distribution system is performed.

Chapter 6 is devoted to the multi-state asymptotic reliability analysis of port and shipyard transportation systems. Theoretical results of this chapter and Chapter 5 are applied to the reliability evaluation and the risk functions determination of some selected port transportation systems. The results of the asymptotic approach to reliability evaluation of non-homogeneous multi-state series-parallel systems are applied to the transportation system used in the Baltic Grain Terminal of the Port of Gdynia for transporting grain from its elevator to the rail carriages. The results of the asymptotic approach to the reliability evaluation of the non-homogeneous multi-state series-parallel systems are applied to the piping transportation system used in the Oil Terminal in Debogorze. This transportation system is set up to take the oil from the tankers that deliver it to the unloading pier located at the breakwater of the Port of Gdynia. The results of the asymptotic approach to reliability evaluation of non-homogeneous multi-state series-parallel and series systems are applied to the transportation system used in the Baltic Bulk Terminal of the Port of Gdynia for loading bulk cargo on the ships. The results of this chapter and Chapter 5 are also applied to reliability evaluation and risk function determination of the shipyard transportation system. Namely, the results of the asymptotic approach to reliability evaluation of homogeneous multi-state parallel-series systems are applied to the ship-rope transportation system used in the Naval Shipyard of Gdynia for docking ships coming for repair. The reliability analysis performed on the considered systems in this chapter is based on the data concerned with the operation processes and reliability of their components coming from experts, from component technical norms and from their producer's certificates.

In Chapter 7 the classes of possible limit reliability functions are fixed for the considered systems in the case where their components have exponential reliability functions. Theoretical results are represented in the form of a very useful guide containing algorithms placed in tables and giving sequential steps for proceeding in the reliability evaluation in each of the possible cases of the considered system shapes. The application of these algorithms for reliability evaluation of the multi-state non-homogeneous series transportation system, the multi-state model homogeneous series-parallel, the multi-state non-homogeneous series-parallel pipeline transportation system and the multi-state non-homogeneous parallel-series bus transportation system is illustrated. The evaluation of reliability functions, risk functions, mean values of sojourn times in subsets of states and mean values of sojourn times in particular states for these systems is carried out. The calculations are performed using a computer program based on the algorithms, so allowing automatic evaluation of the reliability of large real technical systems.

In Chapter 8 the open problems related to the topics considered in the book are presented. The domains of attraction for previously fixed limit reliability functions of the series, parallel, "*m* out of *n*", series-parallel and parallel-series systems are introduced. More exactly, there are formulated theorems giving conditions which reliability functions of the components of the system have to satisfy in order that the system limit reliability function is one of the functions from the system class of all limit reliability functions. Some examples of the result application for series systems are also illustrated. The practically very important problem of the speed of convergence of system reliability function sequences to their limit reliability functions is investigated as well. An exemplary theorem is presented, which allows the differences between the system limit reliability functions and the members of their reliability function sequences to be estimated. Next, an example of the speed of convergence evaluations of reliability function sequences for a homogeneous series-parallel system is given. Partial results of the investigation on the asymptotic approach to reliability evaluation of "*m* out of *n*"-series, series-"*m* out of *n*" and hierarchical systems and on system reliability improvement are presented. These result applications are illustrated graphically as well.

The analysis of large systems' reliability in their operation processes is given at the end of Chapter 8.

The book is completed by the Summary that contains the evaluation of the presented results, the formulation of open problems concerned with large systems' reliability and the perspective of further investigations on the considered problems.

NOTATIONS

E_i	– components of series, parallel and "m out of n" systems
E_{ij}	– components of series-parallel and parallel-series systems
T_i	– component lifetimes of two-state series, parallel and "m out of n" systems
T_{ij}	– component lifetimes of two-state series-parallel and parallel-series systems
T	– a two-state system lifetime
$R(t)$	– a component reliability function of a two-state homogeneous system
$F(t)$	– a component lifetime distribution function of a two-state homogeneous system
$R^{(i)}(t)$	– component reliability functions of two-state non-homogeneous series, parallel and "m out of n" systems
$F^{(i)}(t)$	– component lifetime distribution functions of two-state non-homogeneous series, parallel and "m out of n" systems
$R^{(i,j)}(t)$	– component reliability functions of two-state non-homogeneous series-parallel and parallel-series systems
$F^{(i,j)}(t)$	– component lifetime distribution functions of two-state non-homogeneous series-parallel and parallel-series systems
$\overline{R}_n(t)$	– a reliability function of a two-state homogeneous series system
$\overline{R}'_n(t)$	– a reliability function of a two-state non-homogeneous series system
$R_n(t)$	– a reliability function of a two-state homogeneous parallel system
$R'_n(t)$	– a reliability function of a two-state non-homogeneous parallel system
$R_n^{(m)}(t)$	– a reliability function of a two-state homogeneous "m out of n" system
$R'^{(m)}_n(t)$	– a reliability function of a two-state non-homogeneous "m out of n" system
$\overline{R}_{k_n l_n}(t)$	– a reliability function of a two-state homogeneous parallel-series system

$\bar{R}'_{k_n l_n}(t)$ — a reliability function of a two-state non-homogeneous parallel-series system

$R_{k_n l_n}(t)$ — a reliability function of a two-state homogeneous series-parallel system

$R'_{k_n l_n}(t)$ — a reliability function of a two-state non-homogeneous series-parallel system

$\bar{\mathcal{R}}(t)$ — a limit reliability function of two-state homogeneous series and parallel-series systems

$\bar{\mathcal{R}}'(t)$ — a limit reliability function of two-state non-homogeneous series and parallel-series systems

$\mathcal{R}(t)$ — a limit reliability function of two-state homogeneous parallel and series-parallel systems

$\mathcal{R}'(t)$ — a limit reliability function of two-state non-homogeneous parallel and series-parallel systems

$\mathcal{R}^{(0)}(t)$ — a limit reliability function of a two-state homogeneous "m out of n" system

$\mathcal{R}^{(\mu)}(t)$ — a limit reliability function of a two-state homogeneous "m out of n" system

$\bar{\mathcal{R}}^{(1)}(t,\cdot)$ — a limit reliability function of a two-state homogeneous "m out of n" system

$E(T)$ — a mean lifetime of a two-state system

$\sigma(T)$ — a lifetime standard deviation of a two-state system

z — a number of reliability states of a multi-state component and a multi-state system

$T_i(u)$ — multi-state component lifetimes of series, parallel and "m out of n" systems in a state subset

$T_{ij}(u)$ — multi-state component lifetimes of series-parallel and parallel-series systems in a state subset

$T(u)$ — a multi-state system lifetime in a state subset

$R(t,\cdot)$ — a multi-state component reliability function of a homogeneous system

$F(t,\cdot)$ — a multi-state component lifetime distribution function of a homogeneous system in a state subset

$R^{(i)}(t,\cdot)$ — multi-state component reliability functions of homogeneous series, parallel and "m out of n" systems

$F^{(i)}(t,\cdot)$ — multi-state component lifetime distribution functions of homogeneous series, parallel and "m out of n" systems in a state subset

$R^{(i,j)}(t,\cdot)$	– multi-state component reliability functions of homogeneous series-parallel and parallel-series systems
$F^{(i,j)}(t,\cdot)$	– multi-state component lifetime distribution functions of homogeneous series-parallel and parallel-series systems in a state subset
$\overline{R}_n(t,\cdot)$	– a reliability function of a multi-state homogeneous series system
$\overline{R}'_n(t,\cdot)$	– a reliability function of a multi-state non-homogeneous series system
$R_n(t,\cdot)$	– a reliability function of a multi-state homogeneous parallel system
$R'_n(t,\cdot)$	– a reliability function of a multi-state non-homogeneous parallel system
$R_n^{(m)}(t,\cdot)$	– a reliability function of a multi-state homogeneous "m out of n" system
$\overline{R}_n^{(\overline{m})}(t,\cdot)$	– a reliability function of a multi-state homogeneous "m out of n" system
$R'_n^{(m)}(t,\cdot)$	– a reliability function of a multi-state non-homogeneous "m out of n" system
$\overline{R}'_n^{(\overline{m})}(t,\cdot)$	– a reliability function of a multi-state non-homogeneous "m out of n" system
$\overline{R}_{k_n,l_n}(t,\cdot)$	– a reliability function of a multi-state homogeneous parallel-series system
$\overline{R}'_{k_n,l_n}(t,\cdot)$	– a reliability function of a multi-state non-homogeneous parallel-series system
$R_{k_n,l_n}(t,\cdot)$	– a reliability function of a multi-state homogeneous series-parallel system
$R'_{k_n,l_n}(t,\cdot)$	– a reliability function of a multi-state non-homogeneous series-parallel system
$\overline{\mathcal{R}}(t,\cdot)$	– a limit reliability function of multi-state homogeneous series and parallel-series systems
$\overline{\mathcal{R}}'(t,\cdot)$	– a limit reliability function of multi-state non-homogeneous series and parallel-series systems
$\mathcal{R}(t,\cdot)$	– a limit reliability function of multi-state homogeneous parallel and series-parallel systems
$\mathcal{R}'(t,\cdot)$	– a limit reliability function of multi-state non-homogeneous parallel and series-parallel systems
$\mathcal{R}^{(0)}(t,\cdot)$	– a limit reliability function of a multi-state homogeneous "m out of n" system

$\mathcal{R}^{(\mu)}(t,\cdot)$	– a limit reliability function of a multi-state homogeneous "m out of n" system
$\overline{\mathcal{R}}^{(1)}(t,\cdot)$	– a limit reliability function of a multi-state homogeneous "m out of n" system
r	– a critical reliability state of a system
$r(t)$	– a risk function of a multi-state system
$M_i(u)$	– a multi-state component mean lifetime in a state subset
$\sigma_i(u)$	– a multi-state component lifetime standard deviation in a state subset
$\overline{M}_i(u)$	– a multi-state component mean lifetime in a state
$M(u)$	– a multi-state system mean lifetime in a state subset
$\sigma(u)$	– a multi-state system lifetime standard deviation in a state subset
$\overline{M}(u)$	– a multi-state system mean lifetime in a state
δ	– a permitted level of a multi-state system risk function
τ	– a moment of exceeding a permitted multi-state system risk level
$D_{\overline{\mathcal{R}}_i}$	– domains of attraction of limit reliability functions $\overline{\mathcal{R}}_i(t)$ of two-state homogeneous series system
$R_{k_n,l_1,l_2,\dots,l_{k_n}}^{(m)}(t)$	– a reliability function of a homogeneous two-state series-"m out of k_n" system
$\overline{R}_{k_n,l_1,l_2,\dots,l_{k_n}}^{(\overline{m})}(t)$	– a reliability function of a homogeneous two-state series-"m out of k_n" system
$R_{k_n,l_n}^{(m)}(t)$	– a reliability function of a homogeneous and regular two-state series-"m out of k_n" system
$\overline{R}_{k_n,l_n}^{(\overline{m})}(t)$	– a reliability function of a homogeneous and regular two-state series-"m out of k_n" system
$R_{k_n,l_1,l_2,\dots,l_{k_n}}^{(m_1,m_2,\dots,m_{k_n})}(t)$	– a reliability function of a two-state "m_i out of l_i"-series system
$\overline{R}_{k_n,l_1,l_2,\dots,l_{k_n}}^{(\overline{m}_1,\overline{m}_2,\dots,\overline{m}_{k_n})}(t)$	– a reliability function of a two-state "m_i out of l_i"-series system
$R_{k_n,l_n}^{(m)}(t)$	– a reliability function of a homogeneous and regular two-state "m out of k_n"-series system
$\overline{R}_{k_n,l_n}^{(\overline{m})}(t)$	– a reliability function of a homogeneous and regular two-state "m out of k_n"-series system
$\mathcal{R}_i^{(m)}(t)$	– a limit reliability function of a homogeneous and regular two-state series-"m out of k_n" system
$\mathcal{R}_i^{(\overline{m})}(t)$	– a limit reliability function of a homogeneous and regular two-state series-"m out of k_n" system

$\overline{\mathcal{R}_i^{(m)}}(t)$	– a limit reliability function of a homogeneous and regular two-state "*m* out of k_n"-series system
$\overline{\overline{\mathcal{R}_i^{(\overline{m})}}}(t)$	– a limit reliability function of a homogeneous and regular two-state "*m* out of k_n"-series system
$R_{r,k_n,l_n}(t)$	– a reliability function of a two-state series-parallel system of order *r*
$\mathcal{R}_i(t)$	– a limit reliability function of a two-state series-parallel system of order *r*
$\overline{R}_{r,k_n,l_n}(t)$	– a reliability function of a two-state parallel-series system of order *r*
$\overline{\mathcal{R}}_i(t)$	– a limit reliability function of a two-state parallel-series system of order *r*
ρ	– a factor reducing a component failure rate
$R_n^{(1)}(t)$	– a reliability function of a two-state series system with components improved by reducing their failure rates by a factor ρ
$R_n^{(2)}(t)$	– a reliability function of a two-state series system with a single hot reservation of its components
$R_n^{(3)}(t)$	– a reliability function of a two-state series system with a single cold reservation of its components
$R_n^{(4)}(t)$	– a reliability function of a two-state series system with a single mixed reservation of its components
$R_n^{(5)}(t)$	– a reliability function of a two-state series system with its single hot reservation
$R_n^{(6)}(t)$	– a reliability function of a two-state series system with its single cold reservation
$\mathcal{R}^{(1)}(t)$	– a limit reliability function of a two-state series system with components improved by reducing their failure rates by a factor ρ
$\mathcal{R}^{(2)}(t)$	– a limit reliability function of a two-state series system with a single hot reservation of its components
$\mathcal{R}^{(3)}(t)$	– a limit reliability function of a two-state series system with a single cold reservation of its components
$\mathcal{R}^{(4)}(t)$	– a limit reliability function of a two-state series system with a single mixed reservation of its components
$\mathcal{R}^{(5)}(t)$	– a limit reliability function of a two-state series system with its single hot reservation
$\mathcal{R}^{(6)}(t)$	– a limit reliability function of a two-state series system with its single cold reservation

$T^{(1)}$ – a lifetime mean value of a two-state series system with components improved by reducing their failure rates by a factor ρ

$T^{(2)}$ – a lifetime mean value of a two-state series system with a single hot reservation of its components

$T^{(3)}$ – a lifetime mean value of a two-state series system with a single cold reservation of its components

$T^{(4)}$ – a lifetime mean value of a two-state series system with a single mixed reservation of its components

$T^{(5)}$ – a lifetime mean value of a two-state series system with its single hot reservation

$T^{(6)}$ – a lifetime mean value of a two-state series system with its single cold reservation.

z^k – a system operational state

$Z(t)$ – a process of changing system operational states

θ^{kl} – conditional sojourn times of a process $Z(t)$ at operational states

$[H^{kl}(t)]_{v \times v}$ – a matrix of conditional distribution functions of sojourn times θ^{kl}

$[p^k(0)]_{1 \times v}$ – a vector of probabilities of process $Z(t)$ initial states

$E[\theta^{kl}]$ – mean values of sojourn times θ^{kl}

θ^k – unconditional sojourn times of process $Z(t)$ at states z^k

$H^k(t)$ – unconditional distribution functions of sojourn times θ^k

$E[\theta^k]$ – mean values of unconditional sojourn times θ^k

M^k – mean values of unconditional sojourn times θ^k

$p^k(t)$ – transient probabilities of process $Z(t)$ at states z^k

p^k – limit values of transient probabilities $p^k(t)$

$R^{(k)}(t)$ – conditional reliability functions of a two-state system at operational states z^k

$R(t)$ – an unconditional reliability function of a two-state system

$T_{ij}^{(k)}$ – conditional lifetimes of system components E_{ij} of a non-homogeneous two-state series-parallel system at operational states z^k

$[R^{(i,j)}(t)]^{(k)}$ — conditional reliability functions of system components E_{ij} of a non-homogeneous two-state series-parallel system at operational states z^k

$T^{(k)}$ — conditional lifetimes of a non-homogeneous two-state series-parallel system at operational states z^k

$\boldsymbol{R}^{(k)}_{k_n, l_n}(t)$ — conditional reliability functions of a non-homogeneous two-state series-parallel system at operational states z^k

T — an unconditional lifetime of a non-homogeneous two-state series-parallel system

$\boldsymbol{R}(t)$ — unconditional reliability functions of a non-homogeneous two-state series-parallel system

m — an unconditional mean value of a non-homogeneous two-state series-parallel system lifetime

σ^2 — an unconditional variance of a non-homogeneous two-state series-parallel system lifetime

LIST OF FIGURES

LIST OF TABLES

CHAPTER 1

BASIC NOTIONS

Basic notions and agreements, which are necessary to further considerations, are introduced. The asymptotic approach to the system reliability investigation and the system limit reliability function is defined.

Considering the reliability of two-state systems we assume that the distributions of the component and the system lifetimes T do not necessarily have to be concentrated in the interval $<0,\infty)$. It means that a reliability function

$$R(t) = P(T > t), \quad t \in (-\infty, \infty),$$

does not have to satisfy the usually demanded condition

$$R(t) = 1 \quad \text{for } t \in (-\infty, 0).$$

This is a generalisation of the normally used concept of a reliability function. This generalisation is convenient in the theoretical considerations. At the same time, from the achieved results on the generalised reliability functions, for particular cases, the same properties of the normally used reliability functions appear.

From that assumption it follows that between a reliability function $R(t)$ and a distribution function

$$F(t) = P(T \leq t), t \in (-\infty, \infty),$$

there exists a relationship given by

$$R(t) = 1 - F(t) \quad \text{for } t \in (-\infty, \infty).$$

Thus, the following corollary is obvious.

Corollary 1.1
A reliability function $R(t)$ is non-increasing, right-continuous and moreover

$$R(-\infty) = 1, R(+\infty) = 0.$$

Definition 1.1
A reliability function $R(t)$ is called degenerate if there exists $t_0 \in (-\infty, \infty)$, such that

$$R(t) = \begin{cases} 1, & t < t_0 \\ 0, & t \geq t_0. \end{cases}$$

Corollary 1.2
A function

$$R(t) = 1 - \exp[-V(t)], \quad t \in (-\infty, \infty),$$

is a reliability function if and only if a function $V(t)$ is non-negative, non-increasing, right continuous,

$$V(-\infty) = \infty, \ V(+\infty) = 0$$

and moreover $V(t)$ can be identically equal to ∞ in an interval.

Corollary 1.3
A function

$$\overline{R}(t) = \exp[-\overline{V}(t)], \quad t \in (-\infty, \infty),$$

is a reliability function if and only if a function $\overline{V}(t)$ is non-negative, non-decreasing, right continuous,

$$\overline{V}(-\infty) = 0, \ \overline{V}(+\infty) = \infty,$$

and moreover $\overline{V}(t)$ can be identically equal to ∞ in an interval.

Corollary 1.4
A function

$$R^{(0)}(t) = 1 - \sum_{i=0}^{m-1} \frac{[V(t)]^i}{i!} \exp[-V(t)], \ t \in (-\infty, \infty), \ m \in N,$$

is a reliability function if and only if a function $V(t)$ is non-negative, non-increasing, right continuous,

$$V(-\infty) = \infty, \ V(+\infty) = 0,$$

and moreover $V(t)$ can be identically equal to ∞ in an interval.

Corollary 1.5
A function

$$R^{(\mu)}(t) = 1 - \frac{1}{\sqrt{2\pi}} \int_{-\infty}^{-v(t)} e^{-\frac{x^2}{2}} \, dx, \ t \in (-\infty, \infty), \ 0 < \mu < 1,$$

is a reliability function if and only if a function $v(t)$ is non-increasing, right continuous,

$$v(-\infty) = +\infty, \ v(+\infty) = -\infty$$

and moreover $v(t)$ can be identically equal to $-\infty$ or equal to $+\infty$ in an interval.

Corollary 1.6
A function

$$\overline{R}^{(1)}(t) = \sum_{i=0}^{\overline{m}} \frac{[\overline{V}(t)]^i}{i!} \exp[-\overline{V}(t)], \ t \in (-\infty, \infty), \ \overline{m} \in N,$$

is a reliability function if and only if a function $\overline{V}(t)$ is non-negative, non-decreasing, right continuous,

$$\overline{V}(-\infty) = 0, \ \overline{V}(+\infty) = \infty$$

and moreover $\overline{V}(t)$ can be identically equal to ∞ in an interval.

Agreement 1.1
In further considerations if we use symbols $V(t)$ and $\overline{V}(t)$ we always mean functions of properties given in Corollaries 1.2–1.6. If $V(t)$ and $\overline{V}(t)$ are identically equal to ∞ we assume that

$$\exp[-V(t)] = 0, \ \exp[-\overline{V}(t)] = 0, \ [V(t)]^i \exp[-V(t)] = 0 \ \text{and} \ [\overline{V}(t)]^i \exp[\overline{V}(t)] = 0.$$

If we say that $V(t)$, $v(t)$ and $\overline{V}(t)$ are non-negative, non-increasing or non-decreasing and right-continuous we mean the intervals where

$$V(t) \neq \infty, \ v(t) \neq \infty \ \text{and} \ \overline{V}(t) \neq \infty.$$

Moreover, we denote the set of continuity points of a reliability function $R(t)$ by C_R and the set of continuity points of a function $V(t)$ and points such that $V(t) = \infty$ by C_V. We denote the set of continuity points of a reliability function $\overline{R}(t)$ by $C_{\overline{R}}$ and the set of continuity points of a function $\overline{V}(t)$ and points such that $\overline{V}(t) = \infty$ by $C_{\overline{V}}$. Similarly, we denote the set of continuity points of reliability functions $R^{(0)}(t)$, $R^{(\mu)}(t)$ and $\overline{R}^{(1)}(t)$ by $C_{R^{(0)}}$, $C_{R^{(\mu)}}$ and $C_{\overline{R}^{(1)}}$ respectively and the set of continuity points of a function $v(t)$ and points such that $v(t) = -\infty$ or points such that $v(t) = +\infty$ by C_v.

According to Definition 1.1, Corollaries 1.2–1.6 and Agreement 1.1, we assume the next definitions.

Definition 1.2
A function $V(t)$ defined for $t \in (-\infty, \infty)$, non-negative, non-increasing, right-continuous and such that

$$V(-\infty) = \infty, \; V(+\infty) = 0$$

is called degenerate if there exists $t_0 \in (-\infty, \infty)$ such that

$$V(t) = \begin{cases} \infty, \; t < t_0 \\ 0, \; t \ge t_0. \end{cases}$$

Definition 1.3
A function $v(t)$ defined for $t \in (-\infty, \infty)$, non-increasing, right-continuous and such that

$$v(-\infty) = +\infty, \; v(+\infty) = -\infty$$

is called degenerate if there exists $t_0 \in (-\infty, \infty)$ such that

$$v(t) = \begin{cases} +\infty, \; t < t_0 \\ -\infty, \; t \ge t_0. \end{cases}$$

Definition 1.4
A function $\overline{V}(t)$ defined for $t \in (-\infty, \infty)$, non-negative, non-decreasing, right-continuous and such that

$$\overline{V}(-\infty) = 0, \; \overline{V}(\infty) = \infty$$

is called degenerate if there exists $t_0 \in (-\infty, \infty)$ such that

$$\overline{V}(t) = \begin{cases} 0, & t < t_0 \\ \infty, & t \geq t_0. \end{cases}$$

Under those definitions the following corollaries are clear.

Corollary 1.7
A reliability function

$$R(t) = 1 - \exp[-V(t)], \; t \in (-\infty, \infty),$$

is degenerate if and only if a function $V(t)$ is degenerate.

Corollary 1.8
A reliability function

$$\overline{R}(t) = \exp[-\overline{V}(t)], \;\; t \in (-\infty, \infty),$$

is degenerate if and only if a function $\overline{V}(t)$ is degenerate.

Corollary 1.9
A reliability function

$$R^{(0)}(t) = 1 - \sum_{i=0}^{m-1} \frac{[V(t)]^i}{i!} \exp[-V(t)], \; t \in (-\infty, \infty), \;\; m \in N,$$

is degenerate if and only if a function $V(t)$ is degenerate.

Corollary 1.10
A reliability function

$$R^{(\mu)}(t) = 1 - \frac{1}{\sqrt{2\pi}} \int_{-\infty}^{-v(t)} e^{-\frac{x^2}{2}} dx, \; t \in (-\infty, \infty), \;\; 0 < \mu < 1,$$

is degenerate if and only if a function $v(t)$ is degenerate.

Corollary 1.11
A reliability function

$$\overline{R}^{(1)}(t) = \sum_{i=0}^{\overline{m}} \frac{[\overline{V}(t)]^i}{i!} \exp[-\overline{V}(t)], \; t \in (-\infty, \infty), \;\; \overline{m} \in N,$$

is degenerate if and only if a function $\overline{V}(t)$ is degenerate.

The asymptotic approach to the reliability of two-state systems depends on the investigation of limit distributions of a standardised random variable

$$(T - b_n) / a_n,$$

where T is the lifetime of a system and $a_n > 0$, $b_n \in (-\infty, \infty)$ are suitably chosen numbers called normalising constants.
Since

$$P((T - b_n) / a_n > t) = P(T > a_n t + b_n) = R_n(a_n t + b_n),$$

where $R_n(t)$ is a reliability function of a system composed of n components, then the following definition becomes natural.

Definition 1.5
A reliability function $\mathcal{R}(t)$ is called a limit reliability function or an asymptotic reliability function of a system having a reliability function $R_n(t)$ if there exist normalising constants $a_n > 0$, $b_n \in (-\infty, \infty)$ such that

$$\lim_{n \to \infty} R_n(a_n t + b_n) = \mathcal{R}(t) \text{ for } t \in C_{\mathcal{R}}.$$

Thus, if the asymptotic reliability function $\mathcal{R}(t)$ of a system is known, then for sufficiently large n, the approximate formula

$$R_n(t) \cong \mathcal{R}((t - b_n)/a_n), \ t \in (-\infty, \infty). \tag{1.1}$$

may be used instead of the system exact reliability function $R_n(t)$.
From the condition

$$\lim_{n \to \infty} R_n(a_n t + b_n) = \mathcal{R}(t) \text{ for } t \in C_{\mathcal{R}},$$

it follows that setting

$$\alpha_n = a a_n, \ \beta_n = b a_n + b_n,$$

where $a > 0$ and $b \in (-\infty, \infty)$, we get

$$\lim_{n \to \infty} R_n(\alpha_n t + \beta_n) = \lim_{n \to \infty} R_n(a_n(at + b) + b_n) = \mathcal{R}(at + b) \text{ for } t \in C_{\mathcal{R}}.$$

Hence, if $\mathcal{R}(t)$ is the limit reliability function of a system, then $\mathcal{R}(at + b)$ with arbitrary $a > 0$ and $b \in (-\infty, \infty)$ is also its limit reliability function. That fact, in a natural way, yields the concept of a type of limit reliability function.

Definition 1.6
The limit reliability functions $\mathscr{R}_0(t)$ and $\mathscr{R}(t)$ are said to be of the same type if there exist numbers $a > 0$ and $b \in (-\infty, \infty)$ such that

$$\mathscr{R}_0(t) = \mathscr{R}(at + b) \text{ for } t \in (-\infty, \infty).$$

Agreement 1.2
In further considerations we assume the following notations:

$x(n) << y(n)$ or $x(n) = o(y(n))$, where $x(n)$ and $y(n)$ are positive functions, means that $x(n)$ is of order much less than $y(n)$ in a sense

$$\lim_{n \to \infty} x(n) / y(n) = 0,$$

$x(n) \approx y(n)$ or $x(n) = r(y(n))$, where $x(n)$ and $y(n)$ are either positive or negative functions, means that $x(n)$ is of order $y(n)$ in a sense

$$\lim_{n \to \infty} x(n) / y(n) = 1,$$

$x(n) >> y(n)$ or $x(n) = O(y(n))$, where $x(n)$ and $y(n)$ are positive functions, means that $x(n)$ is of order much greater than $y(n)$ in a sense

$$\lim_{n \to \infty} x(n) / y(n) = \infty.$$

CHAPTER 2

TWO-STATE SYSTEMS

Two-state homogeneous and non-homogeneous series, parallel, "m out of n", series-parallel and parallel-series systems are defined. Their exact reliability functions are determined.

We assume that

$$E_i, \ i = 1,2,...,n, \ n \in N,$$

are two-state components of the system having reliability functions

$$R_i(t) = P(T_i > t), \ t \in (-\infty, \infty),$$

where

$$T_i, \ i = 1,2,...,n,$$

are independent random variables representing the lifetimes of components E_i with distribution functions

$$F_i(t) = P(T_i \le t), \ t \in (-\infty, \infty).$$

The simplest two-state reliability structures are series and parallel systems. We define these first.

Definition 2.1
A two-state system is called series if its lifetime T is given by

$$T = \min_{1 \le i \le n} \{T_i\}.$$

The scheme of a series system is given in Figure 2.1.

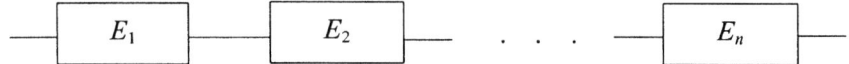

Fig. 2.1. The scheme of a series system

The above definition means that the series system is not failed if and only if all its components are not failed, and therefore its reliability function is given by

$$\overline{R}_n(t) = \prod_{i=1}^{n} R_i(t), \ t \in (-\infty, \infty).$$

Definition 2.2
A two-state series system is called homogeneous if its component lifetimes T_i have an identical distribution function

$$F(t) = P(T_i \leq t), \ t \in (-\infty, \infty), \ i = 1, 2, ..., n,$$

i.e. if its components E_i have the same reliability function

$$R(t) = 1 - F(t), \ t \in (-\infty, \infty).$$

The above definition results in the following simplified formula

$$\overline{R}_n(t) = [R(t)]^n, \ t \in (-\infty, \infty), \tag{2.1}$$

for the reliability function of the homogeneous two-state series system.

Definition 2.3
A two-state system is called parallel if its lifetime T is given by

$$T = \max_{1 \leq i \leq n} \{T_i\}.$$

The above definition means that the parallel system is failed if and only if all its components are failed and therefore its reliability function is given by

$$R_n(t) = 1 - \prod_{i=1}^{n} F_i(t), \ t \in (-\infty, \infty).$$

The scheme of a parallel system is given in Figure 2.2.

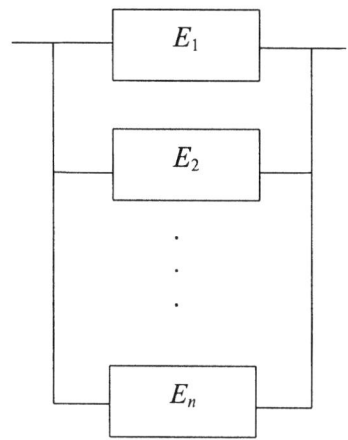

Fig. 2.2. The scheme of a parallel system

Definition 2.4
A two-state parallel system is called homogeneous if its component lifetimes T_i have an identical distribution function

$$F(t) = P(T_i \le t),\ t \in (-\infty, \infty),\ i = 1,2,...,n,$$

i.e. if its components E_i have the same reliability function

$$R(t) = 1 - F(t),\ t \in (-\infty, \infty).$$

Under this definition we get the following formula

$$\mathbf{R}_n(t) = 1 - [F(t)]^n,\ t \in (-\infty, \infty), \tag{2.2}$$

for the reliability function of the homogeneous two-state parallel system.

Definition 2.5
A two-state system is called an "*m* out of *n*" system if its lifetime T is given by

$$T = T_{(n-m+1)},\ m = 1,2,...,n,$$

where $T_{(n-m+1)}$ is the mth maximal order statistic in the sequence of component lifetimes $T_1, T_2,..., T_n$.

The scheme of an "*m* out of *n*" system is given in Figure 2.3, where $i_1, i_2, \ldots, i_n \in \{1,2,\ldots,n\}$ and $i_j \neq i_k$ for $j \neq k$.

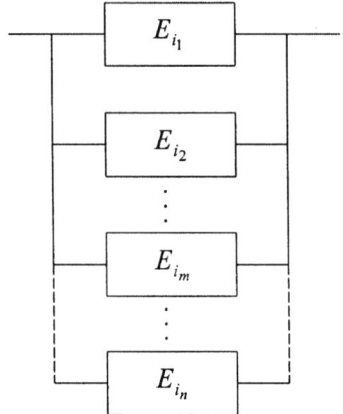

Fig. 2.3. The scheme of an "*m* out of *n*" system

The above definition means that the two-state "*m* out of *n*" system is not failed if and only if at least *m* out of its *n* components are not failed. The two-state "*m* out of *n*" system becomes a parallel system if $m = 1$, whereas it becomes a series system if $m = n$. The reliability function of the two-state "*m* out of *n*" system is given either by

$$\boldsymbol{R}_n^{(m)}(t) = 1 - \sum_{\substack{r_1,r_2,\ldots,r_n=0 \\ r_1+r_2+\ldots+r_n \le m-1}}^{1} \prod_{i=1}^{n} [R_i(t)]^{r_i} [F_i(t)]^{1-r_i} , \ t \in (-\infty,\infty),$$

or by

$$\overline{\boldsymbol{R}}_n^{(\overline{m})}(t) = \sum_{\substack{r_1,r_2,\ldots,r_n=0 \\ r_1+r_2+\ldots+r_n \le \overline{m}}}^{1} \prod_{i=1}^{n} [F_i(t)]^{r_i} [R_i(t)]^{1-r_i} , \ t \in (-\infty,\infty), \ \overline{m} = n - m.$$

Definition 2.6
A two-state "*m* out of *n*" system is called homogeneous if its component lifetimes T_i have an identical distribution function

$$F(t) = P(T_i \le t), \ t \in (-\infty,\infty), \ i = 1,2,\ldots,n,$$

i.e. if its components E_i have the same reliability function

$$R(t) = 1 - F(t), \ t \in (-\infty,\infty).$$

The reliability function of the homogeneous two-state "*m* out of *n*" system is given either by

$$R_n^{(m)}(t) = 1 - \sum_{i=0}^{m-1} \binom{n}{i}[R(t)]^i[F(t)]^{n-i}, \; t \in (-\infty, \infty). \tag{2.3}$$

or by

$$\overline{R}_n^{(\overline{m})}(t) = \sum_{i=0}^{\overline{m}} \binom{n}{i}[F(t)]^i[R(t)]^{n-i}, \; t \in (-\infty, \infty), \; \overline{m} = n - m. \tag{2.4}$$

Other basic, a bit more complex, two-state reliability structures are series-parallel and parallel-series systems. To define them, we assume that

$$E_{ij}, \; i = 1,2,...,k_n, j = 1,2,...,l_i, \; k_n, l_1, l_2,..., l_{k_n} \in N,$$

are two-state components of the system having reliability functions

$$R_{ij}(t) = P(T_{ij} > t), \; t \in (-\infty, \infty),$$

where

$$T_{ij}, \; i = 1,2,...,k_n, j = 1,2,...,l_i,$$

are independent random variables representing the lifetimes of components E_{ij} with distribution functions

$$F_{ij}(t) = P(T_{ij} \leq t), \; t \in (-\infty, \infty).$$

Definition 2.7
A two-state system is called series-parallel if its lifetime *T* is given by

$$T = \max_{1 \leq i \leq k_n} \{ \min_{1 \leq j \leq l_i} \{T_{ij}\} \}.$$

The scheme of a series-parallel system is given in Figure 2.4.
By joining the formulae for the reliability functions of two-state series and parallel systems it is easy to conclude that the reliability function of the two-state series-parallel system is given by

$$R_{k_n, l_1, l_2,..., l_{k_n}}(t) = 1 - \prod_{i=1}^{k_n}[1 - \prod_{j=1}^{l_i} R_{ij}(t)], \; t \in (-\infty, \infty),$$

where k_n is the number of series subsystems linked in parallel and l_i are the numbers of components in the series subsystems.

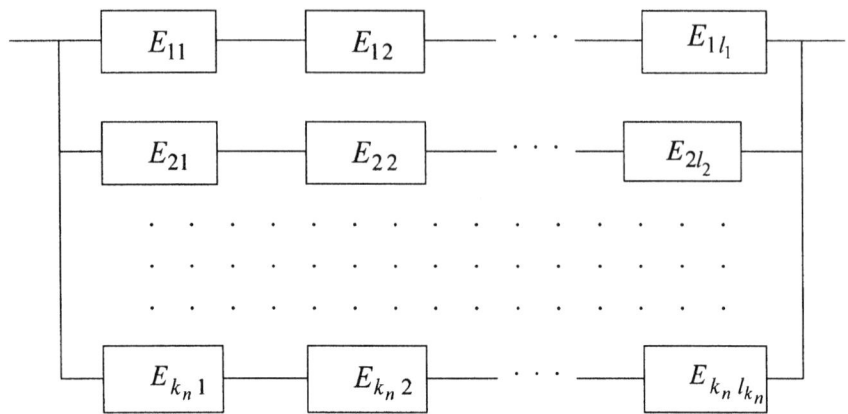

Fig. 2.4. The scheme of a series-parallel system

Definition 2.8
A two-state series-parallel system is called homogeneous if its component lifetimes T_{ij} have an identical distribution function

$$F(t) = P(T_{ij} \leq t), \ i = 1,2,...,k_n, \ j = 1,2,...,l_i,$$

i.e. if its components E_{ij} have the same reliability function

$$R(t) = 1 - F(t), \ t \in (-\infty, \infty).$$

Definition 2.9
A two-state series-parallel system is called regular if

$$l_1 = l_2 = \ldots = l_{k_n} = l_n, \ l_n \in N,$$

i.e. if the numbers of components in its series subsystems are equal.

The scheme of a regular series-parallel system is given in Figure 2.5.
The reliability function of the homogeneous regular two-state series-parallel system is given by

$$\boldsymbol{R}_{k_n,l_n}(t) = 1 - [1 - [R(t)]^{l_n}]^{k_n}, \ t \in (-\infty, \infty), \tag{2.5}$$

where k_n is the number of series subsystems linked in parallel and l_n is the number of components in the series subsystems.

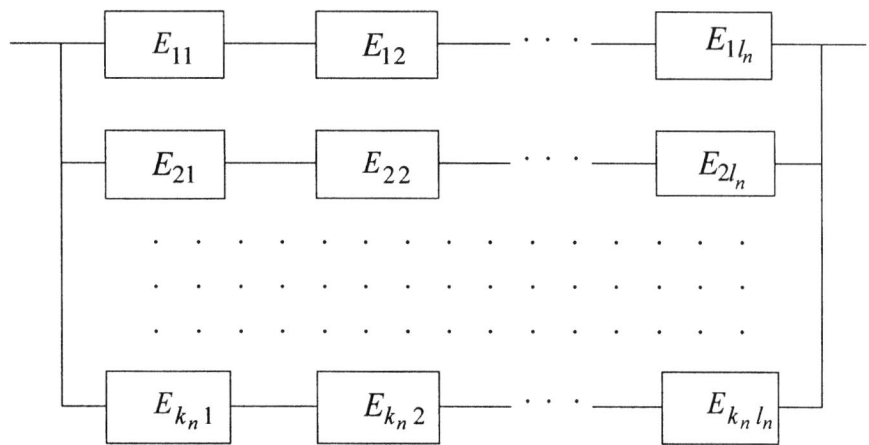

Fig. 2.5. The scheme of a regular series-parallel system

Definition 2.10

A two-state system is called parallel-series if its lifetime T is given by

$$T = \min_{1 \le i \le k_n} \{\max_{1 \le j \le l_i} \{T_{ij}\}\}.$$

The scheme of a parallel-series system is given in Figure 2.6.

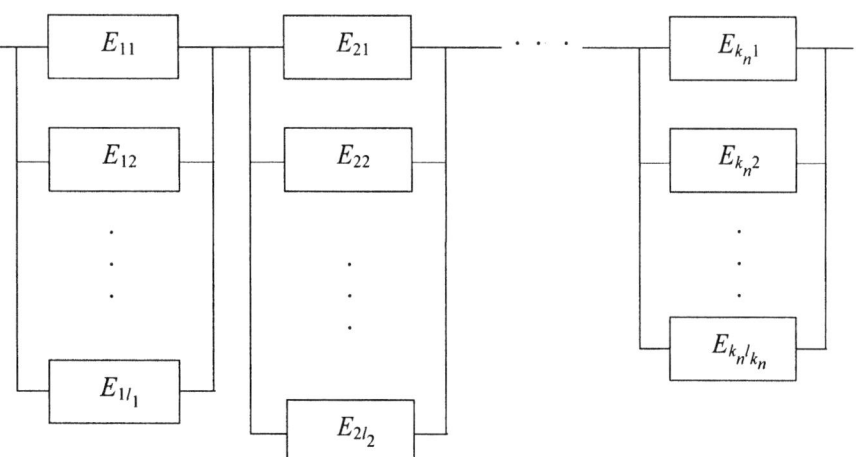

Fig. 2.6. The scheme of a parallel-series system

By superposition of the formulae for the reliability functions of two-state parallel and series systems it is easy to conclude that the reliability function of the two-state parallel-series system is given by

$$\overline{R}_{k_n,l_1,l_2,\ldots,l_{k_n}}(t) = \prod_{i=1}^{k_n}[1 - \prod_{j=1}^{l_i} F_{ij}(t)], \; t \in (-\infty,\infty),$$

where k_n is the number of parallel subsystems linked in series and l_i are the numbers of components in the parallel subsystems.

Definition 2.11
A two-state parallel-series system is called homogeneous if its component lifetimes T_{ij} have an identical distribution function

$$F(t) = P(T_{ij} \leq t), \; i = 1,2,\ldots,k_n, j = 1,2,\ldots,l_i,$$

i.e. if its components E_{ij} have the same reliability function

$$R(t) = 1 - F(t), \;\; t \in (-\infty,\infty).$$

Definition 2.12
A two-state parallel-series system is called regular if

$$l_1 = l_2 = \ldots = l_{k_n} = l_n, l_n \in N,$$

i.e. if the numbers of components in its parallel subsystems are equal.

The scheme of a regular parallel-series system is given in Figure 2.7.

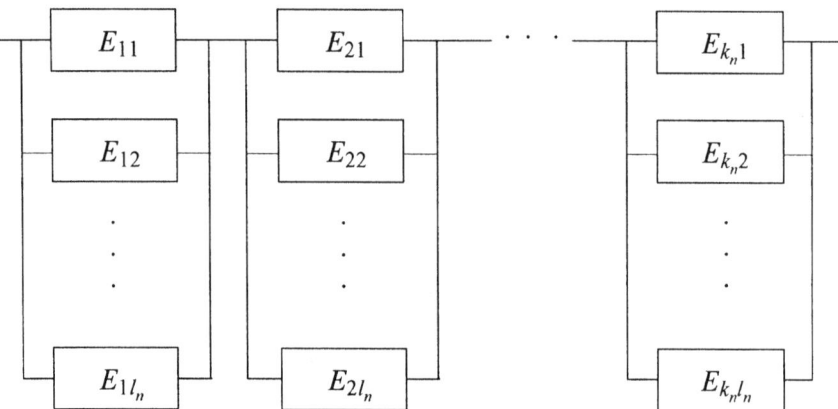

Fig. 2.7. The scheme of a regular parallel-series system

The reliability function of the homogeneous regular two-state parallel-series system is given by

$$\overline{R}_{k_n l_n}(t) = [1-[F(t)]^{l_n}]^{k_n}, \quad t \in (-\infty, \infty),$$ (2.6)

where k_n is the number of parallel subsystems linked in series and l_n is the number of components in the parallel subsystems.

Definition 2.13
A two-state series system is called non-homogeneous if it is composed of a, $1 \leq a \leq n$, different types of components and the fraction of the ith type component in the system is equal to q_i, where $q_i > 0$, $\sum_{i=1}^{a} q_i = 1$. Moreover

$$R^{(i)}(t) = 1 - F^{(i)}(t), \quad t \in (-\infty, \infty), \quad i = 1, 2, ..., a,$$ (2.7)

is the reliability function of the ith type component.

The scheme of a non-homogeneous series system is given in Figure 2.8.

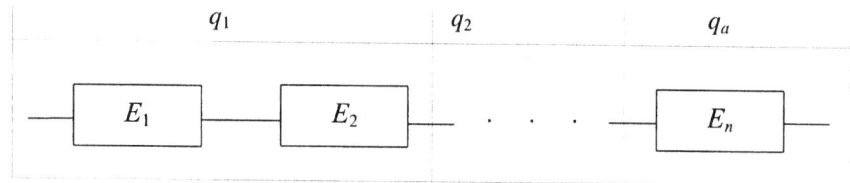

Fig. 2.8. The scheme of a non-homogeneous series system

It is easy to show that the reliability function of the non-homogeneous two-state series system is given by

$$\overline{R}'_{k_n l_n}(t) = \prod_{i=1}^{a} (R^{(i)}(t))^{q_i n}, \quad t \in (-\infty, \infty).$$ (2.8)

Definition 2.14
A two-state parallel system is called non-homogeneous if it is composed of a, $1 \leq a \leq n$, different types of components and the fraction of the ith type component in the system is equal to q_i, where $q_i > 0$, $\sum_{i=1}^{a} q_i = 1$. Moreover

$$R^{(i)}(t) = 1 - F^{(i)}(t), \quad t \in (-\infty, \infty), \quad i = 1, 2, ..., a,$$ (2.9)

is the reliability function of the ith type component.

The scheme of a non-homogeneous parallel system is given in Figure 2.9.

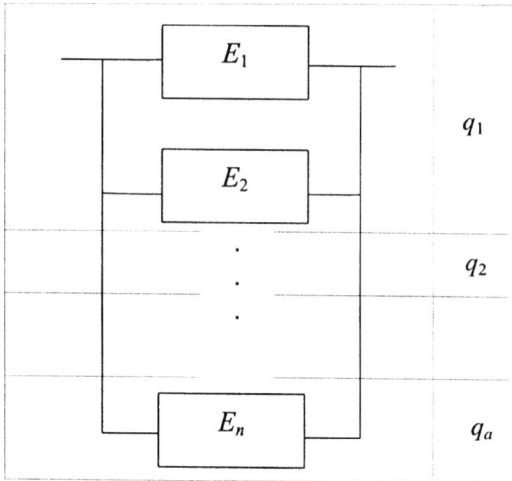

Fig. 2.9. The scheme of a non-homogeneous parallel system

It is possible to work out that the reliability function of the non-homogeneous two-state parallel system is given by

$$R'_n(t) = 1 - \prod_{i=1}^{a} (F^{(i)}(t))^{q_i n} , \quad t \in (-\infty, \infty). \tag{2.10}$$

Definition 2.15
A two-state "*m* out of *n*" system is called non-homogeneous if it is composed of a, $1 \le a \le n$, different types of components and the fraction of the ith type component in the system is equal to q_i, where $q_i > 0$, $\sum_{i=1}^{a} q_i = 1$. Moreover

$$R^{(i)}(t) = 1 - F^{(i)}(t), \quad t \in (-\infty, \infty), \ i = 1, 2, ..., a, \tag{2.11}$$

The scheme of a non-homogeneous "*m* out of *n*" system is given in Figure 2.10, where $i_1, i_2, ..., i_n \in \{1, 2, ..., n\}$ and $i_j \ne i_k$ for $j \ne k$.
The reliability function of the non-homogeneous two-state "*m* out of *n*" system is given either by

$$R'^{(m)}_n(t) = 1 - \sum_{\substack{0 \le r_i \le q_i n \\ r_1 + r_2 + ... + r_a \le m-1}} \prod_{i=1}^{a} \binom{q_i n}{r_i} [R^{(i)}(t)]^{r_i} [F^{(i)}(t)]^{q_i n - r_i} , \quad t \in (-\infty, \infty), \tag{2.12}$$

or by

$$\overline{R}'^{(\overline{m})}_n(t) = \sum_{\substack{0 \le r_i \le q_i n \\ r_1+r_2+...+r_a \le \overline{m}}} \prod_{i=1}^{a} \binom{q_i n}{r_i} [F^{(i)}(t)]^{r_i} [R^{(i)}(t)]^{q_i n - r_i} , \ t \in (-\infty, \infty), \qquad (2.13)$$

where $\overline{m} = n - m$.

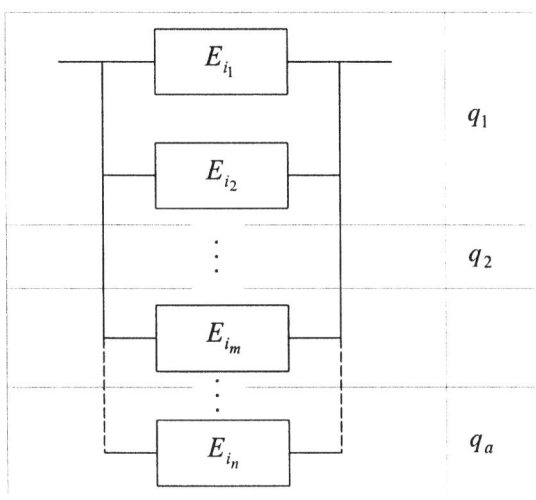

Fig. 2.10. The scheme of a non-homogeneous "*m* out of *n*" system

Definition 2.16
A two-state regular series-parallel system is called non-homogeneous if it is composed of a, $1 \le a \le k_n$, $k_n \in N$, different types of series subsystems and the fraction of the ith type series subsystem is equal to q_i, where $q_i > 0$, $\sum_{i=1}^{a} q_i = 1$. Moreover, the ith type series subsystem consists of e_i, $1 \le e_i \le l_n$, $l_n \in N$, types of components with reliability functions

$$R^{(i,j)}(t) = 1 - F^{(i,j)}(t), \ t \in (-\infty, \infty), \ j = 1,2,...,e_i,$$

and the fraction of the jth type component in this subsystem is equal to p_{ij}, where $p_{ij} > 0$ and $\sum_{j=1}^{e_i} p_{ij} = 1$.

The scheme of a regular non-homogeneous series-parallel system is shown in Figure 2.11.

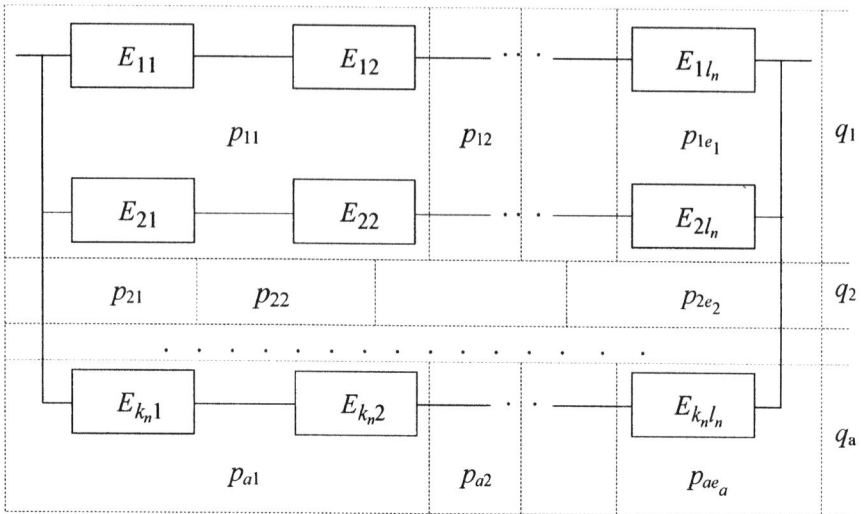

Fig. 2.11. The scheme of a regular non-homogeneous series-parallel system

The reliability function of the regular non-homogeneous two-state series-parallel system is given by

$$R'_{k_n l_n}(t) = 1 - \prod_{i=1}^{a} [1 - (R^{(i)}(t))^{l_n}]^{q_i k_n} , \quad t \in (-\infty, \infty),$$ (2.14)

where

$$R^{(i)}(t) = \prod_{j=1}^{e_i} (R^{(i,j)}(t))^{p_{ij}} , \quad i = 1, 2, ..., a.$$ (2.15)

Definition 2.17
A two-state regular parallel-series system is called non-homogeneous if it is composed of a, $1 \le a \le k_n$, $k_n \in N$, different types of parallel subsystems and the fraction of the ith type parallel subsystem is equal to q_i, where $q_i > 0$, $\sum_{i=1}^{a} q_i = 1$. Moreover, the ith type parallel subsystem consists of e_i, $1 \le e_i \le l_n$, $l_n \in N$, types of components with reliability functions

$$R^{(i,j)}(t) = 1 - F^{(i,j)}(t), \quad t \in (-\infty, \infty), \ j = 1, 2, ..., e_i,$$

and the fraction of the jth type component in this subsystem is equal to p_{ij}, where $p_{ij} > 0$ and $\sum_{j=1}^{e_i} p_{ij} = 1$.

The scheme of a regular non-homogeneous parallel-series system is shown in Figure 2.12.

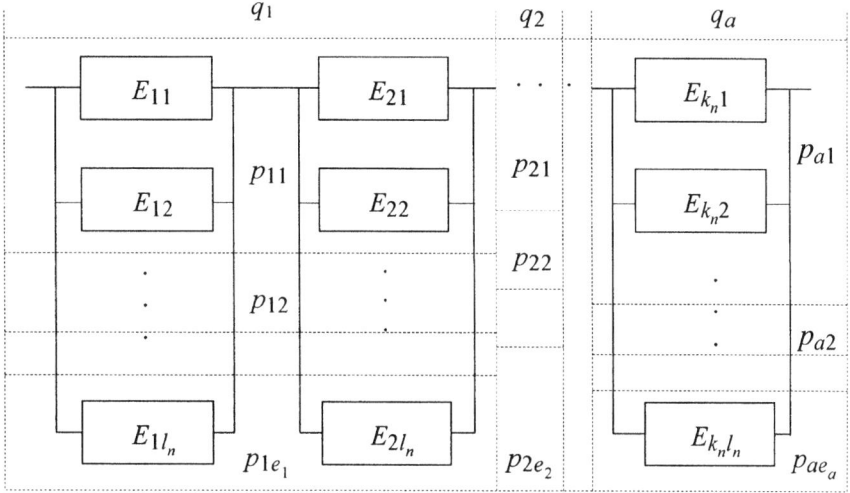

Fig. 2.12. The scheme of a regular non-homogeneous parallel-series system

The reliability function of the regular non-homogeneous two-state parallel-series system is given by

$$\overline{R}'_{k_n,l_n}(t) = \prod_{i=1}^{a}[1-(F^{(i)}(t))^{l_n}]^{q_ik_n} , \ t \in (-\infty,\infty), \tag{2.16}$$

where

$$F^{(i)}(t) = \prod_{j=1}^{e_i}(F^{(i,j)}(t))^{p_{ij}} , \ i = 1,2,...,a. \tag{2.17}$$

Remark 2.1

In our further considerations we suppose that n, k_n and l_n are positive real numbers and we investigate the families of the reliability functions $\overline{R}_n(t)$, $\overline{R}'_n(t)$, $R_n(t)$, $R'_n(t)$, $R_n^{(m)}(t)$, $R'^{(m)}_n(t)$, $\overline{R}^{(\overline{m})}_{k_n,l_n}(t)$, $R'^{(\overline{m})}_n(t)$ for $n \in (0,\infty)$ and the families of the reliability functions $R_{k_n,l_n}(t)$, $R'_{k_n,l_n}(t)$, $\overline{R}_{k_n,l_n}(t)$, $\overline{R}'_{k_n,l_n}(t)$ corresponding to the pair (k_n,l_n), where $k_n \in (0,\infty)$, $l_n \in (0,\infty)$. This assumption is necessary in proving theorems that are cited in the next parts of the book. However, from the practical point of view it is important that n, k_n and l_n should be natural numbers. The return to natural numbers is trivial since the positive real number can be represented by the sum

of its natural part and its real remaining part r. Then, the expression of the form $[R(t)]^r$
exists in the formulae for the reliability functions of the considered systems. Since this
expression is a reliability function, it means that the system or subsystem is composed
of one component that has a reliability function different from the reliability function
$R(t)$ of the remaining components. Such single components do not have a significant
influence on the limit reliability functions of the considered systems, which are
composed of large numbers of components.

CHAPTER 3

MULTI-STATE SYSTEMS

Basic notions of the system multi-state reliability analysis are introduced. The multi-state homogeneous and non-homogeneous series, parallel, "m out of n", series-parallel and parallel-series systems with degrading components are defined and their exact reliability functions are determined. The multi-state limit reliability function of the system, its risk function and other multi-state system reliability characteristics are introduced and determined.

In the multi-state reliability analysis to define systems with degrading (ageing) components we assume that:

– E_i, $i = 1,2,...,n$, are components of a system,
– all components and a system under consideration have the state set $\{0,1,...,z\}$, $z \geq 1$,
– the state indexes are ordered, the state 0 is the worst and the state z is the best,
– $T_i(u)$, $i = 1,2,...,n$, are independent random variables representing the lifetimes of components E_i in the state subset $\{u,u+1,...,z\}$, while they were in the state z at the moment $t = 0$,
– $T(u)$ is a random variable representing the lifetime of a system in the state subset $\{u,u+1,...,z\}$ while it was in the state z at the moment $t = 0$,
– the system state degrades with time t without repair,
– $e_i(t)$ is a component E_i state at the moment t, $t \in (-\infty, \infty)$, given that it was in the state z at the moment $t = 0$,
– $s(t)$ is a system state at the moment t, $t \in (-\infty, \infty)$, given that it was in the state z at the moment $t = 0$.

The above assumptions mean that the states of the system with degrading components may be changed in time only from better to worse. The way in which the components and the system states change is illustrated in Figure 3.1.

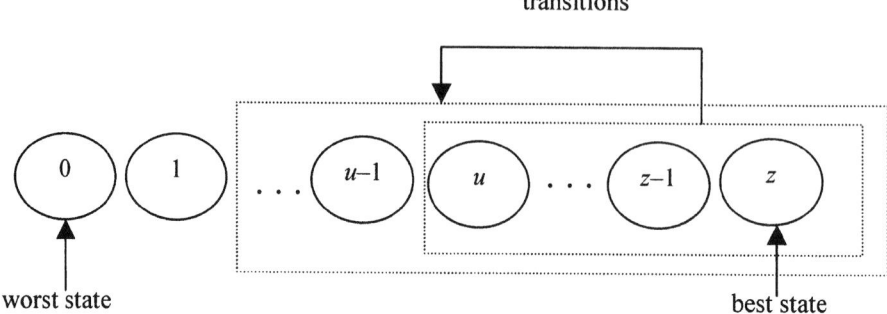

Fig. 3.1. Illustration of states changing in system with ageing components

Definition 3.1
A vector

$$R_i(t,\cdot) = [R_i(t,0), R_i(t,1),...,R_i(t,z)], \quad t \in (-\infty,\infty), \quad i = 1,2,...,n,$$

where

$$R_i(t,u) = P(e_i(t) \geq u \mid e_i(0) = z) = P(T_i(u) > t), \quad t \in (-\infty,\infty), \quad u = 0,1,...,z, \qquad (3.1)$$

is the probability that the component E_i is in the state subset $\{u, u+1,..., z\}$ at the moment t, $t \in (-\infty,\infty)$, while it was in the state z at the moment $t = 0$, is called the multi-state reliability function of a component E_i.

Under this definition we have

$$R_i(t,0) \geq R_i(t,1) \geq \ldots \geq R_i(t,z), \quad t \in (-\infty,\infty), \quad i = 1,2,...,n.$$

Further, if we denote by

$$p_i(t,u) = P(e_i(t) = u \mid e_i(0) = z), \quad t \in (-\infty,\infty), \quad u = 0,1,...,z,$$

the probability that the component E_i is in the state u at the moment t, while it was in the state z at the moment $t = 0$, then by (3.1)

$$R_i(t,0) = 1, \ R_i(t,z) = p_i(t,z), \quad t \in (-\infty,\infty), \quad i = 1,2,...,n, \qquad (3.2)$$

and

$$p_i(t,u) = R_i(t,u) - R_i(t,u+1), \quad u = 0,1,..., z-1, \quad t \in (-\infty,\infty), \quad i = 1,2,...,n. \qquad (3.3)$$

Moreover, if

$$R_i(t,u) = 1 \text{ for } t \le 0, u = 1,2,...,z, i = 1,2,...,n,$$

then

$$M_i(u) = \int_0^\infty R_i(t,u)dt, \ u = 1,2,...,z, i = 1,2,...,n, \tag{3.4}$$

is the mean lifetime of the component E_i in the state subset $\{u, u+1,..., z\}$,

$$\sigma_i(u) = \sqrt{N_i(u) - [M_i(u)]^2}, \ u = 1,2,...,z, i = 1,2,...,n, \tag{3.5}$$

where

$$N_i(u) = 2\int_0^\infty tR_i(t,u)dt, \ u = 1,2,...,z, i = 1,2,...,n, \tag{3.6}$$

is the standard deviation of the component E_i lifetime in the state subset $\{u, u+1,..., z\}$ and

$$\overline{M}_i(u) = \int_0^\infty p_i(t,u)dt, u = 1,2,...,z, i = 1,2,...,n, \tag{3.7}$$

is the mean lifetime of the component E_i in the state u, in the case when the integrals defined by (3.4), (3.6) and (3.7) are convergent.
Next, according to (3.2), (3.3), (3.4) and (3.7), we have

$$\overline{M}_i(u) = M_i(u) - M_i(u+1), \ u = 0,1,..., z-1, \ \overline{M}_i(z) = M_i(z), i = 1,2,...,n. \tag{3.8}$$

Definition 3.2
A vector

$$\boldsymbol{R}_n(t, \cdot) = [\boldsymbol{R}_n(t,0), \boldsymbol{R}_n(t,1),..., \boldsymbol{R}_n(t,z)], \ t \in (-\infty, \infty),$$

where

$$\boldsymbol{R}_n(t,u) = P(s(t) \ge u \mid s(0) = z) = P(T(u) > t), t \in (-\infty,\infty), u = 0,1,...,z, \tag{3.9}$$

is the probability that the system is in the state subset $\{u, u+1,..., z\}$ at the moment t, $t \in (-\infty, \infty)$, while it was in the state z at the moment $t = 0$, is called the multi-state reliability function of a system.

Under this definition we have

$$R_n(t,0) \geq R_n(t,1) \geq \ldots \geq R_n(t,z), \ t \in (-\infty, \infty),$$

and if

$$p(t,u) = P(s(t) = u \mid s(0) = z), \ t \in (-\infty, \infty), \ u = 0,1,\ldots,z, \tag{3.10}$$

is the probability that the system is in the state u at the moment t, $t \in (-\infty, \infty)$, while it was in the state z at the moment $t = 0$, then

$$R_n(t,0) = 1, \ R_n(t,z) = p(t,z), \ t \in (-\infty, \infty), \tag{3.11}$$

and

$$p(t,u) = R_n(t,u) - R_n(t, u+1), \ u = 0,1,\ldots, z-1, \ t \in (-\infty, \infty). \tag{3.12}$$

Moreover, if

$$R_n(t,u) = 1 \text{ for } t \leq 0, \ u = 1,2,\ldots,z,$$

then

$$M(u) = \int_0^\infty R_n(t,u)dt, \ u = 1,2,\ldots,z, \tag{3.13}$$

is the mean lifetime of the system in the state subset $\{u, u+1,\ldots, z\}$,

$$\sigma(u) = \sqrt{N(u) - [M(u)]^2}, \ u = 1,2,\ldots,z, \tag{3.14}$$

where

$$N(u) = 2\int_0^\infty t \, R_n(t,u)dt, \ u = 1,2,\ldots,z, \tag{3.15}$$

is the standard deviation of the system sojourn time in the state subset $\{u, u+1,\ldots, z\}$ and moreover

$$\bar{M}(u) = \int_0^\infty p(t,u)dt, \ u = 1,2,\ldots,z, \tag{3.16}$$

is the mean lifetime of the system in the state u while the integrals (3.13), (3.14) and (3.15) are convergent.

Additionally, according to (3.11), (3.12), (3.13) and (3.16), we get the following relationship

$$\overline{M}(u) = M(u) - M(u+1), \quad u = 0,1,...,z-1, \quad \overline{M}(z) = M(z). \tag{3.17}$$

Definition 3.3
A probability

$$r(t) = P(s(t) < r \mid s(0) = z) = P(T(r) \le t), \quad t \in (-\infty, \infty),$$

that the system is in the subset of states worse than the critical state r, $r \in \{1,...,z\}$ while it was in the state z at the moment $t = 0$ is called a risk function of the multi-state system or, in short, a risk.

Under this definition, from (3.1), we have

$$r(t) = 1 - P(s(t) \ge r \mid s(0) = z) = 1 - R_n(t,r), \quad t \in (-\infty, \infty). \tag{3.18}$$

and if τ is the moment when the risk exceeds a permitted level δ, then

$$\tau = r^{-1}(\delta), \tag{3.19}$$

where $r^{-1}(t)$, if it exists, is the inverse function of the risk function $r(t)$.

Definition 3.4
A multi-state system is called series if its lifetime $T(u)$ in the state subset $\{u, u+1,..., z\}$ is given by

$$T(u) = \min_{1 \le i \le n} \{T_i(u)\}, \quad u = 1,2,...,z.$$

The above definition means that a multi-state series system is in the state subset $\{u, u+1,..., z\}$ if and only if all its components are in this subset of states.
It is easy to work out that the reliability function of the multi-state series system is given by

$$\overline{R}_n(t,\cdot) = [1, \overline{R}_n(t,1),..., \overline{R}_n(t,z)],$$

where

$$\overline{R}_n(t,u) = \prod_{i=1}^{n} R_i(t,u), \quad t \in (-\infty, \infty), \quad u = 1,2,...,z.$$

Definition 3.5

A multi-state series system is called homogeneous if its component lifetimes $T_i(u)$ in the state subsets have an identical distribution function

$$F_i(t,u) = F(t,u), \ u = 1,2,...,z, \ t \in (-\infty, \infty), \ i = 1,2,...,n,$$

i.e. if its components E_i have the same reliability function

$$R_i(t,u) = R(t,u) = 1 - F(t,u), \ u = 1,2,...,z, \ t \in (-\infty, \infty), \ i = 1,2,...,n.$$

The reliability function of the homogeneous multi-state series system is given by

$$\overline{R}_n(t,\cdot) = [1, \overline{R}_n(t,1),..., \overline{R}_n(t,z)], \tag{3.20}$$

where

$$\overline{R}_n(t,u) = [R(t,u)]^n, \ t \in (-\infty, \infty), \ u = 1,2,...,z. \tag{3.21}$$

Definition 3.6

A multi-state system is called parallel if its lifetime $T(u)$ in the state subset $\{u, u+1,..., z\}$ is given by

$$T(u) = \max_{1 \le i \le n}\{T_i(u)\}, \ u = 1,2,...,z.$$

The above definition means that the multi-state parallel system is in the state subset $\{u, u+1,..., z\}$ if and only if at least one of its components is in this subset of states. The reliability function of the multi-state parallel system is given by

$$R_n(t,\cdot) = [1,R_n(t,1),...,R_n(t,z)],$$

where

$$R_n(t,u) = 1 - \prod_{i=1}^{n} F_i(t,u), \ t \in (-\infty, \infty), \ u = 1,2,...,z.$$

Definition 3.7

A multi-state parallel system is called homogeneous if its component lifetimes $T_i(u)$ in the state subsets have an identical distribution function

$$F_i(t,u) = F(t,u), \ u = 1,2,...,z, \ t \in (-\infty, \infty), \ i = 1,2,...,n,$$

i.e. if its components E_i have the same reliability function

$$R_i(t,u) = R(t,u) = 1 - F(t,u), \ u = 1,2,...,z, \ t \in (-\infty, \infty), \ i = 1,2,...,n.$$

The reliability function of the homogeneous multi-state parallel system is given by

$$\boldsymbol{R}_n(t,\cdot) = [1, R_n(t,1), ..., R_n(t,z)], \tag{3.22}$$

where

$$R_n(t,u) = 1 - [F(t,u)]^n, \ t \in (-\infty, \infty), \ u = 1,2,...,z. \tag{3.23}$$

Definition 3.8
A multi-state system is called an "*m* out of *n*" system if its lifetime $T(u)$ in the state subset $\{u, u+1,..., z\}$ is given by

$$T(u) = T_{(n-m+1)}(u), \ m = 1,2,...,n, \ u = 1,2,...,z,$$

where $T_{(n-m+1)}(u)$ is the *m*th maximal order statistic in the sequence of the component lifetimes $T_1(u), T_2(u),..., T_n(u)$.

The above definition means that the multi-state „*m* out of *n*" system is in the state subset $\{u, u+1,..., z\}$ if and only if at least *m* out of its *n* components are in this state subset; and it is a multi-state parallel system if $m = 1$ and it is a multi-state series system if $m = n$.
It can be simply shown that the reliability function of the multi-state "*m* out of *n*" system is given either by

$$\boldsymbol{R}_n^{(m)}(t,\cdot) = [1, \boldsymbol{R}_n^{(m)}(t,1),...,\boldsymbol{R}_n^{(m)}(t,z)],$$

where

$$R_n^{(m)}(t,u) = 1 - \sum_{\substack{r_1,r_2,...,r_n=0 \\ r_1+r_2+...+r_n \leq m-1}}^{1} [R_i(t,u)]^{r_i}[F_i(t,u)]^{1-r_i}, \ t \in (-\infty, \infty), \ u = 1,2,...,z,$$

or by

$$\overline{\boldsymbol{R}}_n^{(\overline{m})}(t,\cdot) = [1, \overline{\boldsymbol{R}}_n^{(\overline{m})}(t,1),..., \overline{\boldsymbol{R}}_n^{(\overline{m})}(t,z)],$$

where

$$\overline{\boldsymbol{R}}_n^{(\overline{m})}(t,u) = \sum_{\substack{r_1,r_2,...,r_n=0 \\ r_1+r_2+...+r_n \leq \overline{m}}}^{1} [F_i(t,u)]^{r_i}[R_i(t,u)]^{1-r_i}, \ t \in (-\infty, \infty), \ \overline{m} = n-m, \ u = 1,2,...,z.$$

Definition 3.9
A multi-state "*m* out of *n*" system is called homogeneous if its component lifetimes $T_i(u)$ in the state subsets have an identical distribution function

$$F_i(t,u) = F(t,u), \ u = 1,2,...,z, \ t \in (-\infty, \infty), \ i = 1,2,...,n,$$

i.e. if its components E_i have the same reliability function

$$R_i(t,u) = R(t,u) = 1 - F(t,u), \ u = 1,2,...,z, \ t \in (-\infty, \infty), \ i = 1,2,...,n.$$

The reliability function of the homogeneous multi-state "*m* out of *n*" system is given either by

$$\boldsymbol{R}_n^{(m)}(t,\cdot) = [1, \boldsymbol{R}_n^{(m)}(t,1),..., \boldsymbol{R}_n^{(m)}(t,z)], \tag{3.24}$$

where

$$\boldsymbol{R}_n^{(m)}(t,u) = 1 - \sum_{i=0}^{m-1} \binom{n}{i} [R(t,u)]^i [F(t,u)]^{n-i}, \ t \in (-\infty, \infty), \ u = 1,2,...,z, \tag{3.25}$$

or by

$$\overline{\boldsymbol{R}}_n^{(\overline{m})}(t,\cdot) = [1, \overline{\boldsymbol{R}}_n^{(\overline{m})}(t,1),..., \overline{\boldsymbol{R}}_n^{(\overline{m})}(t,z)], \tag{3.26}$$

where

$$\overline{\boldsymbol{R}}_n^{(\overline{m})}(t,u) = \sum_{i=0}^{\overline{m}} \binom{n}{i} [F(t,u)]^i [R(t,u)]^{n-i}, \ t \in (-\infty, \infty), \ \overline{m} = n - m, \ u = 1,2,...,z. \tag{3.27}$$

Other basic multi-state reliability structures with components degrading in time are series-parallel and parallel-series systems. To define them, we assume that:
– E_{ij}, $i = 1,2,...,k_n$, $j = 1,2,...,l_i$, k_n, l_1, $l_2,...,l_{k_n} \in N$, are components of a system,
– all components E_{ij} have the same state set as before $\{0,1,...,z\}$,
– $T_{ij}(u)$, $i = 1,2,...,k_n$, $j = 1,2,...,l_i$, k_n, l_1, $l_2,..., l_{k_n} \in N$, are independent random variables
   representing the lifetimes of components E_{ij} in the state subset $\{u, u+1,..., z\}$, while
   they were in the state z at the moment $t = 0$,
– $e_{ij}(t)$ is a component E_{ij} state at the moment t, $t \in (-\infty, \infty)$, while they were in the state
   z at the moment $t = 0$.

Definition 3.10
A vector

$$R_{ij}(t,\cdot) = [R_{ij}(t,0), R_{ij}(t,1),..., R_{ij}(t,z)], \ t \in (-\infty, \infty), \ i = 1,2,...,k_n, j = 1,2,...,l_i,$$

where

$$R_{ij}(t,u) = P(e_{ij}(t) \geq u \mid e_{ij}(0) = z) = P(T_{ij}(u) > t), \; t \in (-\infty, \infty), \; u = 0,1,...,z,$$

is the probability that the component E_{ij} is in the state subset $\{u, u+1,..., z\}$ at the moment t, $t \in (-\infty, \infty)$, while it was in the state z at the moment $t = 0$, is called the multi-state reliability function of a component E_{ij}.

Definition 3.11
A multi-state system is called series-parallel if its lifetime $T(u)$ in the state subset $\{u, u+1,..., z\}$ is given by

$$T(u) = \max_{1 \leq i \leq k_n} \{ \min_{1 \leq j \leq l_i} \{T_{ij}(u)\} , \; u = 1,2,...,z.$$

The reliability function of the multi-state series-parallel system is given by

$$\boldsymbol{R}_{k_n,l_1,l_2,...,l_n}(t,\cdot) = [1, \boldsymbol{R}_{k_n,l_1,l_2,...,l_n}(t,1),...,\boldsymbol{R}_{k_n,l_1,l_2,...,l_n}(t,z)],$$

and

$$\boldsymbol{R}_{k_n,l_1,l_2,...,l_{k_n}}(t,u) = 1 - \prod_{i=1}^{k}[1 - \prod_{j=1}^{l_i} R_{ij}(t,u)], \; t \in (-\infty, \infty), \; u = 1,2,...,z,$$

where k_n is the number of series subsystems linked in parallel and l_i are the numbers of components in the series subsystems.

Definition 3.12
A multi-state series-parallel system is called homogeneous if its lifetimes $T_{ij}(u)$ in the state subset have an identical distribution function

$$F_{ij}(t,u) = F(t,u), u = 1,2,...,z, \; t \in (-\infty, \infty), \; i = 1,2,...,k_n, j = 1,2,...,l_i,$$

i.e. if its components E_{ij} have the same reliability function

$$R_{ij}(t,u) = R(t,u) = 1 - F(t,u), u = 1,2,...,z, \; t \in (-\infty, \infty), \; i = 1,2,...,k_n, j = 1,2,...,l_i.$$

Definition 3.13
A multi-state series-parallel system is called regular if

$$l_1 = l_2 = \ldots = l_{k_n} = l_n, l_n \in N.$$

The reliability function of the multi-state homogeneous regular series-parallel system is given by

$$\boldsymbol{R}_{k_n,l_n}(t,\cdot) = [1, \boldsymbol{R}_{k_n,l_n}(t,1), ..., \boldsymbol{R}_{k_n,l_n}(t,z)], \tag{3.28}$$

and

$$\boldsymbol{R}_{k_n,l_n}(t,u) = 1 - [1 - [R(t,u)]^{l_n}]^{k_n}, \ t \in (-\infty, \infty), \ u = 1,2,...,z, \tag{3.29}$$

where k_n is the number of series subsystems linked in parallel and l_n is the number of components in the series subsystems.

Definition 3.14
A multi-state system is called parallel-series if its lifetime $T(u)$ in the state subset $\{u, u+1, ..., z\}$ is given by

$$T(u) = \min_{1 \le i \le k_n} \{ \max_{1 \le j \le l_i} \{T_{ij}(u)\}, \ u = 1,2,...,z.$$

The reliability function of the multi-state parallel-series system is given by

$$\overline{\boldsymbol{R}}_{k_n,l_1,l_2,...,l_{k_n}}(t,\cdot) = [1, \overline{\boldsymbol{R}}_{k_n,l_1,l_2,...,l_{k_n}}(t,1), ..., \overline{\boldsymbol{R}}_{k_n,l_1,l_2,...,l_{k_n}}(t,z)],$$

and

$$\overline{\boldsymbol{R}}_{k_n,l_1,l_2,...,l_{k_n}}(t,u) = \prod_{i=1}^{k_n}[1 - \prod_{j=1}^{l_i} F_{ij}(t,u)], \ t \in (-\infty, \infty), \ u = 1,2,...,z,$$

where k_n is the number of its parallel subsystems linked in series and l_i are the numbers of components in the parallel subsystems.

Definition 3.15
A multi-state parallel-series system is called homogeneous if its lifetimes $T_{ij}(u)$ in the state subset have an identical distribution function

$$F_{ij}(t,u) = F(t,u), \ u = 1,2,...,z, \ t \in (-\infty, \infty), \ i = 1,2,...,k_n, j = 1,2,...,l_i,$$

i.e. if its components E_{ij} have the same reliability function

$$R_{ij}(t,u) = R(t,u) = 1 - F(t,u), \ u = 1,2,...,z, \ t \in (-\infty, \infty), \ i = 1,2,...,k_n, j = 1,2,...,l_i.$$

Definition 3.16
A multi-state parallel-series system is called regular if

$l_1 = l_2 = \ldots = l_{k_n} = l_n,\ l_n \in N.$

The reliability function of the multi-state homogeneous regular parallel-series system is given by

$$\overline{R}_{k_n,l_n}(t,\cdot) = [1, \overline{R}_{k_n,l_n}(t,1), \ldots, \overline{R}_{k_n,l_n}(t,z)] \tag{3.30}$$

and

$$\overline{R}_{k_n,l_n}(t,u) = [1-[F(t,u)]^{l_n}]^{k_n},\quad t\in(-\infty,\infty),\ u=1,2,\ldots,z, \tag{3.31}$$

where k_n is the number of its parallel subsystems linked in series and l_n is the number of components in the parallel subsystems.

Definition 3.17
A multi-state series system is called non-homogeneous if it is composed of a, $1 \le a \le n$, different types of components and the fraction of the ith type component in the system is equal to q_i, where $q_i > 0$, $\sum\limits_{i=1}^{a} q_i = 1$. Moreover

$$R^{(i)}(t,u) = 1- F^{(i)}(t,u),\ t\in(-\infty,\infty),\ i=1,2,\ldots,a,\ u=1,2,\ldots,z, \tag{3.32}$$

is the reliability function of the ith type component.

It can be easily proved that the reliability function of the non-homogeneous multi-state series system is given by

$$\overline{R}'_n(t,\cdot) = [1, \overline{R}'_n(t,1), \ldots, \overline{R}'_n(t,z)], \tag{3.33}$$

where

$$\overline{R}'_n(t,u) = \prod_{i=1}^{a}[R^{(i)}(t,u)]^{q_i n},\ t\in(-\infty,\infty),\ u=1,2,\ldots,z. \tag{3.34}$$

Definition 3.18
A multi-state parallel system is called non-homogeneous if it is composed of a, $1 \le a \le n$, different types of components and the fraction of the ith type component in the system is equal to q_i, where $q_i > 0$, $\sum\limits_{i=1}^{a} q_i = 1$. Moreover

$$R^{(i)}(t,u) = 1- F^{(i)}(t,u),\ t\in(-\infty,\infty),\ i=1,2,\ldots,a,\ u=1,2,\ldots,z, \tag{3.35}$$

is the reliability function of the ith type component.

It can be easily proved that the reliability function of the non-homogeneous multi-state parallel system is given by

$$\boldsymbol{R'}_n (t, \cdot) = [1, \boldsymbol{R'}_n (t, 1), ..., \boldsymbol{R'}_n (t, z)],$$ (3.36)

where

$$\boldsymbol{R'}_n (t, u) = 1 - \prod_{i=1}^{a} [F^{(i)}(t, u)]^{q_i n}, \quad t \in (-\infty, \infty), \quad u = 1, 2, ..., z.$$ (3.37)

Definition 3.19
A multi-state "*m* out of *n*" system is called non-homogeneous if it is composed of a, $1 \le a \le n$, different types of components and the fraction of the ith type component in the system is equal to q_i, where $q_i > 0$, $\sum_{i=1}^{a} q_i = 1$. Moreover

$$R^{(i)}(t, u) = 1 - F^{(i)}(t, u), \quad t \in (-\infty, \infty), \quad i = 1, 2, ..., a, \quad u = 1, 2, ..., z,$$ (3.38)

is the reliability function of the ith type component.

It can be easily proved that the reliability function of the non-homogeneous multi-state "*m* out of *n*" system is given either by

$$\boldsymbol{R'}_n^{(m)} (t, \cdot) = [1, \boldsymbol{R'}_n^{(m)} (t, 1), ..., \boldsymbol{R'}_n^{(m)} (t, z)],$$ (3.39)

where

$$\boldsymbol{R'}_n^{(m)} (t, u) = 1 - \sum_{\substack{0 \le r_i \le q_i n \\ r_1 + r_2 + ... + r_a \le m-1}} \prod_{i=1}^{a} \binom{q_i n}{r_i} [R^{(i)}(t, u)]^{r_i} [F^{(i)}(t, u)]^{q_i n - r_i}$$ (3.40)

for $t \in (-\infty, \infty)$, $u = 1, 2, ..., z$ or by

$$\overline{\boldsymbol{R}'}_n^{(\overline{m})} (t, \cdot) = [1, \overline{\boldsymbol{R}'}_n^{(\overline{m})} (t, 1), ..., \overline{\boldsymbol{R}'}_n^{(\overline{m})} (t, z)],$$ (3.41)

where

$$\overline{\boldsymbol{R}'}_n^{(\overline{m})} (t, u) = \sum_{\substack{0 \le r_i \le q_i n \\ r_1 + r_2 + ... + r_a \le \overline{m}}} \prod_{i=1}^{a} \binom{q_i n}{r_i} [F^{(i)}(t, u)]^{r_i} [R^{(i)}(t, u)]^{q_i n - r_i}, \quad \overline{m} = n - m,$$ (3.42)

for $t \in (-\infty, \infty)$, $u = 1, 2, ..., z.$

Definition 3.20
A multi-state regular series-parallel system is called non-homogeneous if it is composed of a, $1 \leq a \leq k_n$, $k_n \in N$, different types of series subsystems and the fraction of the ith type series subsystem is equal to q_i, where $q_i > 0$, $\sum_{i=1}^{a} q_i = 1$. Moreover, the ith type series subsystem consists of e_i, $1 \leq e_i \leq l_n$, $l_n \in N$, types of components with reliability functions

$$R^{(i,j)}(t,u) = 1 - F^{(i,j)}(t,u), \ t \in (-\infty, \infty), \ j = 1,2,...,e_i, \ u = 1,2,...,z,$$

and the fraction of the jth type component in this subsystem is equal to p_{ij}, where $p_{ij} > 0$ and $\sum_{j=1}^{e_i} p_{ij} = 1$.

The reliability function of the multi-state non-homogeneous regular series-parallel system is given by

$$\boldsymbol{R'}_{k_n,l_n} (t,\cdot) = [1, \boldsymbol{R'}_{k_n,l_n} (t,1) ,..., \boldsymbol{R'}_{k_n,l_n} (t,z)], \tag{3.43}$$

where

$$\boldsymbol{R'}_{k_n,l_n} (t,u) = 1 - \prod_{i=1}^{a}[1 - [R^{(i)}(t,u)]^{l_n}]^{q_i k_n} , \ t \in (-\infty, \infty), \ u = 1,2,...,z, \tag{3.44}$$

and

$$R^{(i)}(t,u) = \prod_{j=1}^{e_i}[R^{(i,j)}(t,u)]^{p_{ij}} , \ i = 1,2,...,a. \tag{3.45}$$

Definition 3.21
A multi-state regular parallel-series system is called non-homogeneous if it is composed of a, $1 \leq a \leq k_n$, $k_n \in N$, different types of parallel subsystems and the fraction of the ith type parallel subsystem is equal to q_i, where $q_i > 0$, $\sum_{i=1}^{a} q_i = 1$. Moreover, the ith type parallel subsystem consists of e_i, $1 \leq e_i \leq l_n$, $l_n \in N$, types of components with reliability functions

$$R^{(i,j)}(t,u) = 1 - F^{(i,j)}(t,u), \ t \in (-\infty, \infty), \ j = 1,2,...,e_i, \ u = 1,2,...,z,$$

and the fraction of the jth type component in this subsystem is equal to p_{ij}, where $p_{ij} > 0$ and $\sum_{j=1}^{e_i} p_{ij} = 1$.

The reliability function of the multi-state non-homogeneous regular parallel-series system is given by

$$\overline{\boldsymbol{R}}'_{k_n,l_n}(t,\cdot) = [1, \overline{\boldsymbol{R}}'_{k_n,l_n}(t,1),\dots, \overline{\boldsymbol{R}}'_{k_n,l_n}(t,z)],\tag{3.46}$$

where

$$\overline{\boldsymbol{R}}'_{k_n,l_n}(t,u) = \prod_{i=1}^{a}[1-[F^{(i)}(t,u)]^{l_n}]^{q_i k_n}, \ t \in (-\infty, \infty), \ u = 1,2,\dots,z,\tag{3.47}$$

and

$$F^{(i)}(t,u) = \prod_{j=1}^{e_i}[F^{(i,j)}(t,u)]^{p_{ij}}, \ i = 1,2,\dots,a.\tag{3.48}$$

In the asymptotic approach to multi-state system reliability analysis we are interested in the limit distributions of a standardised random variable

$$(T(u) - b_n(u))/a_n(u), \ u = 1,2,\dots,z,$$

where $T(u)$ is the lifetime of the system in the state subset $\{u, u+1,\dots, z\}$ and

$$a_n(u) > 0, \ b_n(u) \in (-\infty, \infty), \ u = 1,2,\dots,z,$$

are some suitably chosen numbers, called normalising constants. And, since

$$P((T(u) - b_n(u))/a_n(u) > t) = P(T(u) > a_n(u)t + b_n(u))$$

$$= \boldsymbol{R}_n(a_n(u)t + b_n(u),u), \ u = 1,2,\dots,z,$$

where

$$\boldsymbol{R}_n(t,\cdot) = [\boldsymbol{R}_n(t,0),\boldsymbol{R}_n(t,1),\dots,\boldsymbol{R}_n(t,z)], \ t \in (-\infty, \infty),$$

is the multi-state reliability function of the system composed of n components, then we assume the following definition.

Definition 3.22
A vector

$$\mathfrak{R}(t,\cdot) = [1,\mathfrak{R}(t,1),\dots,\mathfrak{R}(t,z)], \ t \in (-\infty, \infty),$$

is called the limit multi-state reliability function of the system with reliability function $\boldsymbol{R}_n(t,\cdot)$ if there exist normalising constants $a_n(u) > 0, b_n(u) \in (-\infty, \infty)$ such that

$$\lim_{n\to\infty} \; R_n(a_n(u)t + b_n(u),u) = \mathfrak{R}(t,u) \text{ for } t \in C_{\mathfrak{R}(u)}, \; u = 1,2,...,z,$$

where $C_{\mathfrak{R}(u)}$ is the set of continuity points of $\mathfrak{R}(t,u)$.

Knowing the system limit reliability function allows us, for sufficiently large n, to apply the following approximate formula

$$R_n(t,\cdot) \cong \mathfrak{R}(\frac{t-b_n(u)}{a_n(u)}, \cdot \,),$$

i.e.

$$[1,R_n(t,1),...,R_n(t,z)] \cong [1,\mathfrak{R}\frac{t-b_n(1)}{a_n(1)},1),...,\mathfrak{R}\frac{t-b_n(z)}{a_n(z)},z)], \; t \in (-\infty,\infty). \qquad (3.49)$$

CHAPTER 4

RELIABILITY OF LARGE TWO-STATE SYSTEMS

Auxiliary theorems on limit reliability functions of large two-state systems, which are necessary for their approximate reliability evaluation, are formulated. The classes of limit reliability functions for homogeneous and non-homogeneous series, parallel, series-parallel and parallel-series systems and for a homogeneous "m out of n" system are fixed. Applications of the asymptotic approach to reliability evaluations of model systems are presented. Six corollaries are formulated and proved on the basis of the auxiliary theorems and applied to finding limit reliability functions of the considered systems and approximate evaluations of their reliability functions, lifetime mean values and lifetime standard deviations. The reliability evaluation is done for the following systems: the model non-homogeneous series system, the homogeneous parallel system of an energetic cable, the "16 out of 35" lighting system, the homogeneous regular series-parallel gas distribution system, the non-homogeneous regular series-parallel water supply system and the model homogeneous regular parallel-series system. The accuracy of the performed evaluations is illustrated in tables and figures. The reliability data of components are assumed either to be arbitrary or to come from experts. These reliability evaluations of the considered systems' characteristics are an illustration of the possibility of applying the asymptotic approach in system reliability analysis of large real technical systems.

4.1. Reliability evaluation of two-state series systems

The investigations of limit reliability functions of homogeneous two-state series systems are based on the following auxiliary theorem.

Lemma 4.1
If

(i) $\overline{\mathscr{R}}(t) = \exp[-\overline{V}(t)]$ is a non-degenerate reliability function,

(ii) $\overline{R}_n(t)$ is the reliability function of a homogeneous two-state series system defined
     by (2.1),

(iii) $a_n > 0, \ b_n \in (-\infty, \infty),$

then

$$\lim_{n \to \infty} \overline{R}_n (a_n t + b_n) = \overline{\mathfrak{R}}(t) \ \text{for} \ t \in C_{\overline{\mathfrak{R}}} \tag{4.1}$$

if and only if

$$\lim_{n \to \infty} nF(a_n t + b_n) = \overline{V}(t) \ \text{for} \ t \in C_{\overline{V}} \tag{4.2}$$

Lemma 4.1 is an essential tool in finding limit reliability functions of two-state series
systems. Its various proofs may be found in [7], [36] and [71]. It also is the basis for
fixing the class of all possible limit reliability functions of these systems. This class is
determined by the following theorem proved in [7], [36] and [71].

Theorem 4.1
The only non-degenerate limit reliability functions of the homogeneous two-state series
system are:

$$\overline{\mathfrak{R}}_1(t) \ = \ \exp[-(-t)^{-\alpha}] \ \text{for} \ t < 0, \ \overline{\mathfrak{R}}_1(t) = 0 \ \text{for} \ t \geq 0, \ \alpha > 0, \tag{4.3}$$

$$\overline{\mathfrak{R}}_2(t) \ = 1 \ \text{for} \ t < 0, \ \overline{\mathfrak{R}}_2(t) \ = \ \exp[-t^{\alpha}] \ \text{for} \ t \geq 0, \ \alpha > 0, \tag{4.4}$$

$$\overline{\mathfrak{R}}_3(t) = \ \exp[-\exp[t]] \ \text{for} \ t \in (-\infty, \infty). \tag{4.5}$$

The next auxiliary theorem is an extension of Lemma 4.1 to non-homogeneous two-
state series systems.

Lemma 4.2
If

(i) $\overline{\mathfrak{R}}'(t) = \exp[-\overline{V}'(t)]$ is a non-degenerate reliability function,

(ii) $\overline{R}'_n(t)$ is the reliability function of a non-homogeneous two-state series system
     defined by (2.8),

(iii) $a_n > 0, \ b_n \in (-\infty, \infty),$

then

$$\lim_{n\to\infty} \overline{R}'_n(a_n t + b_n) = \overline{\mathscr{R}}'(t) \text{ for } t \in C_{\overline{\mathscr{R}}'},$$

if and only if

$$\lim_{n\to\infty} n \sum_{i=1}^{a} q_i F^{(i)}(a_n t + b_n) = \overline{V}'(t) \text{ for } t \in C_{\overline{V}'}.$$

The proof of Lemma 4.2 is given in [56] and [71]. This lemma is a particular case of Lemma 1 proved in [60]. In [60] Lemma 2 is also proved. From the latest lemma, as a particular case, it is possible to derive the next auxiliary theorem that is a more convenient tool than Lemma 4.2 for finding limit reliability functions of non-homogeneous series systems and the starting point for fixing limit reliability functions for these systems.

Lemma 4.3
If

(i) $\overline{\mathscr{R}}'(t) = \exp[-\overline{V}'(t)]$ is a non-degenerate reliability function,

(ii) $\overline{R}'_n(t)$ is the reliability function of a non-homogeneous two-state series system defined by (2.8),

(iii) $a_n > 0$, $b_n \in (-\infty, \infty)$,

(iv) $F(t)$ is one of the distribution functions $F^{(1)}(t)$, $F^{(2)}(t)$,...,$F^{(a)}(t)$ defined by (2.7) such that

(v) $\exists N \forall n > N \ F(a_n t + b_n) = 0$ for $t < t_0$ and $F(a_n t + b_n) \neq 0$ for $t \geq t_0$, where $t_0 \in <-\infty, \infty)$,

(vi) $\lim_{n\to\infty} \dfrac{F^{(i)}(a_n t + b_n)}{F(a_n t + b_n)} \leq 1$ for $t \geq t_0$, $i = 1,2,...,a$,

and moreover there exists a non-decreasing function

(vii) $\overline{d}(t) = \begin{cases} 0 & \text{for } t < t_o \\ \lim\limits_{n\to\infty} \sum\limits_{i=1}^{a} q_i \overline{d}_i(a_n t + b_n) & \text{for } t \geq t_o, \end{cases}$ (4.6)

where

(viii) $\bar{d}_i(a_n t + b_n) = \dfrac{F^{(i)}(a_n t + b_n)}{F(a_n t + b_n)}$,

then

$$\lim_{n \to \infty} \overline{R}'_n(a_n t + b_n) = \overline{\mathscr{R}}'(t) \ \text{ for } t \in C_{\overline{\mathscr{R}}'} \tag{4.7}$$

if and only if

$$\lim_{n \to \infty} n F(a_n t + b_n)\bar{d}(t) = \overline{V}'(t) \ \text{ for } t \in C_{\overline{V}'}. \tag{4.8}$$

On the basis of Theorem 4.1 and Lemma 4.3 in [56] and [71] the class of limit reliability functions for non-homogeneous two-state series systems has been fixed. The members of this class are specified in the following theorem ([71]).

Theorem 4.2
The only non-degenerate limit reliability functions of the non-homogeneous two-state series system, under the assumptions of Lemma 4.3, are:

$$\overline{\mathscr{R}}'_1(t) = \exp[-\bar{d}(t)(-t)^{-\alpha}] \text{ for } t < 0, \ \overline{\mathscr{R}}'_1(t) = 0 \text{ for } t \geq 0, \ \alpha > 0, \tag{4.9}$$

$$\overline{\mathscr{R}}'_2(t) = 1 \text{ for } t < 0, \ \overline{\mathscr{R}}'_2(t) = \exp[-\bar{d}(t)t^{\alpha}] \text{ for } t \geq 0, \ \alpha > 0, \tag{4.10}$$

$$\overline{\mathscr{R}}'_3(t) = \exp[-\bar{d}(t)\exp[t]] \text{ for } t \in (-\infty, \infty), \tag{4.11}$$

where $\bar{d}(t)$ is a non-decreasing function dependent on the reliability functions of particular system components and their fractions in the system defined by (4.6).

The above theorem is a particular case of Theorem 2 proved in [60].

Corollary 4.1
If the *i*th type components of the non-homogeneous two-state series system have Weibull reliability functions

$$R^{(i)}(t) = 1 \text{ for } t < 0, \ R^{(i)}(t) = \exp[-\beta_i t^{\alpha_i}] \text{ for } t \geq 0, \ \alpha_i > 0, \ \beta_i > 0, \ i = 1,2,...,a, \tag{4.12}$$

and

$$a_n = 1/(\beta n)^{1/\alpha}, \ b_n = 0, \tag{4.13}$$

where

$$\alpha = \min_{1 \le i \le a}\{\alpha_i\}, \quad \beta = \max\{\beta_i : \alpha_i = \alpha\}, \tag{4.14}$$

then

$$\overline{\mathscr{R}}'_2 (t) = 1 \text{ for } t < 0, \quad \overline{\mathscr{R}}'_2 (t) = 1 - \exp[-\overline{d}(t)t^\alpha] \text{ for } t \ge 0,$$

where

$$\overline{d}(t) = \sum_{(i:\alpha_i=\alpha)} q_i\beta_i / \beta, \tag{4.15}$$

is its limit reliability function.

Motivation: Since, from (4.13), we have

$$a_n t + b_n = (\beta l_n)^{-1/\alpha} t \to 0^- \text{ for } t < 0$$

and

$$a_n t + b_n = (\beta l_n)^{-1/\alpha} t \to 0^+ \text{ for } t \ge 0 \text{ as } n \to \infty,$$

then according to (4.12) for each $i = 1,2,...,a$, we get

$$R^{(i)}(a_n t + b_n) = 1 \text{ for } t < 0$$

and

$$R^{(i)}(a_n t + b_n) = \exp[-\beta_i(a_n t)^{\alpha_i}] \text{ for } t \ge 0.$$

Assuming

$$R(t) = 1 \text{ for } t < 0 \text{ and } R(t) = \exp[-\beta t^\alpha] \text{ for } t \ge 0, \tag{4.16}$$

where α and β are defined by (4.14), for all $i = 1,2,...,a$ and $t \ge t_0 = 0$, we obtain

$$\lim_{n\to\infty} \frac{F^{(i)}(a_n t + b_n)}{F(a_n t + b_n)} = \lim_{n\to\infty} \frac{1 - \exp[-\beta_i(a_n t)^{\alpha_i}]}{1 - \exp[-\beta(a_n t)^\alpha]}$$

$$= \lim_{n\to\infty} \frac{\beta_i}{\beta}(a_n t)^{\alpha_i-\alpha} \le 1.$$

The above means that condition (vii) of Lemma 4.3 holds with $t_0 = 0$. And, due to condition (viii) of Lemma 4.3

$$\lim_{n \to \infty} d_i(a_n t + b_n) = \beta_i / \beta$$

for i such that $\alpha_i = \alpha$ and

$$\lim_{n \to \infty} d_i(a_n t + b_n) = 0$$

otherwise.

Thus, from (4.6), $\overline{d}(t)$ is given by (4.15) for $t \geq 0$ and equal to 0 for $t < 0$. Moreover, according to (4.16), we get

$$\overline{V}'(t) = \lim_{n \to \infty} nF(a_n t + b_n)\overline{d}(t) = 0 \text{ for } t < 0$$

and

$$\overline{V}'(t) = \lim_{n \to \infty} nF(a_n t + b_n)\overline{d}(t)$$

$$= \lim_{n \to \infty} n[1 - \exp[-\beta(a_n t)^\alpha]]\overline{d}(t)$$

$$= \lim_{n \to \infty} n[\beta(a_n t)^\alpha - o(\frac{1}{n})]\overline{d}(t)$$

$$= t^\alpha \overline{d}(t) \text{ for } t \geq 0,$$

which from Lemma 4.3 completes the proof. □

Example 4.1 (*a gas piping system*)
Let us consider a gas piping line composed of $n = 100$ pipe segments of four types linked in series. In the system there are:
40 segments with reliability functions

$$R^{(1)}(t) = \exp[-0.025t] \text{ for } t \geq 0,$$

20 segments with reliability functions

$$R^{(2)}(t) = \exp[-0.020t] \text{ for } t \geq 0,$$

10 segments with reliability functions

$$R^{(3)}(t) = \exp[-0.0015t^2] \text{ for } t \geq 0,$$

and 30 segments with reliability functions

$R^{(4)}(t) = \exp[-0.001t^2]$ for $t \geq 0$.

According to Definition 2.13 the gas piping is a non-homogeneous series system with parameters

$n = 100$, $a = 4$,

$q_1 = 0.4$, $q_2 = 0.2$, $q_3 = 0.1$, $q_4 = 0.3$

and from (2.8) its exact reliability function is given by

$$\overline{R}'_{100}(t) = \prod_{i=1}^{4}[R^{(i)}(t)]^{q_i 100}$$

$$= \exp[-t - 0.4t - 0.015t^2 - 0.03t^2] \text{ for } t \geq 0.$$

Since

$$\alpha_1 = 1, \beta_1 = 0.025, \alpha_2 = 1, \beta_2 = 0.02, \alpha_3 = 2, \beta_3 = 0.0015, \alpha_4 = 2, \beta_4 = 0.001,$$

then

$$\alpha = \min\{1,1,2,2\} = 1, \quad \beta = \max\{0.025, 0.02\} = 0.025.$$

Assuming normalising constants

$a_n = 1/\beta n = 0.4$, $b_n = 0$

and according to (4.15), after determining

$$d(t) = \sum_{(i:\alpha_i = \alpha)} q_i \frac{\beta_i}{\beta} = 0.56,$$

from Corollary 4.1 it follows that the gas piping system limit reliability function is

$$\overline{\mathfrak{R}}'_2(t) = \exp[-0.56t] \text{ for } t \geq 0.$$

Hence, according to (1.1), the exact reliability of the considered system may be approximated by the formula

$$\overline{R}'_{100}(t) \cong \exp[-0.56(t/0.4)]$$

$$= \exp[-1.4t] \text{ for } t > 0. \tag{4.17}$$

The mean values of particular system component lifetimes in years are as follows:

$E(T_1) = 1/0.025 = 40$, $E(T_2) = 1/0.020 = 50$,

$E(T_3) = \Gamma(3/2)(0.0015)^{-1/2} \cong 23$, $E(T_4) = \Gamma(3/2)(0.001)^{-1/2} \cong 28$.

The approximate mean value of the gas piping lifetime and its standard deviation calculated on the base of the formula (4.17), are:

$E(T) \cong 1/1.4 \cong 0.71$ years, $\sigma(T) \cong 1/1.4 \cong 0.71$ years.

The behaviour of the exact and approximate reliability functions of the gas piping is illustrated in Table 4.1 and Figure 4.1. Moreover, in Table 4.1, the differences between the values of these functions are also given. These differences testify that the approximation of the system's exact reliability function by its limit reliability function is good enough.

Table 4.1. The values and differences between the exact and limit reliability functions of the gas piping system

t	$\overline{R}'_{100}(t)$	$\overline{\mathfrak{R}}'_2((t-b_n)/a_n)$	$\Delta = \overline{R}'_{100} - \overline{\mathfrak{R}}'_2$
0.00	1.000	1.000	0.000
0.10	0.869	0.869	0.000
0.20	0.754	0.756	−0.001
0.30	0.654	0.657	−0.003
0.40	0.567	0.571	−0.004
0.50	0.491	0.497	−0.006
0.60	0.425	0.432	−0.007
0.70	0.367	0.375	−0.008
0.80	0.317	0.326	−0.009
0.90	0.274	0.284	−0.010
1.00	0.236	0.247	−0.011
1.20	0.175	0.186	**−0.012**
1.40	0.129	0.141	**−0.012**
1.60	0.095	0.106	**−0.012**
1.80	0.070	0.080	−0.011

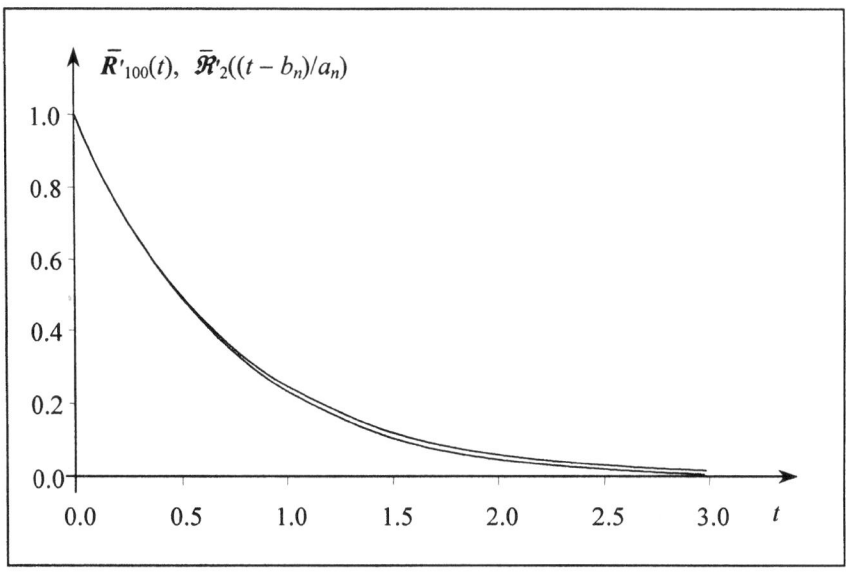

Fig. 4.1. The graphs of the exact and limit reliability functions of the gas piping system

4.2. Reliability evaluation of two-state parallel systems

The class of limit reliability functions for homogeneous two-state parallel systems may be determined on the basis of the following auxiliary theorem proved for instance in [7], [36] and [71].

Lemma 4.4

If $\overline{\mathfrak{R}}(t)$ is the limit reliability function of a homogeneous two-state series system with reliability functions of particular components $\overline{R}(t)$, then

$$\mathfrak{R}(t) = 1 - \overline{\mathfrak{R}}(-t) \text{ for } t \in C_{\overline{\mathfrak{R}}}$$

is the limit reliability function of a homogeneous two-state parallel system with reliability functions of particular components

$$R(t) = 1 - \overline{R}(-t) \text{ for } t \in C_{\overline{R}}.$$

At the same time, if (a_n, b_n) is a pair of normalising constants in the first case, then $(a_n, -b_n)$ is such a pair in the second case.

Applying the above lemma it is possible to prove an equivalent of Lemma 4.1 that allows us to justify facts on limit reliability functions for homogeneous parallel systems. Its form is as follows ([7], [36], [71]).

Lemma 4.5
If

(i) $\mathfrak{R}(t) = 1 - \exp[-V(t)]$ is a non-degenerate reliability function,

(ii) $R_n(t)$ is the reliability function of a homogeneous two-state parallel system defined by (2.2),

(iii) $a_n > 0,\ b_n \in (-\infty, \infty)$,

then

$$\lim_{n \to \infty} R_n(a_n t + b_n) = \mathfrak{R}(t) \text{ for } t \in C_{\mathfrak{R}}, \tag{4.18}$$

if and only if

$$\lim_{n \to \infty} nR(a_n t + b_n) = V(t) \text{ for } t \in C_V. \tag{4.19}$$

By applying Lemma 4.5 and proceeding in an analogous way to the case of homogeneous series systems it is possible to fix the class of limit reliability functions for homogeneous two-state parallel systems. However, it is easier to obtain this result using Lemma 4.4 and Theorem 4.1. Their application immediately results in the following issue.

Theorem 4.3
The only non-degenerate limit reliability functions of the homogeneous parallel system are:

$$\mathfrak{R}_1(t) = 1 \text{ for } t \leq 0,\ \mathfrak{R}_1(t) = 1 - \exp[-t^{-\alpha}] \text{ for } t > 0,\ \alpha > 0, \tag{4.20}$$

$$\mathfrak{R}_2(t) = 1 - \exp[-(-t)^{\alpha}] \text{ for } t < 0,\ \mathfrak{R}_2(t) = 0 \text{ for } t \geq 0,\ \alpha > 0, \tag{4.21}$$

$$\mathfrak{R}_3(t) = 1 - \exp[-\exp[-t]] \text{ for } t \in (-\infty, \infty). \tag{4.22}$$

Corollary 4.2
If components of the homogeneous two-state parallel system have Weibull reliability functions

$$R(t) = 1 \text{ for } t < 0,\ R(t) = \exp[-\beta t^{\alpha}] \text{ for } t \geq 0,\ \alpha > 0,\ \beta > 0,$$

and

$$a_n = b_n/(\alpha\log n), \; b_n = (\log n/\beta)^{1/\alpha},$$

then

$$\mathcal{R}_3(t) = 1 - \exp[-\exp[-t]], \; t \in (-\infty,\infty),$$

is its limit reliability function.

Motivation: Since for sufficiently large n and all $t \in (-\infty,\infty)$ we have

$$a_n t + b_n = b_n(t/(\alpha\log n) + 1) > 0,$$

then

$$R(a_n t + b_n) = \exp[-\beta(a_n t + b_n)^\alpha] \text{ for } t \in (-\infty,\infty).$$

Hence

$$n \, R(a_n t + b_n) = n \, \exp[-\beta(a_n t + b_n)^\alpha]$$

$$= n \, \exp[-\beta(b_n)^\alpha(t/(\alpha\log n) + 1)^\alpha]$$

$$= n \, \exp[-\log n(t/(\alpha\log n) + 1)^\alpha].$$

Further, applying the equality

$$(t/(\alpha\log n) +1)^\alpha = 1 + t/\log n + o(1/\log n) \text{ for } t \in (-\infty,\infty),$$

we obtain

$$V(t) = \lim_{n\to\infty} n \, R \, (a_n t + b_n)$$

$$= \lim_{n\to\infty} n \, \exp[-\log n - t - o(1)]$$

$$= \lim_{n\to\infty} \exp[-t - o(1)] = \exp[-t] \text{ for } t \in (-\infty,\infty),$$

which from Lemma 4.5 completes the proof.

Example 4.2 (*an energetic cable*)

Let us consider an energetic cable composed of 36 wires of the type *A1Si* used in overhead energetic nets and assume that it is able to conduct the current if at least one of its wires is not failed. Under this assumption we may consider the cable as a homogeneous parallel system composed of $n = 36$ basic components. The cross-section of the considered cable is presented in Figure 4.2.

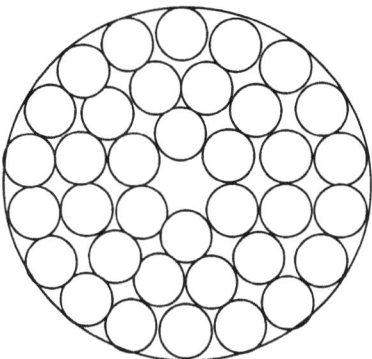

Fig. 4.2. The cross-section of the energetic cable

Further, assuming that the cable wires have Weibull reliability functions with parameters

$$\alpha = 2, \; \beta = (7.07)^{-6},$$

by (2.2), the cable's exact reliability function takes the form

$$\boldsymbol{R}_{36}(t) = 1 \text{ for } t < 0, \; \boldsymbol{R}_{36}(t) = 1 - [1 - \exp[-(7.07)^{-6}t^2]]^{36} \text{ for } t \geq 0.$$

Thus, according to Corollary 4.2, assuming

$$a_n = (7.07)^3/(2\sqrt{\log 36}), \; b_n = (7.07)^3 \sqrt{\log 36},$$

and applying (1.1), we arrive at the approximate formula for the cable reliability function of the form

$$\boldsymbol{R}_{36}(t) \cong \mathfrak{R}_3((t - b_n)/a_n) = 1 - \exp[-\exp[-0.01071t + 7.167]] \text{ for } t \in (-\infty,\infty).$$

The expected value of the cable lifetime T and its standard deviation, in months, calculated on the basis of the above approximate result and according to the formulae ([13])

$$E[T] = Ca_n + b_n, \; \sigma = \pi a_n / \sqrt{6},$$

where $C \cong 0.5772$ is Euler's constant, respectively are:

$$E[T] \cong 723, \; \sigma \cong 120.$$

The values of the exact and approximate reliability functions of the cable are presented in Table 4.2 and Figure 4.3. Moreover, Table 4.2 also shows the differences between those values. The differences are not large, which means that the mistakes in replacing the exact cable reliability function by its approximate form are practically not significant.

Table 4.2. The values of the exact and approximate reliability functions of the energetic cable

t	$R_{36}(t)$	$\mathcal{R}_3\left(\dfrac{t-b_n}{a_n}\right)$	$\Delta = R_{36} - \mathcal{R}_3$
0	1.000	1.000	0.000
400	1.000	1.000	0.000
500	0.995	0.988	−0.003
550	0.965	0.972	−0.007
600	0.874	0.877	−0.003
650	0.712	0.707	0.005
700	0.513	0.513	0.000
750	0.330	0.344	−0.014
800	0.193	0.218	−0.025
900	0.053	0.081	**−0.028**
1000	0.012	0.029	−0.017
1100	0.002	0.010	−0.008
1200	0.000	0.003	−0.003

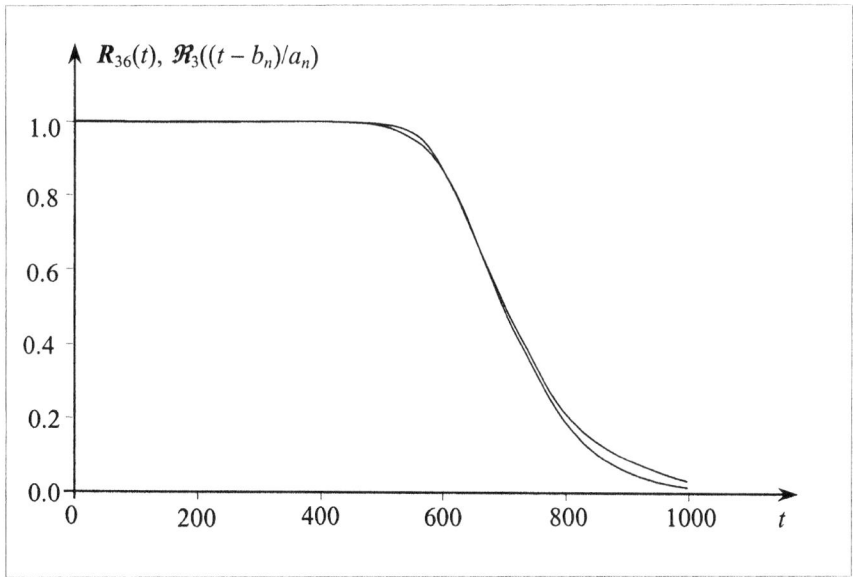

Fig. 4.3. The graphs of the exact and approximate reliability functions of the energetic cable

The next lemma is a slight modification of Lemma 4.5 proved in [56] and [71]. It is also a particular case of Lemma 2, which is proved in [63].

Lemma 4.6

If $\overline{\mathscr{R}}'(t)$ is the limit reliability function of a non-homogeneous two-state series system with reliability functions of particular components

$$\overline{R}^{(i)}(t), \; i = 1,2,...,a,$$

then

$$\mathscr{R}'(t) = 1 - \overline{\mathscr{R}}'(-t) \; \text{ for } t \in C_{\overline{\mathscr{R}}'},$$

is the limit reliability function of a non-homogeneous two-state parallel system with reliability functions of particular components

$$R^{(i)}(t) = 1 - \overline{R}^{(i)}(-t) \text{ for } t \in C_{\overline{R}^{(i)}}, \; i = 1,2,...,a.$$

At the same time, if (a_n, b_n) is a pair of normalising constants in the first case, then $(a_n, -b_n)$ is such a pair in the second case.

Applying the above lemma and Theorem 4.2 it is possible to arrive at the next result ([56], [63], [71]).

Lemma 4.7
If

(i) $\mathscr{R}'(t) = 1 - \exp[-V'(t)]$ is a non-degenerate reliability function,

(ii) $R'_n(t)$ is the reliability function of a non-homogeneous two-state parallel system defined by (2.10),

(iii) $a_n > 0$, $b_n \in (-\infty,\infty)$,

then

$$\lim_{n \to \infty} R'_n(a_n t + b_n) = \mathscr{R}'(t) \; \text{ for } t \in C_{\mathscr{R}'}$$

if and only if

$$\lim_{n \to \infty} n \sum_{i=1}^{a} q_i R^{(i)}(a_n t + b_n) = V'(t) \text{ for } t \in C_{V'}.$$

The next lemma motivated in [56] and [71] that is useful in practical applications is a particular case of Lemma 3 proved in [63].

Lemma 4.8
If

(i) $\mathscr{R}'(t) = 1 - \exp[-V'(t)]$ is a non-degenerate reliability function,

(ii) $R'_n(t)$ is the reliability function of a non-homogeneous two-state parallel system defined by (2.10),

(iii) $a_n > 0$, $b_n \in (-\infty,\infty)$,

(iv) $R(t)$ is one of the reliability functions $R^{(1)}(t)$, $R^{(2)}(t),...,R^{(a)}(t)$ defined by (2.9) such that

(v) $\exists\, N \,\forall\, n > N\ R(a_n t + b_n) \neq 0$ for $t < t_0$ and $R(a_n t + b_n) = 0$ for $t \geq t_0$, where $t_0 \in (-\infty,\infty>$,

(vi) $\lim\limits_{n\to\infty} \dfrac{R^{(i)}(a_n t + b_n)}{R(a_n t + b_n)} \leq 1$ for $t < t_0$, $i = 1,2,..., a$,

and moreover there exists a non-increasing function

(vii) $d(t) = \begin{cases} \lim\limits_{n\to\infty} \sum\limits_{i=1}^{a} q_i d_i (a_n t + b_n) & for\ t < t_0 \\[2mm] 0 & for\ t \geq t_0, \end{cases}$ (4.23)

where

(viii) $d_i(a_n t + b_n) = \dfrac{R^{(i)}(a_n t + b_n)}{R(a_n t + b_n)}$,

then

$$\lim_{n\to\infty} R'_n(a_n t + b_n) = \mathscr{R}'(t)\ \text{ for } t \in C_{\mathscr{R}'}$$ (4.24)

if and only if

$$\lim_{n\to\infty} nR(a_n t + b_n)d(t) - V'(t)\ \text{ for } t \in C_{V'}.$$ (4.25)

Starting from this lemma it is possible to fix the class of possible limit reliability for non-homogeneous two-state parallel systems ([56], [63], [71]).

Theorem 4.4
The only non-degenerate limit reliability functions of the non-homogeneous two-state parallel system, under the assumptions of Lemma 4.8, are:

$$\mathcal{R}'_1(t) = 1 \text{ for } t \leq 0, \ \mathcal{R}'_1(t) = 1 - \exp[-d(t)t^{-\alpha}] \text{ for } t > 0, \ \alpha > 0, \tag{4.26}$$

$$\mathcal{R}'_2(t) = 1 - \exp[-d(t)(-t)^{\alpha}] \text{ for } t < 0, \ \mathcal{R}'_2(t) = 0 \text{ for } t \geq 0, \ \alpha > 0, \tag{4.27}$$

$$\mathcal{R}'_3(t) = 1 - \exp[-d(t)\exp[-t]] \text{ for } t \in (-\infty, \infty), \tag{4.28}$$

where $d(t)$ is a non-increasing function dependent on the reliability functions of particular system components and their fractions in the system defined by (4.23).

Theorem 4.4 is a particular case of Theorem 1 proved in [63].

4.3. Reliability evaluation of two-state "*m* out of *n*" systems

The class of limit reliability function for homogeneous two-state "*m* out of *n*" systems may be established by applying the three following auxiliary theorems proved in [108] and [71].

Lemma 4.9
If

(i) $m = \text{constant } (m/n \to 0 \text{ as } n \to \infty)$,

(ii) $\mathcal{R}^{(0)}(t) = 1 - \sum_{i=0}^{m-1} \frac{[V(t)]^i}{i!} \exp[-V(t)]$ is a non-degenerate reliability function,

(iii) $R_n^{(m)}(t)$ is the reliability function of a homogeneous two-state "*m* out of *n*" system defined by (2.3),

(iv) $a_n > 0, \ b_n \in (-\infty, \infty)$,

then

$$\lim_{n \to \infty} R_n^{(m)}(a_n t + b_n) = \mathcal{R}^{(0)}(t) \text{ for } t \in C_{\mathcal{R}^{(0)}} \tag{4.29}$$

if and only if

$$\lim_{n\to\infty} nR(a_nt + b_n) = V(t) \text{ for } t \in C_V. \tag{4.30}$$

Lemma 4.10
If

(i) $m/n \to \mu$, $0 < \mu < 1$, as $n \to \infty$,

(ii) $\mathcal{R}^{(\mu)}(t) = 1 - \dfrac{1}{\sqrt{2\pi}} \int_{-\infty}^{-v(t)} e^{-\frac{x^2}{2}} dx$ is a non-degenerate reliability function,

(iii) $R_n^{(m)}(t)$ is the reliability function of a homogeneous two-state "m out of n" system defined by (2.3),

(iv) $a_n > 0$, $b_n \in (-\infty,\infty)$,

then

$$\lim_{n\to\infty} R_n^{(m)}(a_nt + b_n) = \mathcal{R}^{(\mu)}(t) \text{ for } t \in C_{\mathcal{R}^{(\mu)}} \tag{4.31}$$

if and only if

$$\lim_{n\to\infty} \frac{(n+1)R(a_nt + b_n) - m}{\sqrt{\dfrac{m(n-m+1)}{n+1}}} = v(t) \text{ for } t \in C_v. \tag{4.32}$$

Lemma 4.11
If

(i) $n - m = \overline{m} = \text{constant}$ $(m/n \to 1 \text{ as } n \to \infty)$,

(ii) $\overline{\mathcal{R}}^{(1)}(t) = \sum_{i=0}^{\overline{m}} \dfrac{[\overline{V}(t)]^i}{i!} \exp[-\overline{V}(t)]$ is a non-degenerate reliability function,

(iii) $\overline{R}_n^{(\overline{m})}(t)$ is the reliability function of a homogeneous two-state "m out of n" system defined by (2.4),

(iv) $a_n > 0$, $b_n \in (-\infty,\infty)$,

then

$$\lim_{n\to\infty} \overline{R}_n^{(\overline{m})}(a_nt + b_n) = \overline{\mathcal{R}}^{(1)}(t) \text{ for } t \in C_{\overline{\mathcal{R}}^{(1)}}, \tag{4.33}$$

if and only if

$$\lim_{n\to\infty} nF(a_n t + b_n) = \overline{V}(t) \text{ for } t \in C_{\overline{V}}. \tag{4.34}$$

The applications of Lemmas 4.9–4.11 allow us to establish the class of possible limit reliability functions for homogeneous two-state "*m* out of *n*" systems pointed out in the following theorem ([71], [108]).

Theorem 4.5
The only non-degenerate limit reliability functions of the homogeneous two-state "*m* out of *n*" system are:

Case 1. $m = \text{constant} \ (m/n \to 0 \text{ as } n \to \infty)$.

$$\mathcal{R}_1^{(0)}(t) = 1 \text{ for } t \le 0, \ \mathcal{R}_1^{(0)}(t) = 1 - \sum_{i=0}^{m-1} \frac{t^{-i\alpha}}{i!} \exp[-t^{-\alpha}] \text{ for } t > 0, \ \alpha > 0, \tag{4.35}$$

$$\mathcal{R}_2^{(0)}(t) = 1 - \sum_{i=0}^{m-1} \frac{(-t)^{i\alpha}}{i!} \exp[-(-t)^{\alpha}] \text{ for } t < 0, \ \mathcal{R}_2^{(0)}(t) = 0 \text{ for } t \ge 0, \ \alpha > 0, \tag{4.36}$$

$$\mathcal{R}_3^{(0)}(t) = 1 - \sum_{i=0}^{m-1} \frac{\exp[-it]}{i!} \exp[-\exp[-t]] \text{ for } t \in (-\infty, \infty). \tag{4.37}$$

Case 2. $m/n = \mu + o(1/\sqrt{n}), \ 0 < \mu < 1, \ (m/n \to \mu \text{ as } n \to \infty)$.

$$\mathcal{R}_4^{(\mu)}(t) = 1 \text{ for } t < 0, \ \mathcal{R}_4^{(\mu)}(t) = 1 - \frac{1}{\sqrt{2\pi}} \int_{-\infty}^{ct^\alpha} e^{-\frac{x^2}{2}} dx \text{ for } t \ge 0, \ c > 0, \alpha > 0, \tag{4.38}$$

$$\mathcal{R}_5^{(\mu)}(t) = 1 - \frac{1}{\sqrt{2\pi}} \int_{-\infty}^{-c|t|^\alpha} e^{-\frac{x^2}{2}} dx \text{ for } t < 0, \ \mathcal{R}_5^{(\mu)}(t) = 0 \text{ for } t \ge 0, \ c > 0, \alpha > 0, \tag{4.39}$$

$$\mathcal{R}_6^{(\mu)}(t) = 1 - \frac{1}{\sqrt{2\pi}} \int_{-\infty}^{-c_1|t|^\alpha} e^{-\frac{x^2}{2}} dx \text{ for } t < 0, \ c_1 > 0, \alpha > 0, \tag{4.40}$$

$$\mathcal{R}_6^{(\mu)}(t) = \frac{1}{2} - \frac{1}{\sqrt{2\pi}} \int_{0}^{c_2 t^\alpha} e^{-\frac{x^2}{2}} dx \text{ for } t \ge 0, \ c_2 > 0, \alpha > 0, \tag{4.41}$$

$$\mathcal{R}_7^{(\mu)}(t) = 1 \text{ for } t < -1, \ \mathcal{R}_7^{(\mu)}(t) = \frac{1}{2} \text{ for } -1 \le t < 1, \ \mathcal{R}_7^{(\mu)}(t) = 0 \text{ for } t \ge 0. \tag{4.42}$$

Case 3. $n - m = \overline{m} = $ constant $(m/n \rightarrow 1$ as $n \rightarrow \infty)$.

$$\overline{\mathcal{R}}_8^{(1)}(t) = \sum_{i=0}^{\overline{m}} \frac{(-t)^{-i\alpha}}{i!} \exp[-(-t)^{-\alpha}] \text{ for } t < 0, \overline{\mathcal{R}}_8^{(1)}(t) = 0 \text{ for } t \geq 0, \alpha > 0, \quad (4.43)$$

$$\overline{\mathcal{R}}_9^{(1)}(t) = 1 \text{ for } t < 0, \overline{\mathcal{R}}_9^{(1)}(t) = \sum_{i=0}^{\overline{m}} \frac{t^{i\alpha}}{i!} \exp[-t^{\alpha}] \text{ for } t \geq 0, \alpha > 0, \quad (4.44)$$

$$\overline{\mathcal{R}}_9^{(1)}(t) = \sum_{i=0}^{\overline{m}} \frac{\exp[it]}{i!} \exp[-\exp[t]] \text{ for } t \in (-\infty,\infty). \quad (4.45)$$

Corollary 4.3
If components of the homogeneous two-state "*m* out of *n*" system have exponential reliability functions

$$R(t) = 1 \text{ for } t < 0, R(t) = \exp[-\lambda t] \text{ for } t \geq 0, \lambda > 0,$$

m tends to infinity in such a way that

$$m/n \rightarrow \mu, \ 0 < \mu < 1, \text{ as } n \rightarrow \infty,$$

and

$$a_n = \frac{1}{\lambda} \sqrt{\frac{n-m+1}{(n+1)m}}, b_n = \frac{1}{\lambda} \log \frac{n+1}{m},$$

then

$$\mathcal{R}_6^{(\mu)}(t) = 1 - \frac{1}{\sqrt{2\pi}} \int_{-\infty}^{t} e^{-\frac{x^2}{2}} dx, t \in (-\infty,\infty),$$

is its limit reliability function.
Motivation: Since for sufficiently large *n* and all $t \in (-\infty,\infty)$ we have

$$a_n t + b_n = \frac{t}{\lambda} \sqrt{\frac{n-m+1}{(n+1)m}} + \frac{1}{\lambda} \log \frac{n+1}{m} > 0,$$

then for sufficiently large *n*

$$R(a_n t + b_n) = \exp[-\lambda (a_n t + b_n)]$$

$$= \exp[- t \sqrt{\frac{n-m+1}{(n+1)m}} - \log\frac{n+1}{m}]$$

$$= [1 - t\sqrt{\frac{n-m+1}{(n+1)m}} + o(\frac{1}{\sqrt{n}})]\frac{m}{n+1} \quad \text{for } t \in (-\infty,\infty).$$

Hence

$$v(t) = \lim_{n\to\infty} \frac{(n+1)R(a_n t+b_n)-m}{\sqrt{\dfrac{m(n-m+1)}{n+1}}}$$

$$= \lim_{n\to\infty} [-t+o(\frac{1}{\sqrt{n}})\sqrt{\frac{(n+1)(n-m+1)}{m}}]$$

$$= -t \text{ for } t \in (-\infty,\infty),$$

which from Lemma 4.10 completes the proof.

Example 4.3 (*a lighting system*)
We consider a lighting system composed of $n = 35$ identical lighting points that is not failed if at least $m = 16$ of the points are not failed. Assuming that the lighting points have exponential reliability functions with the failure rate $\lambda = 1/\text{year}$, from (2.3), the exact system reliability function is given by

$$\boldsymbol{R}_{35}^{(16)}(t) = 1 \text{ for } t < 0, \ \boldsymbol{R}_{35}^{(16)}(t) = 1 - \sum_{i=0}^{15}\binom{35}{i}\exp[-it][1-\exp[-t]]^{35-i} \text{ for } t \geq 0.$$

According to Corollary 4.3, assuming

$$a_n = \frac{\sqrt{5}}{12} \cong 0.1863, \ b_n = \log\frac{9}{4} \cong 0.8109,$$

after applying (1.1), we get the approximate formula for the reliability function of the lighting system in the form

$$\boldsymbol{R}_{35}^{(16)}(t) \cong \mathfrak{R}_6^{(\mu)}((t-b_n)/a_n)$$

$$= 1 - \frac{1}{\sqrt{2\pi}} \int_{-\infty}^{5.37t-4.35} e^{-\frac{x^2}{2}} dx \text{ for } t \in (-\infty,\infty).$$

The mean value of the lighting system lifetime T and its standard deviation, in years, calculated from the above formula are:

$E[T] \cong 0.811$, $\sigma \cong 0.186$.

The values of the exact and approximate reliability functions of the lighting system are presented in Table 4.3 and their graphs in Figure 4.4. The differences between the exact and approximate reliability functions of the system given in Table 4.3 testify good accuracy of the approximation.

Unfortunately, there are no generalisations of Lemmas 4.9–4.11 for the non-homogeneous two-state "m out of n" systems. Each particular case of a non-homogeneous two-state "m out of n" system has to be considered separately and a suitable auxiliary theorem and corollary have to be formulated and proved and then applied to the reliability evaluation of a real system.

Table 4.3. The values of the exact and approximate reliability function of the lighting system

t	$\mathcal{R}_6^{(\mu)}(\dfrac{t-b_n}{a_n})$	$R_{35}^{(16)}(t)$	$\Delta = R_{35}^{(16)} - \mathcal{R}_6^{(\mu)}$
0.0	0.99999	1.00000	0.00001
0.1	0.99994	1.00000	0.00006
0.2	0.99948	1.00000	0.00052
0.3	0.99695	0.99980	0.00285
0.4	0.98629	0.99507	0.00878
0.5	0.95241	0.96253	0.01012
0.6	0.87118	0.86178	−0.00940
0.7	0.72419	0.68205	−0.04214
0.8	0.52339	0.46713	**−0.05626**
0.9	0.31633	0.27703	−0.03929
1.0	0.15513	0.14394	−0.01119
1.1	0.06041	0.06653	0.00612
1.2	0.01840	0.02777	0.00937
1.3	0.00433	0.01062	0.00628
1.4	0.00078	0.00376	0.00298

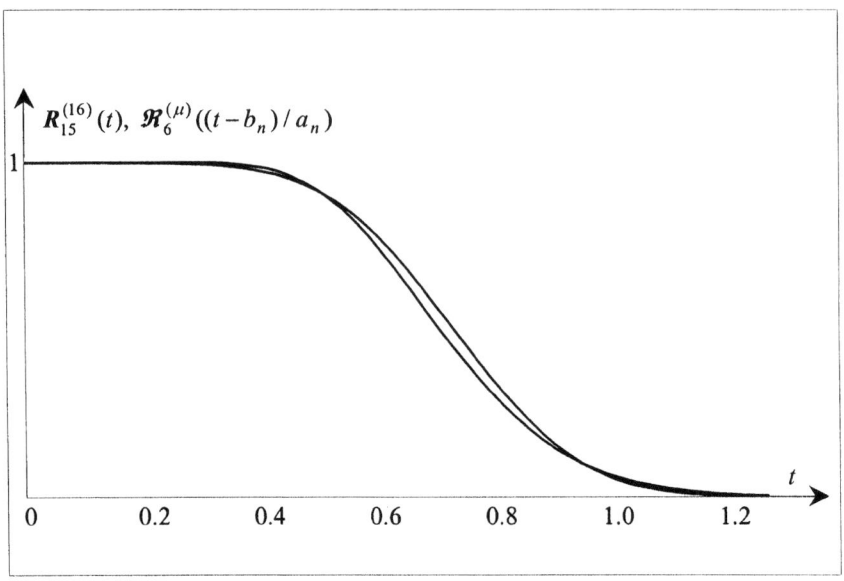

Fig. 4.4. The graphs of the exact and approximate reliability function of the lighting system

4.4. Reliability evaluation of two-state series-parallel systems

Prior to the formulation of the overall results for the classes of limit reliability functions for two-state regular series-parallel systems we introduce some assumptions for all cases of the considered systems shapes. These assumptions distinguish all possible relationships between the number of their series subsystems k_n and the number of components l_n in these subsystems. In the assumptions for two-state regular parallel-series systems, considered in the next section, k_n is the number of parallel subsystems and l_n is the number of components in these subsystems.

Assumption 4.1
Here are considered the relationships between k_n and l_n of the form

$$k_n = n, \; l_n = c(\log n)^{\rho(n)}, n \in (0,\infty), \; c > 0, \tag{4.46}$$

with the following cases distinguished:

Case 1. $k_n = n$, $\left| l_n - c \log n \right| \gg s,\; s > 0,\; c > 0$.

1^0 $l_n \ll c \log n$,

$$\left| \rho(\tau_v) - \rho(n) \right| \underset{\approx}{\ll} \frac{\delta \cdot \log v}{\log n \cdot [\log(\log n)]} \quad \text{for all natural } v > 1,$$

where $0 < \delta \ne 1$ and $\dfrac{\tau_v}{n} = v^{\frac{1}{1-\rho(n)}}$,

2^0 $l_n \approx c \log n$ and $\left| l_n - c \log n \right| \gg \log(\log n)$,

$$\left| \rho(\tau_v) - \rho(n) \right| \underset{\approx}{\ll} \frac{\delta \cdot \log v}{\log n \cdot [\log(\log n)]} \quad \text{for all natural } v > 1,$$

where $0 < \delta \ne 1$ and $\dfrac{\tau_v}{n} = v^{\frac{1}{1-\rho(n)}}$,

3^0 $s \ll \left| l_n - c \log n \right| \underset{\approx}{\ll} C \log(\log n),\; s > 0,\; C > 0$,

$$\left| \rho(\tau_v) - \rho(n) \right| \approx \frac{\delta \cdot \log v}{\log n \cdot [\log(\log n)]} \quad \text{for all natural } v > 1,$$

where $\delta > 0$ and $\dfrac{\tau_v}{n} = v^{\frac{1}{(1-\rho(n)) \cdot \log(\log n)}}$,

4^0 $l_n \gg c \log n$ and $\rho(n) \ll (\log n)^\lambda$ for all $\lambda > 0$,

$$\left| \rho(\tau_v) - \rho(n) \right| \underset{\approx}{\ll} \frac{\delta \cdot \log v}{\log n \cdot [\log(\log n)]} \quad \text{for all natural } v > 1,$$

where $0 < \delta \ne 1$ and $\dfrac{\tau_v}{n} = v^{\frac{1}{1-\rho(n)}}$,

5^0 $\rho(n) \underset{\approx}{\gg} (\log n)^\lambda,\; \lambda > 0$,

$$|\rho(\tau_v) - \rho(n)| \underset{\approx}{<<} \lim_{n\to\infty} \text{ for all natural } v > 1,$$

where $\delta > 0$ and $\dfrac{\tau_v}{n} = v^{\frac{1}{(1-\rho(n))A(t)}}$,

$$A(n) \approx \prod_{i=1}^{v} f_i(\rho(n)),$$

where $f_i(n)$ for $i = 1,2,...,v$, is the ith superposition of a function log n, and v is such that

$$f_{v+1}(\rho(n)) << \log(\log n).$$

Case 2. $k_n = n, \ l_n - c \log n \approx s, \ s \in (-\infty,\infty), \ c > 0.$

Case 3. $k_n \to k, \ k > 0, \ l_n \to \infty.$

The proofs of the theorems on limit reliability functions for homogeneous regular series-parallel systems and methods of finding such functions for individual systems are based on the following essential lemma.

Lemma 4.12
If

    (i) $k_n \to \infty$,
    (ii) $\mathfrak{R}(t) = 1 - \exp[-V(t)]$ is a non-degenerate reliability function,

    (iii) $\boldsymbol{R}_{k_n,l_n}(t)$ is the reliability function of a homogeneous regular two-state series-parallel system defined by (2.5),

  (iv) $a_n > 0, \ b_n \in (-\infty,\infty)$,

then

$$\lim_{n\to\infty} \boldsymbol{R}_{k_n,l_n}(a_n t + b_n) = \mathfrak{R}(t) \text{ for } t \in C_\mathfrak{R}, \qquad (4.47)$$

if and only if

$$\lim_{n\to\infty} k_n[R(a_n t + b_n)]^{l_n} = V(t) \text{ for } t \in C_V. \qquad (4.48)$$

The proof of Lemma 4.12 is given in [53], [56] and [71].

The justification of the next auxiliary theorem that follows from Lemma 4.12 may be found in [56] and [71].

Lemma 4.13
If

(i) $k_n \to k$, $k > 0$, $l_n \to \infty$,

(ii) $\mathcal{R}(t)$ is a non-degenerate reliability function,

(iii) $\boldsymbol{R}_{k_n,l_n}(t)$ is the reliability function of a homogeneous regular two-state series-parallel system defined by (2.5),

(iv) $a_n > 0$, $b_n \in (-\infty,\infty)$,

then

$$\lim_{n\to\infty} \boldsymbol{R}_{k_n,l_n}(a_n t + b_n) = \mathcal{R}(t) \text{ for } t \in C_{\mathcal{R}}, \tag{4.49}$$

if and only if

$$\lim_{n\to\infty} [R(a_n t + b_n)]^{l_n} = \mathcal{R}_0(t) \text{ for } t \in C_{\mathcal{R}_0}, \tag{4.50}$$

where $\mathcal{R}_0(t)$ is a non-degenerate reliability function and moreover

$$\mathcal{R}(t) = 1 - [1 - \mathcal{R}_0(t)]^k \quad \text{for } t \in (-\infty,\infty). \tag{4.51}$$

The results achieved in [53]–[56], [59] and based on Lemma 4.12 and Lemma 4.13 may be formulated in the form of the following theorem ([54], [57], [63]).

Theorem 4.6
The only non-degenerate limit reliability functions of the homogeneous regular two-state series-parallel system are:

Case 1. $k_n = n$, $|l_n - c \log n| \gg s$, $s > 0$, $c > 0$ (under Assumption 4.1).

$$\mathcal{R}_1(t) = 1 \text{ for } t \le 0, \ \mathcal{R}_1(t) = 1 - \exp[-t^{-\alpha}] \text{ for } t > 0, \ \alpha > 0, \tag{4.52}$$

$$\mathcal{R}_2(t) = 1 - \exp[-(-t)^{\alpha}] \text{ for } t < 0, \ \mathcal{R}_2(t) = 0 \text{ for } t \ge 0, \ \alpha > 0, \tag{4.53}$$

$$\mathcal{R}_3(t) = 1 - \exp[-\exp[-t]] \text{ for } t \in (-\infty,\infty), \tag{4.54}$$

Case 2. $k_n = n$, $l_n - c \log n \approx s$, $s \in (-\infty, \infty)$, $c > 0$.

$$\mathcal{R}_4(t) = 1 \text{ for } t < 0, \ \mathcal{R}_4(t) = 1 - \exp[-\exp[-t^\alpha - s/c]] \text{ for } t \geq 0, \ \alpha > 0, \tag{4.55}$$

$$\mathcal{R}_5(t) = 1 - \exp[-\exp[(-t)^\alpha - s/c]] \text{ for } t < 0, \ \mathcal{R}_5(t) = 0 \text{ for } t \geq 0, \ \alpha > 0, \tag{4.56}$$

$$\mathcal{R}_6(t) = 1 - \exp[-\exp[\beta(-t)^\alpha - s/c]] \text{ for } t < 0,$$

$$\mathcal{R}_6(t) = 1 - \exp[-\exp[-t^\alpha - s/c]] \text{ for t } \geq 0, \ \alpha > 0, \ \beta > 0, \tag{4.57}$$

$$\mathcal{R}_7(t) = 1 \text{ for } t < t_1, \ \mathcal{R}_7(t) = 1 - \exp[-\exp[-s/c]] \text{ for } t_1 \leq t < t_2,$$

$$\mathcal{R}_7(t) = 0 \text{ for } t \geq t_2, \ t_1 < t_2, \tag{4.58}$$

Case 3. $k_n \to k$, $k > 0$, $l_n \to \infty$.

$$\mathcal{R}_8(t) = 1 - [1 - \exp[-(-t)^{-\alpha}]]^k \text{ for } t < 0, \ \mathcal{R}_8(t) = 0 \text{ for } t \geq 0, \ \alpha > 0, \tag{4.59}$$

$$\mathcal{R}_9(t) = 1 \text{ for } t < 0, \ \mathcal{R}_9(t) = 1 - [1 - \exp[-t^\alpha]]^k \text{ for } t \geq 0, \ \alpha > 0, \tag{4.60}$$

$$\mathcal{R}_{10}(t) = 1 - [1 - \exp[-\exp t]]^k \text{ for } t \in (-\infty, \infty). \tag{4.61}$$

Corollary 4.4

If components of the homogeneous regular two-state series-parallel system have Weibull reliability functions

$$R(t) = 1 \text{ for } t < 0, \ R(t) = \exp[-\beta t^\alpha] \text{ for } t \geq 0, \ \alpha > 0, \ \beta > 0, \tag{4.62}$$

and

$$k_n \to k, \ l_n > 0, \tag{4.63}$$

$$a_n = 1/(\beta l_n)^{1/\alpha}, \ b_n = 0, \tag{4.64}$$

then

$$\mathcal{R}_9(t) = 1 \text{ for } t < 0, \ \mathcal{R}_9(t) = 1 - [1 - \exp[-t^\alpha]]^k \text{ for } t \geq 0,$$

is its limit reliability function.

Motivation: Since, according to (4.63) and (4.64), we have

$$a_n t + b_n = (\beta l_n)^{-1/\alpha} t < 0 \text{ for } t < 0$$

and

$a_n t + b_n = (\beta l_n)^{-1/\alpha} t \geq 0$ for $t \geq 0$,

then from (4.62) the equalities

$R(a_n t + b_n) = 1$ for $t < 0$

and

$R(a_n t + b_n) = \exp[-\beta(a_n t + b_n)^\alpha] = \exp[-t^\alpha/l_n]$ for $t \geq 0$

are satisfied.
Further, we have

$$\lim_{n \to \infty} [R(a_n t + b_n)]^{l_n} = 1 \text{ for } t < 0$$

and

$$\lim_{n \to \infty} [R(a_n t + b_n)]^{l_n} = \exp[-t^\alpha] \text{ for } t \geq 0.$$

Thus, from Lemma 4.13, $\mathcal{R}_9(t)$ is the limit reliability function of the system.

Example 4.4 (*a gas distribution system*)
The gas distribution system consists of $k_n = 2$ piping lines, each of them composed of $l_n = 1000$ identical pipe segments having Weibull reliability functions

$R(t) = \exp[-0.0002 t^3]$ for $t \geq 0$.

The system is a homogeneous regular series-parallel system and according to (2.5) its exact reliability function is given by

$$R_{2,1000}(t) = 1 - [1 - \exp[-0.2 t^3]]^2 \text{ for } t \geq 0.$$

Assuming, according to (4.64), the normalising constants

$a_n = (0.0002 \cdot 1000)^{-1/3} = 1.71, \, b_n = 0,$

from Corollary 4.4, we conclude that the limit reliability function of the system is given by the formula

$$\mathcal{R}_9(t) = 1 - [1 - \exp[-t^3]]^2 \text{ for } t \geq 0.$$

Thus, according to (1.1), the approximate formula (it is easy to check that it is exact in this case) takes the form

$$R_{2,1000}(t) \cong \mathcal{R}_9((t - b_n)/a_n) = 1 - [1 - \exp[-0.2t^3]]^2 \text{ for } t \geq 0.$$

The expected values of the lifetimes of pipe segments are ([71])

$$E(T_i) = \Gamma(4/3)(0.0002)^{-1/3} \cong 15.3 \text{ years,}$$

while the mean value of the system lifetime is ([71])

$$E(T) = 2\Gamma(4/3)(0.2)^{-1/3} - \Gamma(4/3)(0.04)^{-1/3} \cong 2.4 \text{ years.}$$

The behaviour of the reliability function of the gas distribution system is illustrated in Table 4.4 and Figure 4.5.

Table 4.4. The behaviour of the exact and approximate reliability function of the gas distribution system

t	$R_{2,1000}(t) = \mathcal{R}_9((t - b_n)/a_n)$
0.0	1.0000
0.2	1.0000
0.4	0.9998
0.6	0.9982
0.8	0.9905
1.0	0.9671
1.2	0.9146
1.4	0.8216
1.6	0.6873
1.8	0.5259
2.0	0.3630
2.2	0.2236
2.4	0.1220

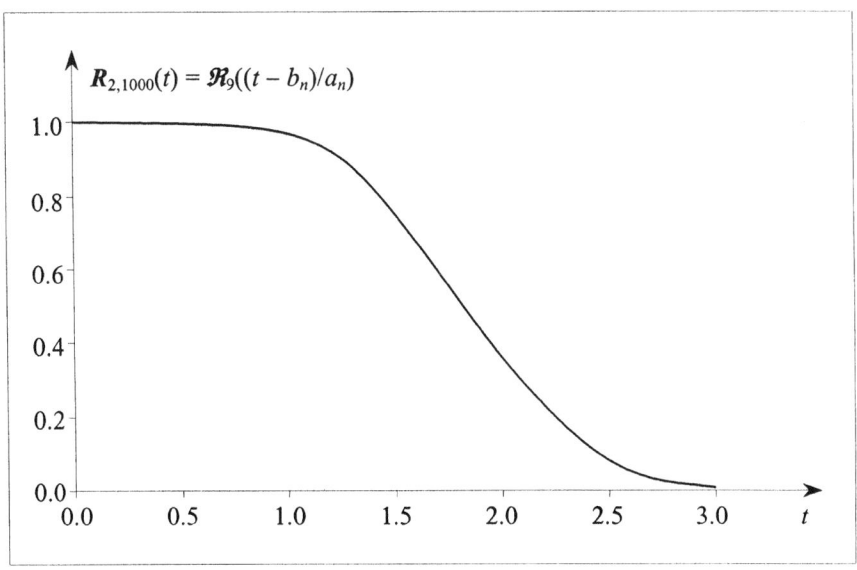

$$R_{2,1000}(t) = \mathcal{R}_9((t - b_n)/a_n)$$

Fig. 4.5. The graphs of the reliability function of the gas distribution system

The proofs of the facts concerned with limit reliability functions of non-homogeneous two-state series-parallel systems are based on the following auxiliary theorems formulated and proved in [56], [60], [63] and [71].

Lemma 4.14
If

(i) $k_n \to \infty$,

(ii) $\mathcal{R}'(t) = 1 - \exp[-V'(t)]$ is a non-degenerate reliability function,

(iii) $R'_{k_n,l_n}(t)$ is the reliability function of a non-homogeneous regular two-state series-parallel system defined by (2.14)–(2.15),

(iv) $a_n > 0, b_n \in (-\infty, \infty)$,

then

$$\lim_{n \to \infty} R'_{k_n,l_n}(a_n t + b_n) = \mathcal{R}'(t) \text{ for } t \in C_{\mathcal{R}'},$$

if and only if

$$\lim_{n\to\infty} k_n \sum_{i=1}^{a} q_i [R^{(i)}(a_n t + b_n)]^{l_n} = V(t) \text{ for } t \in C_{V'}.$$

Lemma 4.15
If

(i) $k_n \to \infty$,

(ii) $\mathscr{R}'(t) = 1 - \exp[-V'(t)]$ is a non-degenerate reliability function,

(iii) $R'_{k_n,l_n}(t)$ is the reliability function of a non-homogeneous regular two-state series-parallel system defined by (2.14)–(2.15),

(iv) $a_n > 0$, $b_n \in (-\infty,\infty)$,

(v) $R(t)$ is one of the reliability functions $R^{(1)}(t)$, $R^{(2)}(t)$,...,$R^{(a)}(t)$ defined by (2.15) such that

(vi) $\exists N \; \forall \, n > N \; R(a_n t + b_n) \neq 0$ for $t < t_0$ and $R(a_n t + b_n) = 0$ for $t \geq t_0$, where $t_0 \in (-\infty,\infty>$,

(vii) $\displaystyle\lim_{n\to\infty} \frac{R^{(i)}(a_n t + b_n)}{R(a_n t + b_n)} \leq 1$ for $t < t_0$, i $= 1,2,...,$a,

      and moreover there exists a non-increasing function

(viii) $d(t) = \begin{cases} \displaystyle\lim_{n\to\infty} \sum_{i=1}^{a} q_i d_i (a_n t + b_n) & \text{for } t < t_0 \\ 0 & \text{for } t \geq t_0, \end{cases}$ (4.65)

      where

$$d_i(a_n t + b_n) = [\frac{R^{(i)}(a_n t + b_n)}{R(a_n t + b_n)}]^{l_n},$$

then

$$\lim_{n\to\infty} R'_{k_n,l_n}(a_n t + b_n) = \mathscr{R}'(t) \text{ for } t \in C_{\mathscr{R}'}$$ (4.66)

if and only if

$$\lim_{n\to\infty} k_n [R(a_n t + b_n)]^{l_n} d(t) = V'(t) \text{ for } t \in C_{V'}.$$ (4.67)

Lemma 4.16
If

(i) $k_n \to k,\ k > 0,\ l_n \to \infty,$

(ii) $\mathscr{R}'(t)$ is a non-degenerate reliability function,

(iii) $R'_{k_n,l_n}(t)$ is the reliability function of a non-homogeneous regular two-state series-parallel system defined by (2.14)–(2.15),

(iv) $a_n > 0,\ b_n \in (-\infty,\infty),$

(v) $R(t)$ is one of reliability functions $R^{(1)}(t),\ R^{(2)}(t),...,R^{(a)}(t)$ defined by (2.15) such that

(vi) $\exists\, N\ \forall\, n > N\ R(a_n t + b_n) \neq 0$ for $t < t_0$ and $R(a_n t + b_n) = 0$ for $t \geq t_0,$ where $t_0 \in (-\infty,\infty>,$

(vii) $\displaystyle\lim_{n\to\infty} \frac{R^{(i)}(a_n t + b_n)}{R(a_n t + b_n)} \leq 1$ for $t < t_0,\ i = 1,2,...,a,$

and moreover there exist non-increasing functions

(viii) $d_i(t) = \begin{cases} \displaystyle\lim_{n\to\infty} d_i(a_n t + b_n) & \text{for } t < t_0 \\[2mm] 0 & \text{for } t \geq t_0, \end{cases}$ (4.68)

where

$$d_i(a_n t + b_n) = [\frac{R^{(i)}(a_n t + b_n)}{R(a_n t + b_n)}]^{l_n},$$

then

$$\lim_{n\to\infty} R'_{k_n,l_n}(a_n t + b_n) = \mathscr{R}'(t) \text{ for } t \in C_{\mathscr{R}'}$$ (4.69)

if and only if

$$\lim_{n\to\infty} [R(a_n t + b_n)]^{l_n} = \mathscr{R}_0(t) \text{ for } t \in C_{\mathscr{R}_0},$$ (4.70)

where $\mathscr{R}_0(t)$ is a non-degenerate reliability function and moreover

$$\mathscr{R}'(t) = 1 - \prod_{i=1}^{a}[1-d_i(t)\,\mathscr{R}_0(t)]^{q_ik}, \; t \in (-\infty,\infty). \tag{4.71}$$

Theorem 4.6, Lemma 4.15 and Lemma 4.16 determine the class of limit reliability functions for non-homogeneous regular series-parallel systems whose members are pointed out in the following theorem ([56]–[57], [60], [71]).

Theorem 4.7
The only non-degenerate limit reliability functions of the non-homogeneous regular two-state series-parallel system are:

Case 1. $k_n = n$, $|l_n - c \log n| \gg s$, $s > 0$, $c > 0$ (under Assumption 4.1 and the assumptions of Lemma 4.15).

$$\mathscr{R}'_1(t) = 1 \text{ for } t \le 0, \; \mathscr{R}'_1(t) = 1 - \exp[-d(t)t^{-\alpha}] \text{ for } t > 0, \; \alpha > 0, \tag{4.72}$$

$$\mathscr{R}'_2(t) = 1 - \exp[-d(t)(-t)^{\alpha}] \text{ for } t < 0, \; \mathscr{R}'_2(t) = 0 \text{ for } t \ge 0, \; \alpha > 0, \tag{4.73}$$

$$\mathscr{R}'_3(t) = 1 - \exp[-d(t)\exp[-t]] \text{ for } t \in (-\infty,\infty), \tag{4.44}$$

Case 2. $k_n = n$, $l_n - c \log n \approx s$, $s \in (-\infty,\infty)$, $c > 0$ (under the assumptions of Lemma 4.15).

$$\mathscr{R}'_4(t) = 1 \text{ for } t < 0, \; \mathscr{R}'_4(t) = 1 - \exp[-d(t)\exp[-t^{\alpha} - s/c]] \text{ for } t \ge 0, \; \alpha > 0, \tag{4.75}$$

$$\mathscr{R}'_5(t) = 1 - \exp[-d(t)\exp[(-t)^{\alpha} - s/c]] \text{ for } t < 0,$$

$$\mathscr{R}'_5(t) = 0 \text{ for } t \ge 0, \; \alpha > 0, \tag{4.76}$$

$$\mathscr{R}'_6(t) = 1 - \exp[-d(t)\exp[\beta(-t)^{\alpha} - s/c]] \text{ for } t < 0,$$

$$\mathscr{R}'_6(t) = 1 - \exp[-d(t)\exp[-t^{\alpha} - s/c]] \text{ for } t \ge 0, \; \alpha > 0, \; \beta > 0, \tag{4.77}$$

$$\mathscr{R}'_7(t) = 1 \text{ for } t < t_1, \; \mathscr{R}'_7(t) = 1 - \exp[-d(t)\exp[-s/c]] \text{ for } t_1 \le t < t_2,$$

$$\mathscr{R}'_7(t) = 0 \text{ for } t \ge t_2, \; t_1 < t_2, \tag{4.78}$$

Case 3. $k_n \to k$, $k > 0$, $l_n \to \infty$ (under the assumptions of Lemma 4.16).

$$\mathcal{R}'_8(t) = 1 - \prod_{i=1}^{a}[1 - d_i(t)\exp[-(-t)^{-\alpha}]]^{q_i k} \text{ for } t < 0,$$

$$\mathcal{R}'_8(t) = 0 \text{ for } t \geq 0, \ \alpha > 0, \tag{4.79}$$

$$\mathcal{R}'_9(t) = 1 \text{ for } t < 0, \ \mathcal{R}'_9(t) = 1 - \prod_{i=1}^{a}[1 - d_i(t)\exp[-t^{\alpha}]]^{q_i k} \text{ for } t \geq 0, \ \alpha > 0, \tag{4.80}$$

$$\mathcal{R}'_{10}(t) = 1 - \prod_{i=1}^{a}[1 - d_i(t)\exp[-\exp t]]^{q_i k} \text{ for } t \in (-\infty,\infty), \tag{4.81}$$

where $d(t)$ and $d_i(t)$ are non-increasing functions dependent on the reliability functions of the system's particular components and their fractions in the system defined by (4.65) and (4.68) respectively.

Corollary 4.5
If components of the non-homogeneous regular two-state series-parallel system have Weibull reliability functions

$$R^{(i,j)}(t) = 1 \text{ for } t < 0, \ R^{(i,j)}(t) = \exp[-\beta_{ij}t^{\alpha_{ij}}] \text{ for } t \geq 0, \ \alpha_{ij} > 0, \ \beta_{ij} > 0, \tag{4.82}$$
$$i = 1,2,...,a, j = 1,2,...,e_i,$$

and

$$k_n \to k, l_n \to \infty, \tag{4.83}$$

$$a_n = 1/(\beta l_n)^{-1/\alpha}, b_n = 0, \tag{4.84}$$

where

$$\alpha_i = \min_{1 \leq j \leq e_i}\{\alpha_{ij}\}, \ \beta_i = \sum_{(j:\alpha_{ij}=\alpha_i)} p_{ij}\beta_{ij}, \tag{4.85}$$

$$\alpha = \max_{1 \leq i \leq a}\{\alpha_i\}, \ \beta = \min\{\beta_i : \alpha_i = \alpha\}, \tag{4.86}$$

then

$$\mathcal{R}'_9(t) = 1 \text{ for } t < 0, \ \mathcal{R}'_9(t) = 1 - \prod_{(i:\alpha_i=\alpha)}[1 - \exp[-\frac{\beta_i}{\beta}t^{\alpha}]]^{q_i k} \text{ for } t \geq 0,$$

is its limit reliability function.
Motivation: Since, according to (4.83) and (4.84), we have

$$a_n t + b_n = (\beta l_n)^{-1/\alpha} t \to 0^- \text{ for } t < 0$$

and

$$a_n t + b_n = (\beta l_n)^{-1/\alpha} t \to 0^+ \text{ for } t \geq 0 \text{ as } n \to \infty,$$

then for every $i = 1,2,...,a$, we get

$$R^{(i)}(a_n t + b_n) = 1 \text{ for } t < 0$$

and applying (2.15), (4.82) and (4.84), we obtain

$$R^{(i)}(a_n t + b_n) = \exp[-\sum_{j=1}^{e_i} p_{ij} \beta_{ij} (a_n t)^{\alpha_{ij}}]$$

$$= \exp[-(a_n t)^{\alpha_i} \sum_{j=1}^{e_i} p_{ij} \beta_{ij} (a_n t)^{\alpha_{ij} - \alpha_i}]$$

$$= \exp[-\beta_i (a_n t)^{\alpha_i} + o(1)] \text{ for } t \geq 0.$$

Letting

$$R(t) = 1 \text{ for } t < 0 \text{ and } R(t) = \exp[-\beta t^{\alpha}] \text{ for } t \geq 0,$$

where α and β are defined by (4.86), for all $i = 1,2,...,a$, we have

$$\lim_{n \to \infty} \frac{R^{(i)}(a_n t + b_n)}{R(a_n t + b_n)} = 1 \text{ for } t < 0$$

and

$$\lim_{n \to \infty} \frac{R^{(i)}(a_n t + b_n)}{R(a_n t + b_n)} = \lim_{n \to \infty} \frac{\exp[-\beta_i (a_n t + b_n)^{\alpha_i}]}{\exp[-\beta (a_n t + b_n)^{\alpha}]}$$

$$= \lim_{n \to \infty} \exp[-\beta(a_n t)^{\alpha} [\frac{\beta_i}{\beta} (a_n t)^{\alpha_i - \alpha} - 1]] \leq 1 \text{ for } t \geq 0.$$

The above means that condition (vii) of Lemma 4.16 holds with $t_0 = \infty$. Further, according to (4.68), we get

$$d_i(t) = \begin{cases} 1 & \text{for } t < 0 \\ \exp[-(\frac{\beta_i}{\beta} - 1)t^{\alpha}] & \text{for } t \geq 0 \end{cases}$$

for i such that $\alpha_i = \alpha$ and

$$d_i(t) = \begin{cases} 1 \text{ for } t < 0 \\ 0 \text{ for } t \geq 0 \end{cases}$$

otherwise.
Moreover, we have

$$\lim_{n \to \infty}[R(a_n t + b_n)])^{l_n} = 1 \text{ for } t < 0$$

and

$$\lim_{n \to \infty}[R(a_n t + b_n)]^{l_n} = \lim_{n \to \infty} \exp[-l_n \beta(a_n t)^{\alpha}]$$

$$= \exp[-t^{\alpha}] \text{ for } t \geq 0,$$

which from Lemma 4.16 completes the proof. □

Example 4.5 (*a water supply system*)
The water supply system consists of $k_n = 3$ lines of segment pipes. Each line is composed of $l_n = 100$ segments. The scheme of the considered system is shown in Figure 4.6.

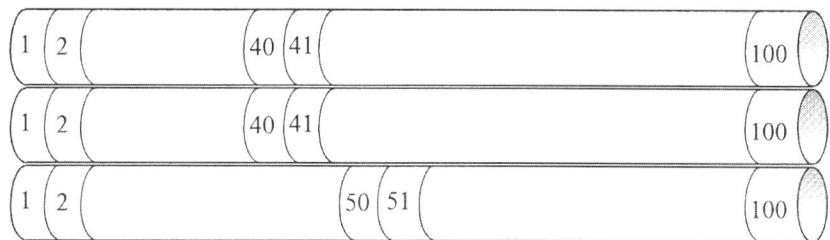

Fig. 4.6. The model of a non-homogeneous regular series-parallel water supply system

In two of the lines there are 40 segment pipes with a reliability function

$$R^{(1,1)}(t) = \exp[-0.05t] \text{ for } t \geq 0,$$

and 60 segment pipes with a reliability function

$$R^{(1,2)}(t) = \exp[-0.0015t^2] \text{ for } t \geq 0,$$

In the third line there are 50 segment pipes with a reliability function

$$R^{(2,1)}(t) = \exp[-0.0007t^3] \text{ for } t \geq 0,$$
and 50 segment pipes with a reliability function

$$R^{(2,2)}(t) = \exp[-0.2 \sqrt{t}] \text{ for } t \geq 0.$$

Thus, according to Definition 2.16, the water supply system is a non-homogeneous regular series-parallel system with the following parameters

$$k_n = k = 3, \, l_n = 100, \, a = 2, \, q_1 = 2/3, \, q_2 = 1/3.$$

Considering (2.14), we have

$$\boldsymbol{R'}_{3,100}(t) = 1 - \prod_{i=1}^{2} [1 - (R^{(i)}(t))^{100}]^{q_i 3}$$

$$= 1 - [1 - (R^{(1)}(t))^{100}]^2 [1 - (R^{(2)}(t))^{100}],$$

where after considering the substitutions:

$$e_1 = 2, \, p_{11} = 0.4, \, p_{12} = 0.6,$$

$$\alpha_{11} = 1, \, \beta_{11} = 0.05, \, \alpha_{12} = 2, \, \beta_{12} = 0.0015,$$

$$e_2 = 2, \, p_{21} = 0.5, \, p_{22} = 0.5,$$

$$\alpha_{21} = 3, \, \beta_{21} = 0.0007, \, \alpha_{22} = 0.5, \, \beta_{22} = 0.2,$$

and the formula (2.15)

$$R^{(1)}(t) = \prod_{j=1}^{e_1} (R^{(1,j)}(t))^{p_{1j}}$$

$$= (R^{(1,1)}(t))^{0.4} (R^{(1,2)}(t))^{0.6} = \exp[-0.02t - 0.0009t^2]$$

and

$$R^{(2)}(t) = \prod_{j=1}^{e_2} (R^{(2,j)}(t))^{p_{2j}}$$

$$= (R^{(2,1)}(t))^{0.5}(R^{(2,2)}(t))^{0.5} = \exp[-0.00035t^3 - 0.1\sqrt{t}\ ].$$

From the above it follows that the exact reliability function of the system is given by

$$R'_{3,100}(t) = 1 - [1 - \exp[-2t - 0.09t^2]]^2 \cdot [1 - \exp[-0.035t^3 - 10\sqrt{t}\ ]] \text{ for } t \geq 0.$$

Further, according to (4.85), (4.86) and (4.84), we have

$\alpha_1 = \min\{\alpha_{11}, \alpha_{12}\} = \min\{1,2\} = 1,$

$\beta_1 = p_{11}\beta_{11} = 0.4 \cdot 0.05 = 0.02,$

$\alpha_2 = \min\{\alpha_{21}, \alpha_{22}\} = \min\{3, 0.5\} = 0.5,$

$\beta_2 = p_{22}\beta_{22} = 0.5 \cdot 0.2 = 0.1,$

$\alpha = \max\{\alpha_1, \alpha_2\} = \max\{1, 0.5\} = 1,$

$\beta = \min\{\beta_1\} = \min\ \{0.02\} = 0.02,$

$a_n = (0.02 \cdot 100)^{-1} = 0.5,\ b_n = 0,$

and from Corollary 4.5 the limit reliability function of the system is given by

$$\mathscr{R}'_9(t) = 1 - [1 - \exp[-t]]^2 \text{ for } t \geq 0.$$

Hence, after considering (1.1), the reliability function of the system is approximately given by

$$R'_{3,100}(t) \cong \mathscr{R}'_9((t - b_n)/a_n) = 1 - [1 - \exp[-2t]]^2 \text{ for } t \geq 0. \tag{4.87}$$

The reliability data of the system components have come from experts. According to their opinions the mean lifetimes of the pipe segments, depending on their types, vary in a range from 10 up to 50 years and are as follows ([71]):

$$E(T_{11}) = 1/0.05 = 20,\ E(T_{12}) = \Gamma(3/2)(0.0015)^{-1/2} \cong 23,$$

$$E(T_{21}) = \Gamma(4/3)(0.0007)^{-1/3} \cong 10,\ E(T_{22}) = \Gamma(3)(0.2)^{-2} \cong 50.$$

The water supply system lifetime and its standard deviation calculated on the basis of the approximate formula (4.87) are:

$$E(T) \cong 0.75 \text{ years},\ \sigma(T) \cong 0.56 \text{ years}.$$

The values of the exact and approximate reliability functions of the system and the differences between them are presented in Table 4.5 and Figure 4.7.

Table 4.5. The behaviour of the exact and approximate reliability functions of the water supply system

t	$R'_{3,100}(t)$	$\mathcal{R}'_9\left(\dfrac{t-b_n}{a_n}\right)$	$\Delta = R'_{3,100} - \mathcal{R}'_9$
0.0	1.0000	1.0000	0.0000
0.2	0.8910	0.8913	−0.0003
0.4	0.6902	0.6968	−0.0066
0.6	0.4984	0.`5117	−0.0133
0.8	0.3449	0.3630	−0.0181
1.0	0.2321	0.2524	**−0.0202**
1.2	0.1530	0.1732	**−0.0202**
1.4	0.0994	0.1179	−0.0186
1.6	0.0637	0.0799	−0.0162
1.8	0.0404	0.0539	−0.0135
2.0	0.0254	0.0363	−0.0109
2.2	0.0158	0.0244	−0.0086
2.4	0.0098	0.0164	−0.0066

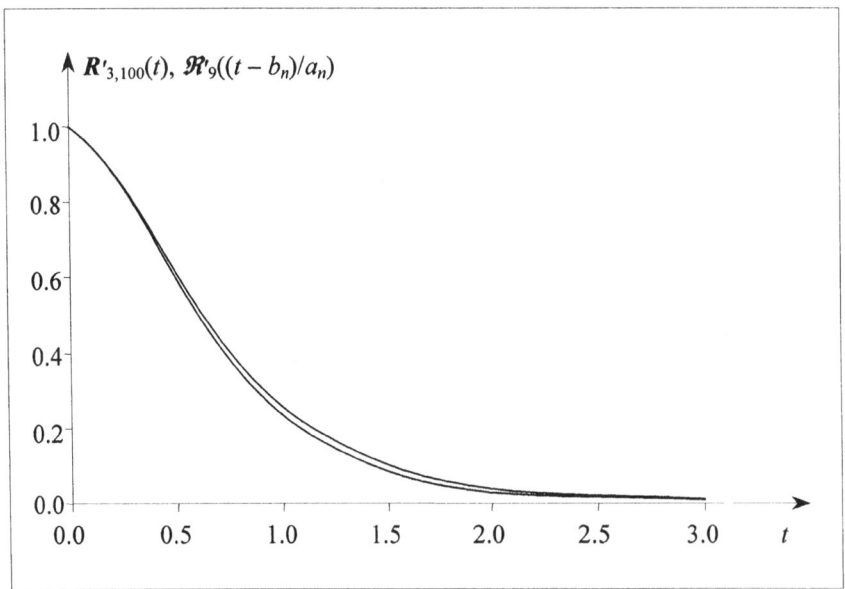

Fig. 4.7. The graphs of the exact and approximate reliability functions of the water supply system

4.5. Reliability evaluation of two-state parallel-series systems

The class of limit reliability functions for homogeneous regular two-state parallel-series systems is successively fixed in ([53]–[56], [59]) on the basis of the following lemmas.

Lemma 4.17
If $\mathcal{R}(t)$ is the limit reliability function of a homogeneous regular two-state series-parallel system composed of components with a reliability function $R(t)$, then

$$\overline{\mathcal{R}}(t) = 1 - \mathcal{R}(-t) \text{ for } t \in C_{\overline{\mathcal{R}}}$$

is the limit reliability function of a homogeneous regular two-state parallel-series system composed of components with a reliability function

$$\overline{R}(t) = 1 - R(-t) \text{ for } t \in C_R.$$

At the same time, if (a_n, b_n) is the pair of normalising constants in the first case, then $(a_n, -b_n)$ is such a pair in the second case.

Lemma 4.18
If

(i) $k_n \to \infty$,

(ii) $\overline{\mathcal{R}}(t) = \exp[-\overline{V}(t)]$ is a non-degenerate reliability function,

(iii) $\overline{R}_{k_n l_n}(t)$ is the reliability function of a homogeneous regular two-state parallel-series system defined by (2.6),

(iv) $a_n > 0, b_n \in (-\infty, \infty)$,

then

$$\lim_{n \to \infty} \overline{R}_{k_n l_n}(a_n t + b_n) = \overline{\mathcal{R}}(t) \text{ for } t \in C_{\overline{\mathcal{R}}} \tag{4.88}$$

if and only if

$$\lim_{n \to \infty} k_n [F(a_n t + b_n)]^{l_n} = \overline{V}(t) \text{ for } t \in C_{\overline{V}}. \tag{4.89}$$

Lemma 4.19
If

(i) $k_n \to k,\ k > 0,\ l_n \to \infty,$

(ii) $\overline{\mathscr{R}}(t)$ is a non-degenerate reliability function,

(iii) $\overline{R}_{k_n l_n}(t)$ is the reliability function of a homogeneous regular two-state parallel-series system defined by (2.6),

(iv) $a_n > 0, b_n \in (-\infty,\infty),$

then

$$\lim_{n\to\infty} \overline{R}_{k_n l_n}(a_n t + b_n) = \overline{\mathscr{R}}(t) \text{ for } t \in C_{\overline{\mathscr{R}}}, \tag{4.90}$$

if and only if

$$\lim_{n\to\infty} [F(a_n t + b_n)]^{l_n} = \mathfrak{I}_0(t) \text{ for } t \in C_{\mathfrak{I}_0}, \tag{4.91}$$

where $\mathfrak{I}_0(t)$ is a non-degenerate distribution function and moreover

$$\overline{\mathscr{R}}(t) = [1 - \mathfrak{I}_0(t)]^k \text{ for } t \in (-\infty,\infty). \tag{4.92}$$

By applying Lemma 4.18 and Lemma 4.19 and proceeding in the same way as in the case of homogeneous regular series-parallel systems it is possible to fix the class of limit reliability functions for homogeneous regular parallel-series systems. This class is presented in [54] and [56] as the successive results given in [53], [55], [59] and [61]. However, it is much easier to obtain this result by applying Lemma 4.17 and Theorem 4.6. Their direct application immediately results in the following theorem ([54], [56], [71]).

Theorem 4.8
The only non-degenerate limit reliability functions of the homogeneous regular two-state parallel-series system are:

Case 1. $k_n = n,\ |l_n - c \log n| >> s, s > 0, c > 0$ (under Assumption 4.1).

$$\overline{\mathscr{R}}_1(t) = \exp[-(-t)^{-\alpha}] \text{ for } t < 0,\ \overline{\mathscr{R}}_1(t) = 0, \text{ for } t \geq 0,\ \alpha > 0, \tag{4.93}$$

$$\overline{\mathscr{R}}_2(t) = 1 \text{ for } t < 0,\ \overline{\mathscr{R}}_2(t) = \exp[-t^{\alpha}], \text{ for } t \geq 0,\ \alpha > 0, \tag{4.94}$$

$$\overline{\mathfrak{R}}_3(t) = \exp[-\exp[t]] \text{ for } t \in (-\infty,\infty), \tag{4.95}$$

Case 2. $k_n = n, \, l_n - c \log n \approx s, \, s \in (-\infty,\infty), \, c > 0;$

$$\overline{\mathfrak{R}}_4(t) = \exp[-\exp[-(-t)^\alpha - s/c]] \text{ for } t < 0, \; \overline{\mathfrak{R}}_4(t) = 0 \text{ for } t \geq 0, \; \alpha > 0, \tag{4.96}$$

$$\overline{\mathfrak{R}}_5(t) = 1 \text{ for } t < 0, \; \overline{\mathfrak{R}}_5(t) = \exp[-\exp[t^\alpha - s/c]] \text{ for } t \geq 0, \; \alpha > 0, \tag{4.97}$$

$$\overline{\mathfrak{R}}_6(t) = \exp[-\exp[-(-t)^\alpha - s/c]] \text{ for } t < 0,$$

$$\overline{\mathfrak{R}}_6(t) = \exp[-\exp[\beta t^\alpha - s/c]] \text{ for } t \geq 0, \; \alpha > 0, \; \beta > 0, \tag{4.98}$$

$$\overline{\mathfrak{R}}_7(t) = 1 \text{ for } t < t_1, \; \overline{\mathfrak{R}}_7(t) = \exp[-\exp[-s/c]] \text{ for } t_1 \leq t < t_2,$$

$$\overline{\mathfrak{R}}_7(t) = 0 \text{ for } t \geq t_2, \; t_1 < t_2, \tag{4.99}$$

Case 3. $k_n \to k, \; k > 0, \, l_n \to \infty.$

$$\overline{\mathfrak{R}}_8(t) = 1 \text{ for } t \leq 0, \; \overline{\mathfrak{R}}_8(t) = [1 - \exp[-t^{-\alpha}]]^k \text{ for } t > 0, \; \alpha > 0, \tag{4.100}$$

$$\overline{\mathfrak{R}}_9(t) = [1 - \exp[-(-t)^\alpha]]^k \text{ for } t < 0, \; \overline{\mathfrak{R}}_9(t) = 0 \text{ for } t \geq 0, \; \alpha > 0, \tag{4.101}$$

$$\overline{\mathfrak{R}}_{10}(t) = [1 - \exp[-\exp[-t]]]^k \text{ for } t \in (-\infty,\infty). \tag{4.102}$$

Corollary 4.6
If components of the homogeneous regular two-state parallel-series system have Weibull reliability functions

$$R(t) = 1 \text{ for } t < 0, \; R(t) = \exp[-\beta t^\alpha] \text{ for } t \geq 0, \; \alpha > 0, \; \beta > 0,$$

and

$$k_n = n, \, l_n - c \log n \gg s, \, c > 0, \, s > 0,$$

$$a_n = b_n/(\alpha\beta(b_n)^\alpha \log n), \; b_n = [(1/\beta)\log(l_n/\log n)]^{1/\alpha},$$

then

$$\overline{\mathfrak{R}}_3(t) = \exp[-\exp[t]], \, t \in (-\infty,\infty),$$

is its limit reliability function.

Motivation: Since for sufficiently large n and all $t \in (-\infty,\infty)$, we have

$$a_n t + b_n > 0$$

and

$$a_n/b_n \to 0 \text{ as } n \to \infty,$$

then

$$F(a_n t + b_n) = 1 - \exp[-\beta(a_n t + b_n)^{\alpha}]$$

$$= 1 - \exp[-\beta(b_n)^{\alpha}(1 + (a_n/b_n)t)^{\alpha}]$$

$$= 1 - \exp[[-\beta(b_n)^{\alpha}(1 + \alpha(a_n/b_n t) + o(a_n/b_n)].$$

Moreover

$$\alpha\beta(b_n)^{\alpha} a_n/b_n = 1/\log n \to 0 \text{ as } n \to \infty,$$

and therefore

$$F(a_n t + b_n) = 1 - \exp[-\log(l_n/\log n) - t/\log n + o(1/\log n)]$$

$$= 1 - (\log n)/l_n(1 - t/\log n + o(1/\log n)$$

$$= 1 - (\log n)/l_n + t/l_n - o(1/l_n) \text{ for } t \in (-\infty,\infty),$$

and

$$\overline{V}(t) = \lim_{n\to\infty} k_n[F(a_n t + b_n)]^{l_n}$$

$$= \lim_{n\to\infty} n[1 - (\log n)/l_n + t/l_n - o(1/l_n)]^{l_n}$$

$$= \lim_{n\to\infty} \exp[t - l_n o(1/l_n)] = \exp[t] \text{ for } t \in (-\infty,\infty),$$

which from Lemma 4.18 completes the proof.

Example 4.6 (*a model parallel-series system*)
If the shape of a homogeneous regular parallel-series system is such that

$$k_n = 30, \, l_n = 60,$$

and its components have Weibull reliability functions with parameters

$\beta = 1/100$, $\alpha = 1$,

then, according to (2.6), its exact reliability function is given by

$$\overline{R}_{30,60}(t) = 1 \text{ for } t < 0, \ \overline{R}_{30,60}(t) = [1 - [1 - \exp[-0.01t]]^{60}]^{30} \text{ for } t \geq 0.$$

Applying Corollary 4.6 with the normalising constants

$$a_n = 100/\log30 \cong 29.4, \ b_n = 100\log(60/\log30) \cong 287,$$

from (1.1), we get the following approximate expression for the reliability function of the system

$$\overline{R}_{30,60}(t) \cong \overline{\mathfrak{R}}_3((t - b_n)/a_n) = \exp[-\exp[0.034t - 9.76]] \text{ for } t \in (-\infty,\infty). \quad (4.103)$$

The mean values of the system component lifetimes T_i are here

$$E[T_i] = 1 / \beta = 100 \text{ h},$$

while the expected value and the standard deviation of the system lifetime calculated from (4.103) are ([13], [71]):

$$E[T] \cong -0.5772a_n + b_n \cong 270 \text{ h}, \ \sigma(T) = \pi a_n / \sqrt{6} = 38 \text{ h}.$$

The behaviour of the exact and approximate reliability functions of the considered system is illustrated in Table 4.6 and Figure 4.8.

Table 4.6. The values of the exact and approximate reliability functions of the model parallel-series system

t	$\overline{R}_{30,60}(t)$	$\overline{\mathfrak{R}}_3((t - b_n)/a_n)$	$\Delta = \overline{R}_{30,60} - \overline{\mathfrak{R}}_3$
0	1.0000	1.0000	0.0000
100	1.0000	0.9983	0.0017
150	1.0000	0.9906	0.0084
200	0.9961	0.9495	0.0456
220	0.9742	0.9028	0.0714
240	0.9049	0.8172	0.0877
260	0.7453	0.6713	0.0742
280	0.4947	0.3973	**0.0974**
300	0.2382	0.2117	0.0265
320	0.0760	0.0467	0.0293
340	0.0151	0.0024	0.0127
360	0.0018	0.0000	0.0018

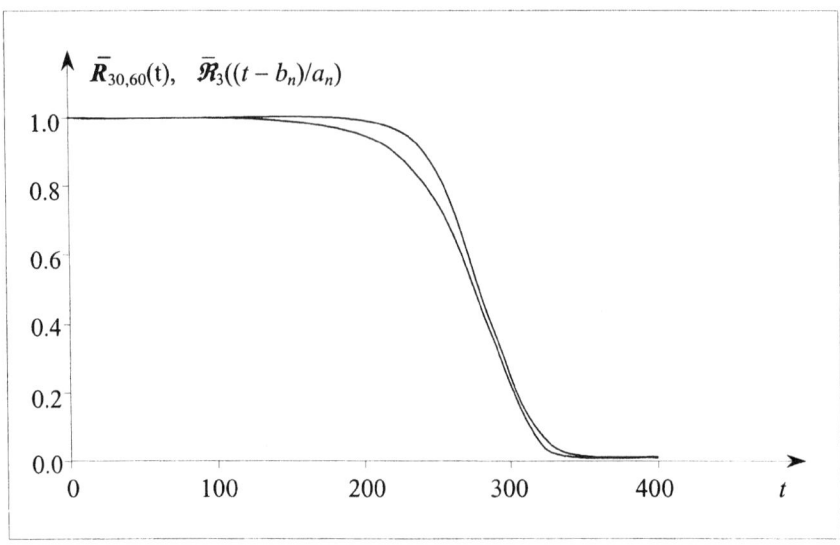

Fig. 4.8. The graphs of the exact and approximate reliability functions of the model parallel-series system

The generalisations of Lemmas 4.18–4.19 are the next lemmas proved in [56], [60]–[61] and [63], giving the way of finding limit reliability functions for non-homogeneous regular parallel-series systems.

Lemma 4.20

If $\overline{\mathfrak{R}}'(t)$ is the limit reliability function of a non-homogeneous regular two-state series-parallel system composed of components with reliability functions

$$R^{(i,j)}(t), \, i = 1,2,...,a, \, j = 1,2,...,e_i,$$

then

$$\overline{\mathfrak{R}}'(t) = 1 - \overline{\mathfrak{R}}'(-t) \text{ for } t \in C_{\mathfrak{R}'}$$

is the limit reliability function of a non-homogeneous regular two-state parallel-series system composed of components with reliability functions

$$\overline{R}^{(i,j)}(t) = 1 - R^{(i,j)}(-t) \text{ for } t \in C_{R^{(i,j)}}, \, i = 1,2,...,a, \, j = 1,2,...,e_i.$$

At the same time, if (a_n, b_n) is the pair of normalising constants in the first case, then $(a_n, -b_n)$ is such a pair in the second case.

Lemma 4.21
If

(i) $\overline{\mathscr{R}}'(t) = \exp[-\overline{V}'(t)]$ is a non-degenerate reliability function,

(ii) $\overline{R}'_{k_n l_n}(t)$ is the reliability function of a non-homogeneous regular two-state parallel-series system defined by (2.16)–(2.17),

(iii) $a_n > 0$, $b_n \in (-\infty, \infty)$,

then

$$\lim_{n \to \infty} \overline{R}'_{k_n l_n}(a_n t + b_n) = \overline{\mathscr{R}}'(t) \text{ for } t \in C_{\overline{\mathscr{R}}'},$$

if and only if

$$\lim_{n \to \infty} k_n \sum_{i=1}^{a} q_i [F^{(i)}(a_n t + b_n)]^{l_n} = \overline{V}'(t) \text{ for } t \in C_{\overline{V}'}.$$

Lemma 4.22
If

(i) $k_n \to \infty$,

(ii) $\overline{\mathscr{R}}'(t) = \exp[-\overline{V}'(t)]$ is a non-degenerate reliability function,

(iii) $\overline{R}'_{k_n l_n}(t)$ is the reliability function of a non-homogeneous regular two-state parallel-series system defined by (2.16)–(2.17),

(iv) $a_n > 0$, $b_n \in (-\infty, \infty)$,

(v) $F(t)$ is one of the distribution functions $F^{(1)}(t)$, $F^{(2)}(t),...,F^{(a)}(t)$ defined by (2.17) such that

(vi) $\exists N \; \forall \; n > N \; F(a_n t + b_n) = 0$ for $t < t_0$ and $F(a_n t + b_n) \neq 0$ for $t \geq t_0$, where $t_0 \in <-\infty, \infty)$,

(vii) $\displaystyle \lim_{n \to \infty} \frac{F^{(i)}(a_n t + b_n)}{F(a_n t + b_n)} \leq 1$ for $t \geq t_0$, $i = 1,2,...,a$,

and moreover there exists a non-decreasing function

(viii) $\bar{d}(t) = \begin{cases} 0 & \text{for } t < t_o \\ \lim\limits_{n \to \infty} \sum\limits_{i=1}^{a} q_i \bar{d}_i(a_n t + b_n) & \text{for } t \geq t_o, \end{cases}$ (4.104)

where

$$\bar{d}_i(a_n t + b_n) = \left[\frac{F^{(i)}(a_n t + b_n)}{F(a_n t + b_n)}\right]^{l_n}$$

then

$$\lim_{n \to \infty} \overline{R}'_{k_n l_n}(a_n t + b_n) = \overline{\mathscr{R}}'(t) \text{ for } t \in C_{\overline{\mathscr{R}}'},$$ (4.105)

if and only if

$$\lim_{n \to \infty} k_n [F(a_n t + b_n)]^{l_n} \bar{d}(t) = \overline{V}'(t) \text{ for } t \in C_{\overline{V}'}.$$ (4.106)

Lemma 4.23
If

(i) $k_n \to k, \ k > 0, \ l_n \to \infty,$

(ii) $\overline{\mathscr{R}}'(t)$ is a non-degenerate reliability function,

(iii) $\overline{R}_{k_n l_n}(t)$ is the reliability function of a non-homogeneous regular two-state parallel-series system defined by (2.16)–(2.17),

(iv) $a_n > 0, \ b_n \in (-\infty, \infty),$

(v) $F(t)$ is one of the distribution functions $F^{(1)}(t), F^{(2)}(t),..., F^{(a)}(t)$ defined by (2.17) such that

(vi) $\exists N \ \forall n > N \ F(a_n t + b_n) = 0$ for $t < t_0$ and $F(a_n t + b_n) \neq 0$ for $t \geq t_0,$
      where $t_0 \in <-\infty, \infty),$

(vii) $\lim\limits_{n \to \infty} \dfrac{F^{(i)}(a_n t + b_n)}{F(a_n t + b_n)} \leq 1$ for $t \geq t_0$, i = 1,2,...,a,

and moreover there exist non-decreasing functions

(viii) $\bar{d}_i(t) = \begin{cases} 0 & \text{for } t < t_0 \\ \lim\limits_{n\to\infty} \bar{d}_i(a_n t + b_n) & \text{for } t \geq t_0, \end{cases}$ (4.107)

where

$$\bar{d}_i(a_n t + b_n) = [\frac{F^{(i)}(a_n t + b_n)}{F(a_n t + b_n)}]^{l_n},$$

then

$$\lim_{n\to\infty} \bar{R}'_{k_n l_n}(a_n t + b_n) = \bar{R}'(t) \text{ for } t \in C_{\bar{R}'},$$ (4.108)

if and only if

$$\lim_{n\to\infty} [F(a_n t + b_n)]^{l_n} = \mathfrak{I}_0(t) \text{ for } t \in C_{\mathfrak{I}_0},$$ (4.109)

where $\mathfrak{I}_0(t)$ is a non-degenerate distribution function and moreover

$$\bar{R}'(t) = \prod_{i=1}^{a}[1 - \bar{d}_i(t)\,\mathfrak{I}_0(t)]^{q_i k}, \; t \in (-\infty,\infty).$$ (4.110)

Theorem 4.8, Lemma 4.22 and Lemma 4.23 determine the class of limit reliability functions for non-homogeneous regular two-state parallel-series system pointed out in the following theorem ([56], [60], [71]).

Theorem 4.9
The only non-degenerate limit reliability functions of the non-homogeneous regular two-state parallel-series system are:

Case 1. $k_n = n$, $|l_n - c \log n| >> s$, $s > 0$, $c > 0$ (under Assumption 4.1 and the assumptions of Lemma 4.22).

$\bar{R}'_1(t) = \exp[-\bar{d}(t)(-t)^{-\alpha}]$ for $t < 0$, $\bar{R}'_1(t) = 0$ for $t \geq 0$, $\alpha > 0$,   (4.111)

$\bar{R}'_2(t) = 1$ for $t < 0$, $\bar{R}'_2(t) = \exp[-\bar{d}(t)t^{\alpha}]$ for $t \geq 0$, $\alpha > 0$,   (4.112)

$\bar{R}'_3(t) = \exp[-\bar{d}(t)\exp[t]]$ for $t \in (-\infty,\infty)$,   (4.113)

Case 2. $k_n = n$, $l_n - c \log n \approx s$, $s \in (-\infty, \infty)$, $c > 0$ (under the assumptions of Lemma 4.22).

$$\overline{\mathcal{R}}'_4(t) = \exp[-\overline{d}(t) \exp[-(-t)^\alpha - s/c]] \text{ for } t < 0,$$

$$\overline{\mathcal{R}}'_4(t) = 0 \text{ for } t \geq 0, \ \alpha > 0, \tag{4.114}$$

$$\overline{\mathcal{R}}'_5(t) = 1 \text{ for } t < 0, \quad \overline{\mathcal{R}}'_5(t) = \exp[-\overline{d}(t) \exp[t^\alpha - s/c]] \text{ for } t \geq 0, \ \alpha > 0, \tag{4.115}$$

$$\overline{\mathcal{R}}'_6(t) = \exp[-\overline{d}(t) \exp[-(-t)^\alpha - s/c]] \text{ for } t < 0,$$

$$\overline{\mathcal{R}}'_6(t) = \exp[-\overline{d}(t) \exp[\beta t^\alpha - s/c]] \text{ for } t \geq 0, \ \alpha > 0, \ \beta > 0, \tag{4.116}$$

$$\overline{\mathcal{R}}'_7(t) = 1 \text{ for } t < t_1, \quad \overline{\mathcal{R}}'_7(t) = \exp[-\overline{d}(t) \exp[-s/c]] \text{ for } t_1 \leq t < t_2,$$

$$\overline{\mathcal{R}}'_7(t) = 0 \text{ for } t \geq t_2, \ t_1 < t_2, \tag{4.117}$$

Case 3. $k_n \to k$, $k > 0$, $l_n \to \infty$ (under the assumptions of Lemma 4.23).

$$\overline{\mathcal{R}}'_8(t) = 1 \text{ for } t \leq 0, \quad \overline{\mathcal{R}}'_8(t) = \prod_{i=1}^{a}[1 - \overline{d}_i(t) \exp[-t^{-\alpha}]]^{q_i k} \text{ for } t > 0, \ \alpha > 0, \tag{4.118}$$

$$\overline{\mathcal{R}}'_9(t) = \prod_{i=1}^{a}[1 - \overline{d}_i(t) \exp[-(-t)^\alpha]]^{q_i k} \text{ for } t < 0,$$

$$\overline{\mathcal{R}}'_9(t) = 0 \text{ for } t \geq 0, \ \alpha > 0, \tag{4.119}$$

$$\overline{\mathcal{R}}'_{10}(t) = \prod_{i=1}^{a}[1 - \overline{d}_i(t) \exp[-\exp(-t)]]^{q_i k} \text{ for } t \in (-\infty, \infty), \tag{4.120}$$

where $\overline{d}(t)$ and $\overline{d}_i(t)$ are non-decreasing functions dependent on the reliability functions of particular system components and their fractions in the system defined by (4.104) and (4.107) respectively.

CHAPTER 5

RELIABILITY OF LARGE MULTI-STATE SYSTEMS

Auxiliary theorems on limit reliability functions of multi-state systems, which are necessary for their approximate reliability evaluation, are formulated and proved. The classes of limit reliability functions for homogeneous and non-homogeneous series, parallel, series-parallel and parallel-series multi-state systems and for a homogeneous multi-state "m out of n" system are fixed. Practical applications of the multi-state asymptotic approach to reliability evaluation of real technical systems are presented. On the basis of auxiliary theorems some corollaries are formulated and proved and then applied to approximate reliability and risk characteristics determination of real technical multi-state systems having series, parallel, "m out of n", series-parallel and parallel-series reliability structures. Evaluations are given of multi-state reliability functions, mean sojourn times in the state subsets and their standard deviations, mean lifetimes in the states, risk functions, and exceeding moments of a permitted risk level for selected real systems. The homogeneous series piping transportation system, the model·homogeneous series telecommunication network, the homogeneous series bus transportation system, the non-homogeneous series piping transportation system, the homogeneous parallel system of an electrical cable, the non-homogeneous parallel rope system, the "10 out of 36" homogeneous steel rope system, the model homogeneous series-parallel system, the homogeneous and non-homogeneous series-parallel pipeline systems and the homogeneous parallel-series electrical energy distribution system are analysed and their reliability characteristics are evaluated. Necessary data on system components reliability and system operation processes come from experts, from trade norms and from certificates of the system component producers. Component reliability and system operation processes data by necessity are approximate and concerned with the components' mean lifetimes in the reliability state subsets and the hypothetical distributions of these lifetimes. The accuracy of the asymptotic approach to the reliability evaluation of the considered systems is illustrated in tables and figures.

5.1. Reliability evaluation of multi-state series systems

In proving facts on limit reliability functions of homogeneous multi-state series systems we apply the following obvious extension of Lemma 4.1 ([74]).

Lemma 5.1
If

(i) $\overline{\mathfrak{R}}(t,u) = \exp[-\overline{V}(t,u)]$, $u = 1,2,...,z$, is a non-degenerate reliability function,

(ii) $\overline{R}_n(t,\cdot) = [1, \overline{R}_n(t,1),..., \overline{R}_n(t,z)]$, $t \in (-\infty,\infty)$, is the reliability function of a homogeneous multi-state series system defined by (3.20)–(3.21),

(iii) $a_n(u) > 0$, $b_n(u) \in (-\infty,\infty)$, $u = 1,2,...,z$,

then

$$\overline{\mathfrak{R}}(t,\cdot) = [1, \overline{\mathfrak{R}}(t,1),..., \overline{\mathfrak{R}}(t,z)], \ t \in (-\infty,\infty),$$

is the multi-state limit reliability function of this system, i.e.

$$\lim_{n\to\infty} \overline{R}_n(a_n(u)t + b_n(u),u) = \overline{\mathfrak{R}}(t,u) \ \text{for} \ t \in C_{\overline{\mathfrak{R}}(u)}, u = 1,2,...,z, \tag{5.1}$$

if and only if

$$\lim_{n\to\infty} nF(a_n(u)t + b_n(u),u) = \overline{V}(t,u) \ \text{for} \ t \in C_{\overline{V}(u)}, u = 1,2,...,z. \tag{5.2}$$

Motivation: For each fixed u, $u = 1,2,...,z$, assumptions (i)–(iii) of Lemma 5.1 are identical to assumptions (i)–(iii) of Lemma 4.1, condition (5.1) is identical to condition (4.1) and moreover condition (5.2) is identical to condition (4.2). Since, from Lemma 4.1, condition (4.1) and condition (4.2) are equivalent, then conditions (5.1) and (5.2) are also equivalent. □

Lemma 5.1 and Theorem 4.1 from Chapter 4 allow us to fix the class of all limit reliability functions for homogeneous multi-state series systems. Their application results in the following theorem ([74]).

Theorem 5.1
The class of limit non-degenerate reliability functions of the homogeneous multi-state series system is composed of 3^z reliability functions of the form

$$\overline{\mathfrak{R}}(t,\cdot) = [1, \overline{\mathfrak{R}}(t,1),..., \overline{\mathfrak{R}}(t,z)], \ t \in (-\infty,\infty), \tag{5.3}$$

where

$$\overline{\mathcal{R}}(t,u) \in \{\,\overline{\mathcal{R}}_1(t)\,,\overline{\mathcal{R}}_2(t)\,,\overline{\mathcal{R}}_3(t)\,\}, u = 1,2,...,z, \tag{5.4}$$

and $\overline{\mathcal{R}}_i(t)$, $i = 1,2,3$, are defined by (4.3)–(4.5).

Motivation: For each fixed u, $u = 1,2,...,z$, the co-ordinate $\overline{\mathcal{R}}(t,u)$ of the vector $\overline{\mathcal{R}}(t,\cdot)$ defined by (5.3), from Theorem 4.1 that is the consequence of Lemma 4.1, may be one of the three types of reliability functions defined by (4.3)–(4.5). Thus the number of different multi-state limit reliability functions of the considered system is equal to the number of z-term variations of the 3-component set (5.4), i.e. 3^z, and they are of the form (5.3). $\square$

Corollary 5.1
If the homogeneous multi-state series system is composed of components having Weibull reliability functions

$$R(t,\cdot) = [1,R(t,1),...,R(t,z)], t \in (-\infty,\infty),$$

where

$$R(t,u) = 1 \text{ for } t < 0, R(t,u) = \exp[-\beta(u)\,t^{\alpha(u)}\,] \text{ for } t \ge 0, \alpha(u) > 0, \beta(u) > 0,$$
$$u = 1,2,...,z,$$

and

$$a_n(u) = (n\beta(u))^{-1/\alpha(u)}, \quad b_n(u) = 0, \quad u = 1,2,...,z,$$

then

$$\overline{\mathcal{R}}_2(t,\cdot) = [1,\overline{\mathcal{R}}_2(t,1),...,\overline{\mathcal{R}}_2(t,z)], t \in (-\infty,\infty),$$

where

$$\overline{\mathcal{R}}_2(t,u) = 1 \text{ for } t < 0, \overline{\mathcal{R}}_2(t,u) = \exp[-t^{\alpha(u)}\,] \text{ for } t \ge 0, u = 1,2,...,z,$$

is its limit reliability function.
Motivation: Since for each fixed u we have

$$a_n(u)t + b_n(u) < 0 \text{ for } t < 0$$

and

$$a_n(u)t + b_n(u) \ge 0 \text{ for } t \ge 0,$$

then

$$F(a_n(u)t + b_n(u),u) = 0 \text{ for } t < 0$$

and

$$F(a_n(u)t + b_n(u),u) = 1 - \exp[-\beta(u)(a_n(u)t + b_n(u))^{\alpha(u)}]$$

$$= 1 - \exp[-t^{\alpha(u)}/n] = t^{\alpha(u)}/n - o(1/n) \text{ for } t \geq 0.$$

Hence

$$\overline{V}(t,u) = \lim_{n \to \infty} nF (a_n(u)t + b_n(u),u) = 0 \text{ for } t < 0$$

and

$$\overline{V}(t,u) = \lim_{n \to \infty} nF (a_n(u)t + b_n(u),u) = t^{\alpha(u)} \text{ for } t \geq 0,$$

which from Lemma 5.1 completes the proof.

Example 5.1 (*a piping transportation system*)
The piping system is composed of $n = 1000$ identical pipe segments with reliability functions

$$R(t,u) = \exp[-0.0002ut^3] \text{ for } t \geq 0, u = 1,2,3,4.$$

Since it is a homogeneous five-state system, then according to (3.20) and (3.21) its multi-state reliability function is given by

$$\overline{R}_{1000}(t, \cdot) = [1,\exp[-0.2t^3],\exp[-0.4t^3],\exp[-0.6t^3],\exp[-0.8t^3]] \text{ for } t \geq 0.$$

Assuming normalising constants

$$a_n(u) = (0.0002 \cdot 1000u)^{1/3}, b_n(u) = 0, u = 1,2,3,4,$$

on the basis of Corollary 5.1, we conclude that the system limit reliability function is

$$\overline{\mathfrak{R}}_2(t,\cdot) = [1,\exp[-t^3],\exp[-t^3],\exp[-t^3],\exp[-t^3]] \text{ for } t \geq 0.$$

Hence, considering (3.49), we arrive at the following approximate formula (it is exact in this case)

$$\overline{R}_{1000}(t, \cdot) \cong \overline{\mathfrak{R}}_2((t - b_n(u))/a_n(u),\cdot)$$

$$= [1, \exp[-0.2t^3], \exp[-0.4t^3], \exp[-0.6t^3], \exp[-0.8t^3]] \text{ for } t \geq 0.$$

The mean values of the sojourn times $T_i(u)$ in the state subsets in years, according to (3.4), are:

$$M_i(u) = E[T_i(u)] = \Gamma(4/3)(0.0002u)^{-1/3}, \ u = 1,2,3,4,$$

i.e.

$$M_i(1) = 15.3, \ M_i(2) = 12.1, \ M_i(3) = 10.6, \ M_i(4) = 9.6,$$

and according to (3.8), their mean sojourn lifetimes in the particular states are:

$$\overline{M}_i(1) = 3.2, \ \overline{M}_i(2) = 1.5, \ \overline{M}_i(3) = 1.0, \ \overline{M}_i(4) = 9.6.$$

The expected values of the system sojourn times $T(u)$ in the state subsets, according to (3.13), are:

$$M(u) = E[T(u)] = \Gamma(4/3)(0.2u)^{-1/3}, \ u = 1,2,3,4,$$

i.e.

$$M(1) \cong 1.53, \ M(2) \cong 1.21, \ M(3) \cong 1.06, \ M(4) \cong 0.96.$$

Thus, from (3.17), the expected values of the system sojourn times in the particular states are:

$$\overline{M}(1) \cong 0.32, \overline{M}(2) \cong 0.15, \overline{M}(3) \cong 0.10, \ \overline{M}(4) \cong 0.96.$$

If the critical state is $r = 2$, then from (3.18), the system risk function is given by

$$r(t) \cong 1 - \exp[-0.4t^3] \text{ for } t \geq 0.$$

The moment when the risk exceeds a permitted level $\delta = 0.05$, calculated according to (3.19), is

$$\tau = r^{-1}(\delta) \cong (-\log(1 - \delta)/0.4)^{1/3} = 0.5 \text{ years.}$$

The graphs of the piping system limit multi-state reliability function and its risk function are plotted using a computer program ([71]) and presented in Figure 5.1.

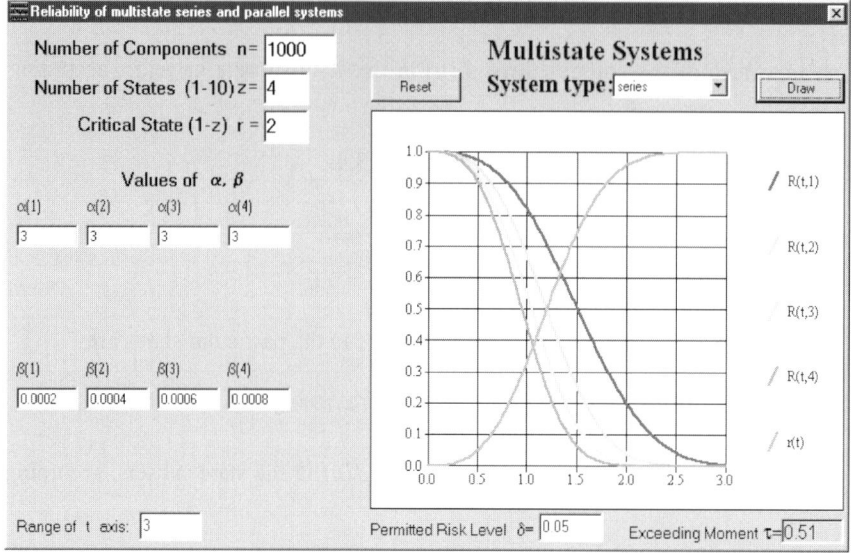

Fig. 5.1. The graphs of the piping system reliability function and risk function

Corollary 5.2
If the homogeneous multi-state series system is composed of components having Erlang's reliability function of order 2 given by

$$R(t,\cdot) = [1, R(t,1),...,R(t,z)], \ t \in (-\infty,\infty),$$

where

$$R(t,u) = 1 \text{ for } t < 0, \ R(t,u) = [1 + \lambda(u)] \exp[-\lambda(u)\,t] \text{ for } t \ge 0, \ \lambda(u) > 0, \ u = 1,2,...,z,$$

and

$$a_n(u) = \frac{\sqrt{2}}{\lambda(u)\sqrt{n}}, \ b_n(u) = 0, \ u = 1,2,...,z,$$

then

$$\overline{\mathcal{R}}_2(t,\cdot) = [1, \overline{\mathcal{R}}_2(t,1),..., \overline{\mathcal{R}}_2(t,z)], \ t \in (-\infty,\infty),$$

where

$$\overline{\mathcal{R}}_2(t,u) = 1 \text{ for } t < 0, \ \overline{\mathcal{R}}_2(t,u) = \exp[-t^2] \text{ for } t \ge 0, \ u = 1,2,...,z,$$

is its limit reliability function.

Motivation: Since for each fixed u, we have

$$a_n(u)t + b_n(u) < 0 \text{ for } t < 0$$

and

$$a_n(u)t + b_n(u) \geq 0 \text{ for } t \geq 0,$$

then

$$F(a_n(u)t + b_n(u),u) = 0 \text{ for } t < 0$$

and

$$F(a_n(u)t + b_n(u),u) = 1 - [1+ \lambda(u)\, a_n(u)t]\exp[-\lambda(u)\,(a_n(u)t)]$$

$$= 1 - [1 + \sqrt{2}t / \sqrt{n}\,]\exp[-\sqrt{2}t / \sqrt{n}\,]$$

$$= 1 - [1 + \sqrt{2}t / \sqrt{n}\,][1 - \sqrt{2}t / \sqrt{n}\, + t^2 / n - o(1/n)]$$

$$= t^2 / n + o(1/n) \text{ for } t \geq 0.$$

Hence

$$\overline{V}(t,u) = \lim_{n\to\infty} nF(a_n(u)t + b_n(u),u) = 0 \text{ for } t < 0$$

and

$$\overline{V}(t,u) = \lim_{n\to\infty} nF(a_n(u)t + b_n(u),u) = \lim_{n\to\infty} n[t^2 / n - o(1/n)] = t^2 \text{ for } t \geq 0,$$

which from Lemma 5.1 completes the proof.

Example 5.2 (*a model telecommunication network*)
The telecommunication network operating for telephone subscribers is composed of $n = 2000$ subscriber terminals, subscriber cables and one head linking subscriber cables with distributing cables. We analyse the reliability of the network part that consists of subscriber cables only. Thus the considered system is composed of $n = 2000$ double cables that consist of one basic cable and one cable in a cold reserve. The cables are five-state ($z = 4$) components of the system having exponential reliability functions with the following transition rates between the state subsets

$$\lambda(u) = \frac{1}{60 - 10u}, \, u = 1,2,3,4.$$

Under this assumption, since the sojourn time of a double cable in the state subsets is the sum of two lifetimes having exponential distributions, then it has Erlang's distribution of order 2, i.e. its reliability function is given by

$$R(t,u) = 1 \text{ for } t < 0, \ R(t,u) = [1 + \frac{1}{60 - 10u} t] \exp[-\frac{1}{60 - 10u} t] \text{ for } t \geq 0, \ u = 1,2,3,4.$$

Thus the considered part of the telecommunication network is a homogeneous five-state series system and according to Corollary 5.2, assuming the normalising constants

$$a_n(u) = 0.6 - 0.1u, \ b_n(u) = 0 \text{ for } u = 1,2,3,4,$$

we conclude that its limit reliability function is

$$\overline{\mathcal{R}}_2(t,\cdot) = [1, \exp[-t^2], \exp[-t^2], \exp[-t^2], \exp[-t^2]] \text{ for } t \geq 0.$$

Hence, from (3.49), we get the following approximate formula

$$\overline{R}_{20000}(t,\cdot) \cong \overline{\mathcal{R}}_2((t - b_n(u))/a_n(u),\cdot)$$

$$= [1, \exp[-0.781t^2], \exp[-1.389t^2], \exp[-3.125t^2], \exp[-12.5t^2]] \text{ for } t \geq 0.$$

The mean lifetimes $T_i(u)$ of the system components in the state subsets in years, according to (3.4), are:

$$M_i(u) = E[T_i(u)] = 60 - 10u, \ u = 1,2,3,4,$$

i.e.

$$M_i(1) = 50, \ M_i(2) = 40, \ M_i(3) = 30, \ M_i(4) = 20,$$

and from (3.8), the system component mean lifetimes in particular states are:

$$\overline{M}_i(1) = 10, \ \overline{M}_i(2) = 10, \ \overline{M}_i(3) = 10, \ \overline{M}_i(4) = 20.$$

The expected values of the network sojourn times $T(u)$ in the state subsets, according to (3.13), are:

$$M(u) = E[T(u)] = \Gamma(3/2)(0.6 - 0.1u), \ u = 1,2,3,4,$$

i.e.

$$M(1) \cong 0.44, \ M(2) \cong 0.35, \ M(3) \cong 0.27, \ M(4) \cong 0.18.$$

Thus, from (3.17), the lifetime expected values of the network in particular states are:

$\overline{M}(1) \cong 0.09, \ \overline{M}(2) \cong 0.08, \ \overline{M}(3) \cong 0.09, \ \overline{M}(4) \cong 0.18.$

If the critical state is $r = 2$, then from (3.18), the network risk function is given by

$r(t) \cong 1 - \exp[-6.25t^2]$ for $t \geq 0$.

The moment when the network risk exceeds an admissible level $\delta = 0.05$, from (3.19), is

$\tau = r^{-1}(\delta) \cong [-\log(1 - \delta)/6.25]^{1/2} = 0.09$ years.

Corollary 5.3
If components of the homogeneous multi-state series system have reliability functions

$R(t,\cdot) = [1,R(t,1),...,R(t,z)], \ t \in (-\infty,\infty),$

where

$R(t,u) = 1$ for $t < 0$, $R(t,u) = r_1 \exp[-\lambda_1(u) t] + r_2 \exp[-\lambda_2(u) t]$ for $t \geq 0$,

$u = 1,2,...,z, \ \lambda_1(u) > 0, \ \lambda_2(u) > 0, \ 0 \leq r_1 \leq 1, r_2 = 1 - r_1,$

and

$a_n(u) = \dfrac{1}{[r_1 \lambda_1(u) + r_2 \lambda_2(u)]n}, \ b_n(u) = 0, \ u = 1,2,...,z,$

then

$\overline{\mathcal{R}}_2(t,\cdot) = [1, \overline{\mathcal{R}}_2(t,1),..., \overline{\mathcal{R}}_2(t,z)], \ t \in (-\infty,\infty),$

where

$\overline{\mathcal{R}}_2(t,u) = 1$ for $t < 0$, $\overline{\mathcal{R}}_2(t,\cdot) = \exp[-t]$ for $t \geq 0, u = 1,2,...,z,$

is its limit reliability function.
Motivation: Since for each fixed u, we have

$a_n(u)t + b_n(u) < 0$ for $t < 0$

and

$a_n(u)t + b_n(u) \geq 0$ for $t \geq 0,$

then

$F(a_n(u)t + b_n(u),u) = 0$ for $t < 0$

and

$F(a_n(u)t + b_n(u),u) = 1 - [r_1\exp[-\lambda_1(u)(a_n(u)t) + r_2\exp[-\lambda_2(u)(a_n(u)t)]$

$$= [r_1\lambda_1(u) + r_2\lambda_2(u)][a_n(u)t - o(a_n(u))]$$

$$= t/n - o(1/n) \text{ for } t \geq 0.$$

Hence

$$\overline{V}(t,u) = \lim_{n\to\infty} nF(a_n(u)t + b_n(u),u) = 0 \text{ for } t < 0$$

and

$$\overline{V}(t,u) = \lim_{n\to\infty} nF(a_n(u)t + b_n(u),u) = \lim_{n\to\infty} n[t/n - o(1/n)] = t \text{ for } t \geq 0,$$

which from Lemma 5.1 completes the proof.

Example 5.3 (*a bus transportation system*)
The city transportation system is composed of n, $n \geq 1$, buses necessary to perform its communication tasks. We assume that the bus lifetimes are independent random variables and that the system is operating in successive cycles (days) $c = 1,2,...$. In each of the cycles the following three operating phases of all components are distinguished:
f_1 – components waiting for inclusion in the operation process, lasting from the moment t_0 up to the moment t_1,
f_2 – components' activation for the operation process, lasting from t_1 up to t_2,
f_3 – components operating, lasting from t_2 up to $t_3 = t_0$.
Each of the system components during the waiting phase may be damaged because of the circumstances at the stoppage place. We assume that the probability that at the end moment t_1 of the first phase the ith component is not failed is equal to $p_i^{(1)}$, where $0 \leq p_i^{(1)} \leq 1$, $i = 1,2,...,n$. Since component lifetimes are independent then the system availability at the end moment t_1 of phase f_1 is given by

$$p^{(1)} = \prod_{i=1}^{n} p_i^{(1)} . \tag{5.5}$$

In the activation phase f_2 system components are prepared for the operation process by the service. They are checked and small flaws are removed. Sometimes the flaws cannot be removed and particular components are not prepared to fulfil their tasks. We assume that the probability that at the end moment t_2 of the first phase the ith component is not failed is equal to $p_i^{(2)}$, where $0 \le p_i^{(2)} \le 1$, $i = 1,2,..., n$. Since component lifetimes are independent then the system availability at the end moment t_2 of the phase f_2 is given by

$$p^{(2)} = \prod_{i=1}^{n} p_i^{(2)}.$$ (5.6)

Thus, finally, the system S availability after two phases is given by

$$p^{(1,2)} = p^{(1)} p^{(2)},$$ (5.7)

where $p^{(1)}$ and $p^{(2)}$ are defined respectively by (5.5) and (5.6).
In the operating phase f_3, during the time $t_4 = t_3 - t_2$, each of the system components is performing one of two tasks:

z_1 – a first task (working at normal communication conditions),

z_2 – a second task (working at a communication peak), with probabilities

respectively equal to r_1 and r_2, where $0 \le r_1 \le 1$, $r_2 = 1 - r_1$.

Let

$$R^{(1)}(t,u) = 1 \text{ for } t < 0, \quad R^{(1)}(t,u) = \exp[-\frac{1}{15-5u}t] \text{ for } t \ge 0, u = 1,2,$$

be the reliability function of the ith component during performance of task z_1 and

$$R^{(2)}(t,u) = 1 \text{ for } t < 0, \quad R^{(2)}(t,u) = \exp[-\frac{1}{10-2u}t] \text{ for } t \ge 0, u = 1,2,$$

be the reliability function of the ith component during performance of task z_2. Then, according to the formula for total probability

$$R^{(1,2)}(t) = 1 \text{ for } t < 0,$$

$$R^{(1,2)}(t,u) = r_1 \exp[-\frac{1}{15-5u}t] + r_2 \exp[-\frac{1}{10-2u}t] \text{ for } t \ge 0, u = 1,2,$$

is the reliability function of the ith component performing two tasks.
Thus the considered transportation system is a homogeneous three-state series system and according to Corollary 5.3, assuming the normalising constants

$$a_n(u) = 1/[[r_1 /(15-5u) + r_2 /(10-2u)]n], \quad b_n(u) = 0 \text{ for } u = 1,2,$$

we conclude that its limit reliability function is

$$\overline{\mathscr{R}}_2(t,\cdot) = [1, \exp[-t], \exp[-t]] \text{ for } t \geq 0.$$

Hence, from (3.49), we get the following approximate formula

$$\overline{R}_n(t,\cdot) \cong \overline{\mathscr{R}}_2((t - b_n(u))/a_n(u),\cdot)$$

$$= [1, \exp[-(r_1/10 + r_2/8)nt], \exp[-(r_1/5 + r_2/6)nt]] \text{ for } t \geq 0.$$

The mean values of the system lifetimes $T(u)$ in the state subsets, according to (3.13), are:

$$M(u) = E[T(u)] = 1/[[r_1/(15 - 5u) + r_2(10 - 2u)]n], u = 1,2.$$

If we assume that

$$n = 30, r_1 = 0.8, r_2 = 0.2,$$

then

$$\overline{R}_{30}(t,\cdot) \cong [1, \exp[-3.15t], \exp[-5.80t]] \text{ for } t \geq 0 \qquad (5.8)$$

and

$$M(1) \cong 0.32, M(2) \cong 0.17.$$

Thus, considering (3.17), the expected values of the sojourn times in the particular states are:

$$\overline{M}(1) \cong 0.15, \overline{M}(2) \cong 0.17.$$

If a critical state is $r = 1$, then according to (3.18), the system risk function is given by

$$r(t) \cong 1 - \exp[-3.15t] \text{ for } t \geq 0.$$

The moment when the system risk exceeds a permitted level $\delta = 0.05$, according to (3.19), is

$$\tau = r^{-1}(\delta) \cong -\log(1 - \delta)/3.15 = 0.016 \text{ years} \cong 6 \text{ days}.$$

At the end moment of the system activation phase, which is simultaneously the starting moment of the system operating phase t_2 the system is able to perform its tasks with the probability $p^{(1,2)}$ defined by (5.7). Therefore, after applying the formula (5.8), we conclude that the system reliability in c cycles, $c = 1,2,...$, is given by the following formula

$$G(c,\cdot) \cong [1, p^{(1,2)} \exp[-3.15ct_4], p^{(1,2)} \exp[-5.80ct_4]],$$

where $t_4 = t_3 - t_2$ is the time duration of the system operating phase f_3. Further, assuming for instance

$$p^{(1,2)} = p^{(1)} p^{(2)} = 0.99 \cdot 0.99 = 0.98, \ t_4 = 18 \text{ hours} = 0.002055 \text{ years}$$

for the number of cycles

$$c = 7 \text{ days} = 1 \text{ week},$$

we get

$$G(7,\cdot) \cong [1, 0.966, 0.902].$$

This result means that during 7 days the considered transportation system will be able to perform its tasks in state not worse than the first state with probability 0.966, whereas it will be able to perform its tasks in the second state with probability 0.902.

In finding the class of limit reliability functions for non-homogeneous multi-state series systems we use an obvious extension of Lemma 4.3 formulated as follows ([74]).

Lemma 5.2
If

(i) $\overline{\mathscr{R}}'(t,u) = \exp[-\overline{V}'(t,u)]$, $u = 1,2,...,z$, is a non-degenerate reliability function,

(ii) $\overline{\boldsymbol{R}}'_n (t,\cdot) = [1, \overline{\boldsymbol{R}}'_n (t,1),..., \overline{\boldsymbol{R}}'_n (t,z)]$, $t \in (-\infty,\infty)$, is the reliability function of a non-homogeneous multi-state series system defined by (3.33)–(3.34),

(iii) $a_n (u) > 0$, $b_n(u) \in (-\infty,\infty)$, $u = 1,2,...,z$,

(iv) $F(t,u)$ for each fixed u, is one of distribution functions $F^{(1)}(t,u), F^{(2)}(t,u),..., F^{(a)}(t,u)$ defined by (3.32) such that

(v) $\exists \ N(u) \ \forall \ n > N(u) \ \ F(a_n(u)t + b_n(u),u) = 0$ for $t < t_0(u)$, $F(a_n(u)t + b_n(u),u) \neq 0$ for $t \geq t_0(u)$, where $t_0(u) \in <-\infty,\infty)$,

(vi) $\displaystyle \lim_{n\to\infty} \frac{F^{(i)}(a_n(u)t + b_n(u),u)}{F(a_n(u)t + b_n(u),u)} \leq 1$ for $t \geq t_0(u)$, $i = 1,2,...,a$, $u = 1,2,...,z$,

and moreover there exist non-decreasing functions

$$
\text{(vii)} \quad \overline{d}(t,u) = \begin{cases} 0 & \text{for } t < t_o(u) \\ \lim_{n \to \infty} \sum_{i=1}^{a} q_i \overline{d}_i(a_n(u)t + b_n(u), u) & \text{for } t \geq t_o(u), \end{cases}
\tag{5.9}
$$

where

$$
\text{(viii)} \quad \overline{d}_i(a_n(u)t + b_n(u), u) = \frac{F^{(i)}(a_n(u)t + b_n(u), u)}{F(a_n(u)t + b_n(u), u)},
\tag{5.10}
$$

then

$$
\overline{\mathscr{R}}'(t,\cdot) = [1, \ \overline{\mathscr{R}}'(t,1), ..., \ \overline{\mathscr{R}}'(t,z)], \ t \in (-\infty, \infty),
$$

is the multi-state limit reliability function of this system, i.e.

$$
\lim_{n \to \infty} \overline{R}'_n(a_n(u)t + b_n(u), u) = \overline{\mathscr{R}}'(t,u) \text{ for } t \in C_{\overline{\mathscr{R}}'(u)}, \ u = 1,2,...,z,
\tag{5.11}
$$

if and only if

$$
\lim_{n \to \infty} nF(a_n(u)t + b_n(u), u) \ \overline{d}(t,u) = \overline{V}'(t,u) \text{ for } t \in C_{\overline{V}'(u)}, \ u = 1,2,...,z.
\tag{5.12}
$$

Motivation: For each fixed u, $u = 1,2,...,z$, assumptions (i)–(viii) of Lemma 5.2 are identical to assumptions (i)–(viii) of Lemma 4.3, condition (5.11) is identical to condition (4.7), and condition (5.12) is identical to condition (4.8). Moreover, since from Lemma 4.3, conditions (4.7) and (4.8) are equivalent, then Lemma 5.2 is valid. $\square$

Lemma 5.2 and Theorem 4.2 from Chapter 4 establish the class of limit reliability functions for non-homogeneous multi-state series systems pointed out in the form of the next theorem ([74]).

Theorem 5.2
The class of limit non-degenerate reliability functions of the non-homogeneous multi-state series system, under the assumptions of Lemma 5.2, is composed of 3^z reliability functions of the form

$$
\overline{\mathscr{R}}'(t,\cdot) = [1, \overline{\mathscr{R}}'(t,1), ..., \overline{\mathscr{R}}'(t,z)], \ t \in (-\infty, \infty),
\tag{5.13}
$$

where

$$
\overline{\mathscr{R}}'(t,u) \in \{ \overline{\mathscr{R}}'_1(t), \overline{\mathscr{R}}'_2(t), \overline{\mathscr{R}}'_3(t) \}, \ u = 1,2,...,z,
\tag{5.14}
$$

and $\overline{\mathfrak{R}}'_i(t)$, $i = 1,2,3$, are given by (4.9)–(4.11) with $\overline{d}(t) = \overline{d}(t,u)$, $u = 1,2,...z$, where $\overline{d}(t,u)$ are defined by (5.9).

Motivation: For each fixed u, $u = 1,2,...,z$, the co-ordinate $\overline{\mathfrak{R}}'(t,u)$ of the vector $\overline{\mathfrak{R}}'(t,\cdot)$ defined by (5.13), from Theorem 4.2 that is the consequence of Lemma 4.3, may be one of the three types of reliability functions given by (4.9)–(4.11) with $\overline{d}(t) = \overline{d}(t,u)$, where $\overline{d}(t,u)$ are defined by (5.9). Therefore, the number of different multi-state limit reliability functions of the considered system is equal to the number of z-term variations of the 3-component set (5.14), i.e. 3^z, and they are of the form (5.13). □

Corollary 5.4
If the non-homogeneous multi-state series system is composed of components having Weibull reliability functions

$$R^{(i)}(t,\cdot) = [1,R^{(i)}(t,1),...,R^{(i)}(t,z)],\ t \in (-\infty,\infty),$$

where

$$R^{(i)}(t,u) = 1 \text{ for } t < 0,\ R^{(i)}(t,u) = \exp[-\beta_i(u)\,t^{\alpha_i(u)}] \text{ for } t \geq 0,\ \alpha_i(u) > 0,\ \beta_i(u) > 0,$$
$$i = 1,2,...,a,\ u = 1,2,...,z,$$

and

$$a_n(u) = (\beta(u)n)^{-1/\alpha(u)},\ b_n(u) = 0 \text{ for } u = 1,2,...,z,$$

where

$$\alpha(u) = \min_{1 \leq i \leq a}\{\alpha_i(u)\},\ \beta(u) = \max_{(i:\alpha_i(u)=\alpha(u))}\{\beta_i(u)\} \text{ for } u = 1,2,...,z,$$

then

$$\overline{\mathfrak{R}}'_2(t,\cdot) = [1,\overline{\mathfrak{R}}'_2(t,1),...,\overline{\mathfrak{R}}'_2(t,z)],$$

where

$$\overline{\mathfrak{R}}'_2(t,u) = 1 \text{ for } t < 0,\ \overline{\mathfrak{R}}'_2(t,u) = \exp[-\overline{d}(t,u)t^{\alpha(u)}] \text{ for } t \geq 0,\ u = 1,2,...,z,$$

and

$$\overline{d}(t,u) = \sum_{(i:\alpha_i(u)=\alpha(u))} q_i\beta_i(u)/\beta(u),$$

is its limit reliability function.

Motivation: Since for all fixed u, we have

$$a_n(u)t + b_n(u) = a_n(u)t < 0 \text{ for } t < 0 \text{ and } a_n(u)t + b_n(u) = a_n(u)t \geq 0 \text{ for } t \geq 0,$$

then

$$F^{(i)}(a_n(u)t + b_n(u), u) = 0 \text{ for } t < 0$$

and

$$F^{(i)}(a_n(u)t + b_n(u), u) = 1 - \exp[-\beta_i(u)(a_n(u)t + b_n(u))\, t^{\alpha_i(u)}] \text{ for } t \geq 0.$$

Assuming

$$F(t,u) = 0 \text{ for } t < 0 \text{ and } F(t,u) = 1 - \exp[-\beta(u)t^{\alpha(u)}] \text{ for } t \geq 0,$$

for all $i = 1, 2, \ldots, a$ and $t \geq t_0(u) = 0$, we have

$$\lim_{n \to \infty} \frac{F^{(i)}(a_n(u)t + b_n(u), u)}{F(a_n(u)t + b_n(u), u)} = \lim_{n \to \infty} \frac{1 - \exp[-\beta_i(u)(a_n(u)t)^{\alpha_i(u)}]}{1 - \exp[-\beta(u)(a_n(u)t)^{\alpha(u)}]}$$

$$= \lim_{n \to \infty} \frac{\beta_i(u)}{\beta(u)} (a_n(u)t)^{\alpha_i(u) - \alpha(u)} \leq 1,$$

which means that condition (vi) of Lemma 5.2 holds. Moreover, from (5.9), we have

$$\bar{d}(t,u) = \begin{cases} 0 & \text{for } t < 0 \\ \sum_{(i:\alpha_i(u) = \alpha(u))} q_i \beta_i(u) / \beta(u) & \text{for } t \geq 0. \end{cases}$$

Therefore

$$\bar{V}'(t,u) = \lim_{n \to \infty} n F(a_n(u)t + b_n(u), u) \bar{d}(t,u) = 0 \text{ for } t < 0, \; u = 1, 2, \ldots, z,$$

and

$$\bar{V}'(t,u) = \lim_{n \to \infty} n F(a_n(u)t + b_n(u), u) \, \bar{d}(t,u)$$

$$= \lim_{n \to \infty} n[1 - \exp[-\beta(u)(a_n(u)t)^{\alpha(u)}]] \, \bar{d}(t,u)$$

$$= \lim_{n\to\infty} n\beta(u)(a_n(u)t)^{\alpha(u)} \, \overline{d}(t,u) = \overline{d}(t,u) \, t^{\alpha(u)} \text{ for } t \geq 0, \, u = 1,2,\dots,z,$$

which from Lemma 5.2 completes the proof. □

Example 5.4 (*a piping transportation system*)
The pipeline system is composed of $n = 80$ five-state pipe segments, i.e. $z = 4$. There are four types of pipe segments in the system, namely:
20 segments with exponential reliability functions

$$R^{(1)}(t,1) = \exp[-0.01t], \, R^{(1)}(t,2) = \exp[-0.0120t],$$

$$R^{(1)}(t,3) = \exp[-0.015t], \, \mathrm{R}^{(1)}(t,4) = \exp[-0.025t],$$

20 segments with exponential reliability functions

$$R^{(2)}(t,1) = \exp[-0.018t], \, R^{(2)}(t,2) = \exp[-0.019t],$$

$$R^{(2)}(t,3) = \exp[-0.020t], \, R^{(2)}(t,4) = \exp[-0.023t],$$

10 segments with Weibull reliability functions

$$R^{(3)}(t,1) = \exp[-0.0005t^2], \, R^{(3)}(t,2) = \exp[-0.0006t^2],$$

$$R^{(3)}(t,3) = \exp[-0.0010t^2], \, R^{(3)}(t,4) = \exp[-0.0015t^2],$$

and 30 segments with Weibull reliability functions

$$R^{(4)}(t,1) = \exp[-0.00061t^3], \, R^{(4)}(t,2) = \exp[-0.00062t^3],$$

$$R^{(4)}(t,3) = \exp[-0.00064t^3], \, R^{(4)}(t,4) = \exp[-0.00070t^3].$$

According to Definition 3.17 the considered system is a non-homogeneous multi-state series system with parameters

$$n = 80, \, a = 4, \, q_1 = 2/8, \, q_2 = 2/8, \, q_3 = 1/8, \, q_4 = 3/8.$$

Thus, from (3.33)–(3.34), we have

$$\overline{R}'_{80} (t,\cdot) = [1, \overline{R}'_{80} (t,1), \overline{R}'_{80} (t,2), \overline{R}'_{80} (t,3), \overline{R}'_{80} (t,4)],$$

where

$$\overline{R}'_{80} (t,1) = \exp[-0.2t - 0.34t - 0.005t^2 - 0.00183t^3],$$

$$\overline{R}'_{80} (t,2) = \exp[-0.24t - 0.38t - 0.006t^2 - 0.00186t^3],$$

$\overline{R}'_{80} (t,3) = \exp[-0.3t - 0.4t - 0.01t^2 - 0.00192t^3]$,

$\overline{R}'_{80} (t,4) = \exp[-0.5t - 0.46t - 0.015t^2 - 0.0021t^3]$ for $t \geq 0$.

Further, from Corollary 5.4, we find

$\alpha(u) = \min\{1,1,2,3\} = 1$ for $u = 1,2,3,4$,

$\beta(1) = \max\{0.01,0.018\} = 0.018$, $\beta(2) = \max\{0.012,0.019\} = 0.019$,

$\beta(3) = \max\{0.015,0.02\} = 0.020$, $\beta(4) = \max\{0.025,0.023\} = 0.025$,

$a_n(1) = 1/(0.018\cdot80) = 0.694$, $a_n(2) = 1/(0.019\cdot80) = 0.658$,

$a_n(3) = 1/(0.020\cdot80) = 0.625$, $a_n(4) = 1/(0.025\cdot80) = 0.500$, $b_n(u) = 0$ for $u = 1,2,3,4$,

$d(t,1) = (2/8)(0.018/0.018) + (2/8)(0.010/0.018) = 0.389$,

$d(t,2) = (2/8)(0.019/0.019) + (2/8)(0.012/0.019) = 0.408$,

$d(t,3) = (2/8)(0.020/0.020) + (2/8)(0.015/0.020) = 0.438$,

$d(t,u) = (2/8)(0.025/0.025) + (2/8)(0.023/0.025) = 0.480$

and we conclude that the system limit reliability function takes the form

$\mathfrak{R}'_2 (t,\cdot) = [\exp[-0.389t],\exp[-0.408t],\exp[-0.438t],\exp[-0.48t]]$ for $t \geq 0$.

Hence, according to (3.49), the approximate formula for $t \geq 0$ is given by

$\overline{R}'_{80} (t,\cdot) \cong \overline{\mathfrak{R}}'_2 ((t - b_n(u))/a_n(u),\cdot)$

$= [\exp[-0.56t],\exp[-0.62t],\exp[-0.7t],\exp[-0.96t]]$.

The mean values of the system sojourn times in the state subsets in years for instance for segments of the second type, after applying (3.4), are:

$M_2(1) = 1/0.018 = 55.56$, $M_2(2) = 1/0.019 = 52.63$,

$M_2(3) = 1/0.020 = 50.00$, $M_2(4) = 1/0.023 = 43.48$,

and their mean times in the particular states, from (3.8), are:

$\overline{M}_2(1) = 2.93$, $\overline{M}_2(2) = 2.63$, $\overline{M}_2(3) = 6.52$, $\overline{M}_2(4) = 43.48$.

The approximate expected values of the pipeline system sojourn times in the state subsets calculated from the approximate reliability function according to (3.13), are:

$M(1) = 1/0.56 = 1.79, M(2) = 1/0.62 = 1.61,$

$M(3) = 1/0.7 = 1.43, M(4) = 1/0.96 = 1.04,$

and from (3.17), the system lifetimes in the particular states are:

$\overline{M}(1) = 0.18, \overline{M}(2) = 0.18, \overline{M}(3) = 0.39, \overline{M}(4) = 1.04.$

If the critical state is $r = 2$, then the system risk function, according to (3.18), is given by

$r(t) = 1 - \exp[-0.62t]$ for $t \geq 0.$

The moment when the system risk function exceeds a permitted level $\delta = 0.05$, from (3.19), is

$\tau = r^{-1}(\delta) = -(1/0.62)\log(1 - \delta) = 0.0827$ years.

The graphs of the piping reliability function and its risk function have been drown by a computer program ([71]) and are presented in Figure 5.2.

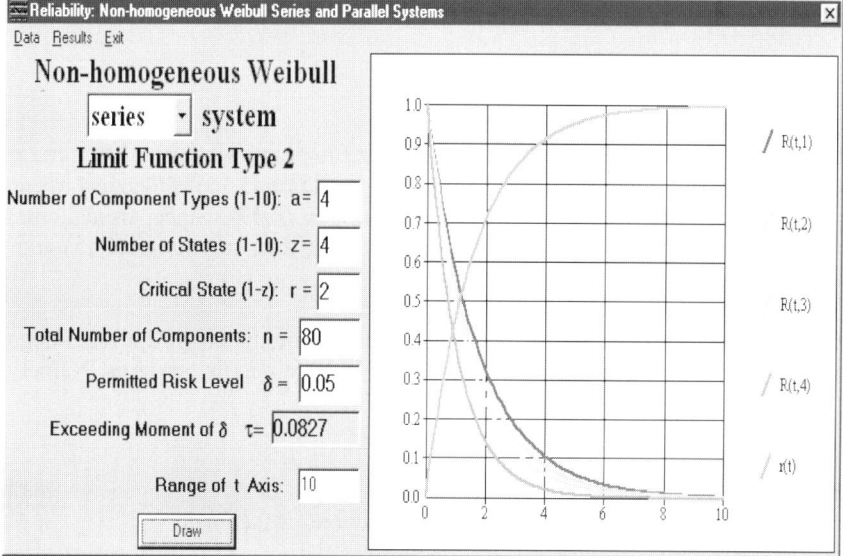

Fig. 5.2. The graphs of multi-state reliability function and risk function of the piping system

5.2. Reliability evaluation of multi-state parallel systems

In proving facts on limit reliability functions of homogeneous parallel multi-state systems the following extension of Lemma 4.5 is useful ([74]).

Lemma 5.3
If

(i) $\mathcal{R}(t,u) = 1 - \exp[-V(t,u)]$, $u = 1,2,...,z$, is a non-degenerate reliability function,

(ii) $\boldsymbol{R}_n(t,\cdot) = [1,R_n(t,1),...,R_n(t,z)]$, $t \in (-\infty,\infty)$, is the reliability function of a homogeneous multi-state parallel system defined by (3.22)–(3.23),

(iii) $a_n(u) > 0$, $b_n(u) \in (-\infty,\infty)$, $u = 1,2,...,z$,

then

$$\mathcal{R}(t,\cdot) = [1,\mathcal{R}(t,1),...,\mathcal{R}(t,z)], \ t \in (-\infty,\infty),$$

is the multi-state limit reliability function of this system, i.e.

$$\lim_{n\to\infty} \boldsymbol{R}_n(a_n(u)t + b_n(u),u) = \mathcal{R}(t,u) \text{ for } t \in C_{\mathcal{R}(u)}, \ u = 1,2,...,z, \tag{5.15}$$

if and only if

$$\lim_{n\to\infty} nR(a_n(u)t + b_n(u),u) = V(t,u) \text{ for } t \in C_{V(u)}, \ u = 1,2,...,z. \tag{5.16}$$

Motivation: For each fixed u, $u = 1,2,...,z$, assumptions (i)–(iii) of Lemma 5.3 are identical to assumptions (i)–(iii) of Lemma 4.5, condition (5.15) is identical to condition (4.18) and condition (5.16) is identical to condition (4.19). Since, from Lemma 4.5, condition (4.18) and condition (4.19) are equivalent, then conditions (5.15) and (5.16) are equivalent. □

Lemma 5.3 and Theorem 4.3 from Chapter 4 determine the class of all non-degenerate limit reliability functions for homogeneous multi-state parallel systems. Namely, their application results in the following theorem ([74]).

Theorem 5.3
The class of limit non-degenerate reliability functions of the homogeneous multi-state parallel system is composed of 3^z reliability functions of the form

$$\mathcal{R}(t,\cdot) = [1,\mathcal{R}(t,1),...,\mathcal{R}(t,z)], \ t \in (-\infty,\infty), \tag{5.17}$$

where

$$\mathcal{R}(t,u) \in \{\mathcal{R}_1(t), \mathcal{R}_2(t), \mathcal{R}_3(t)\}, \ u = 1,2,...,z, \tag{5.18}$$

and $\mathcal{R}_i(t)$, $i = 1,2,3$, are defined by (4.20)–(4.22).

Motivation: For each fixed u, $u = 1,2,...,z$, the co-ordinate $\mathcal{R}(t,u)$ of the vector $\mathcal{R}(t,\cdot)$ defined by (5.17), from Theorem 4.3 that is the consequence of Lemma 4.5, may be one of the three types of reliability functions given by (4.20)–(4.22). It means that the number of different limit multi-state reliability functions of the considered system is equal to the number of z-term variations of the 3-component set (5.18), i.e. 3^z, and they are of the form (5.17). □

Corollary 5.5

If components of the homogeneous multi-state parallel system have Weibull reliability functions

$$R(t,\cdot) = [1,R(t,1),...,R(t,z)], \ t \in (-\infty,\infty),$$

where

$$R(t,u) = 1 \text{ for } t < 0, \ R(t,u) = \exp[-\beta(u)\,t^{\alpha(u)}] \text{ for } t \ge 0, \ \alpha(u) > 0, \ \beta(u) > 0,$$
$$u = 1,2,...,z,$$

and

$$a_n(u) = b_n(u)/(\alpha(u)\log n), \ b_n(u) = (\log n/\beta(u))^{1/\alpha(u)}, \ u = 1,2,...,z,$$

then

$$\mathcal{R}_3(t,\cdot) = [1,\mathcal{R}_3(t,1),...,\mathcal{R}_3(t,z)], \ t \in (-\infty,\infty),$$

where

$$\mathcal{R}_3(t,u) = 1 - \exp[-\exp[-t]] \text{ for } t \in (-\infty,\infty), \ u = 1,2,...,z,$$

is its limit reliability function.

Motivation: Since for each fixed u, sufficiently large n and all $t \in (-\infty,\infty)$, we have

$$a_n(u)t + b_n(u) = b_n(u)(t/(\alpha(u)\log n) + 1) > 0,$$

then

$$R(a_n(u)t + b_n(u),u) = \exp[-\beta(u)(a_n(u)t + b_n(u))^{\alpha(u)}] \text{ for } t \in (-\infty,\infty).$$

Hence

$$nR(a_n(u)t + b_n(u),u) = n \exp[-\beta(u)(a_n(u)t + b_n(u))^{\alpha(u)}]$$

$$= n \exp[-\beta(u)(b_n(u))^{\alpha(u)}(t/(\alpha(u)\log n) +1))^{\alpha(u)}]$$

$$= n \exp[-\log n[t/(\alpha(u)\log n) +1]^{\alpha(u)}].$$

Further, applying the equality

$$[t/(\alpha(u)\log n) +1]^{\alpha(u)} = 1 + t/\log n + o(1/\log n) \text{ for } t \in (-\infty,\infty),$$

we get

$$V(t,u) = \lim_{n \to \infty} n\, R(a_n(u)t + b_n(u),u)$$

$$= \lim_{n \to \infty} n \exp[-\log n - t - o(1)]$$

$$= \lim_{n \to \infty} \exp[-t - o(1)] = \exp[-t] \text{ for } t \in (-\infty,\infty),$$

which from Lemma 5.3 completes the proof.

Example 5.5 (*an energetic cable*)
Let us consider a model energetic network stretched between two poles and composed of three energetic cables, six insulators and two bearers and analyse the reliability of a single cable. The cable consists of 36 identical wires. Assuming that the cable is able to conduct the current if at least one of its wires is not failed we conclude that it is a homogeneous parallel system composed of $n = 36$ basic components. Further, assuming that the wires are four-state components, i.e. $z = 3$, having Weibull reliability functions with parameters

$$\alpha(u) = 2, \ \beta(u) = (7.07)^{2u - 8}, \ u = 1,2,3,$$

after applying Corollary 5.5 with normalising constants

$$a_n(u) = (7.07)^{4 - u}/(2 \sqrt{\log 36}), \ b_n(u) = (7.07)^{4 - u} \sqrt{\log 36}, \ u = 1,2,3,$$

and considering (3.49), we obtain, for $t \in (-\infty,\infty)$, the following approximate formula for the reliability function of the considered energetic cable

$$[R_{36}(t,0),R_{36}(t,1),R_{36}(t,2),R_{36}(t,3)] \cong [1,1 - \exp[-\exp[-0.01071t + 7.167]],$$

$$1 - \exp[-\exp[-0.07572t + 7.167]],$$

$$1- \exp[-\exp[-0.53543t + 7.167]]].$$

The mean values of the cable lifetimes $T(u)$ in the state subsets in months and their standard deviations calculated according to the formulae ([13], [71])

$$M(u) = E[T(u)] = Ca_n(u) + b_n(u), \ \sigma(u) = \pi a_n / \sqrt{6} , u = 1,2,3,$$

where $C \cong 0.5772$ is Euler's constant, are:

$$M(1) \cong 723, M(2) \cong 102, M(3) \cong 14.5, \sigma(1) \cong 120, \sigma(2) \cong 17, \sigma(3) \cong 2.4,$$

while the mean values of the cable lifetimes in the particular states, from (3.17), are:

$$\overline{M}(1) \cong 621, \ \overline{M}(2) \cong 87.5, \ \overline{M}(3) \cong 14.5.$$

If the critical state is $r = 2$, then from (3.18), the cable risk function is given by

$$r(t) \cong \exp[-\exp[-0.07572t + 7.167]].$$

The moment when the cable risk function exceeds a permitted level $\delta = 0.05$, according to (3.19), is

$$\tau = r^{-1}(\delta) \cong [7.167 - \log[-\log \delta]]]/0.07572 = 80 \text{ months.}$$

The values of the cable risk function are given in Table 5.1. The graphs of the cable multi-state reliability function and its risk function plotted by the computer program ([71]) are given in Figure 5.3.

Table 5.1. The values of the energetic cable risk function

t	$r(t)$
0	0.000
50	0.000
60	0.000
70	0.017
80	0.048
90	0.276
100	0.514
110	0.731
120	0.856
130	0.934
140	0.968
150	0.985
160	0.993

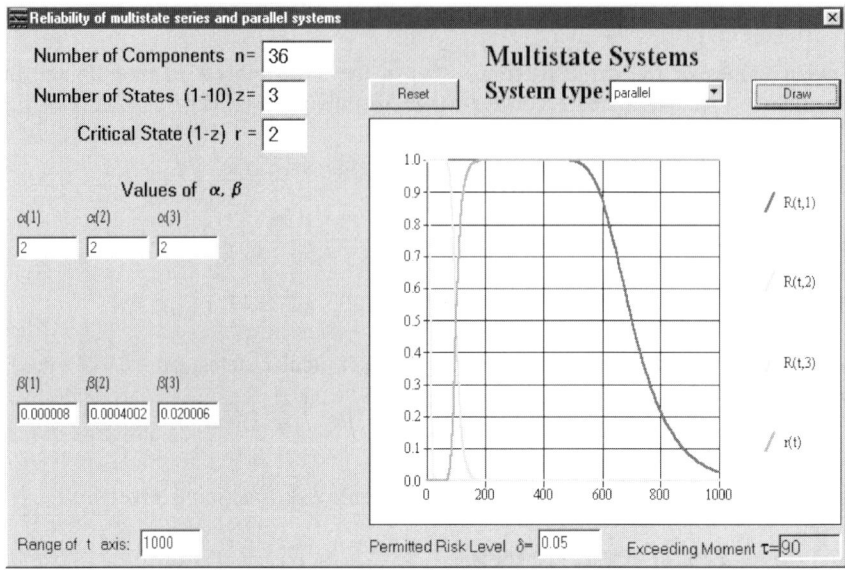

Fig. 5.3. The graphs of the energetic cable reliability function and risk function

In fixing the class of limit reliability functions for non-homogeneous multi-state parallel systems we apply the following extension of Lemma 4.8.

Lemma 5.4
If

(i) $\mathscr{R}'(t,u) = 1 - \exp[-V'(t,u)]$, $u = 1,2,...,z$, is a non-degenerate reliability function,

(ii) $\mathbf{R}'_n(t,\cdot) = [1, \mathbf{R}'_n(t,1),..., \mathbf{R}'_n(t,z)]$, $t \in (-\infty,\infty)$, is the reliability function of a non-homogeneous multi-state parallel system defined by (3.36)–(3.37),

(iii) $a_n(u) > 0$, $b_n(u) \in (-\infty,\infty)$, $u = 1,2,...,z$,

(iv) $R(t,u)$ for each fixed u, is one of the reliability functions $R^{(1)}(t,u)$, $R^{(2)}(t,u)$, ..., $R^{(a)}(t,u)$ defined by (3.35) such that

(v) $\exists\ N(u)\ \forall\ n > N(u)\ R(a_n(u)t + b_n(u),u) \neq 0$ for $t < t_0(u)$, $R(a_n(u)t + b_n(u),u) = 0$ for $t \geq t_0(u)$, where $t_0(u) \in (-\infty,\infty>$,

(vi) $\lim\limits_{n\to\infty} \dfrac{R^{(i)}(a_n(u)t + b_n(u),u)}{R(a_n(u)t + b_n(u),u)} \leq 1$ for $t < t_0(u)$, $i = 1,2,...,a$, $u = 1,2,...,z$,

and moreover there exist non-increasing functions

$$\text{(vii)} \quad d(t,u) = \begin{cases} \lim\limits_{n \to \infty} \sum\limits_{i=1}^{a} q_i d_i (a_n(u)t + b_n(u), u) & \text{for } t < t_0(u) \\ \\ 0 & \text{for } t \geq t_0(u), \end{cases}$$
(5.19)

where

$$\text{(viii)} \quad d_i(a_n(u)t + b_n(u), u) = \frac{R^{(i)}(a_n(u)t + b_n(u), u)}{R(a_n(u)t + b_n(u), u)},$$

then

$$\mathscr{R}'(t, \cdot) = [1, \mathscr{R}'(t,1), ..., \mathscr{R}'(t, z)], \ t \in (-\infty, \infty),$$

is the multi-state limit reliability function of this system, i.e.

$$\lim_{n \to \infty} \mathbf{R'}_n(a_n(u)t + b_n(u), u) = \mathscr{R}'(t, u) \text{ for } t \in C_{\mathscr{R}'(u)}, \ u = 1, 2, ..., z,$$
(5.20)

if and only if

$$\lim_{n \to \infty} nR(a_n(u)t + b_n(u), u) d(t, u) = V'(t, u) \text{ for } t \in C_{V'(u)}, \ u = 1, 2, ..., z.$$
(5.21)

Motivation: For each fixed u, $u = 1, 2, ..., z$, assumptions (i)–(viii) of Lemma 5.4 are identical to assumptions (i)–(viii) of Lemma 4.8, condition (5.20) is identical to condition (4.24) and condition (5.21) is identical to condition (4.25). And, since from Lemma 4.8, condition (4.24) and the condition (4.25) are equivalent, then Lemma 5.4 is valid. □

Lemma 5.4 and Theorem 4.4 from Chapter 4 allow us to fix the class of all possible non-degenerate limit reliability functions of the non-homogeneous multi-state parallel system listed in the following theorem ([74]).

Theorem 5.4
The class of limit non-degenerate reliability functions of the non-homogeneous multi-state parallel system, under the assumptions of Lemma 5.4, is composed of 3^z reliability functions of the form

$$\mathscr{R}'(t, \cdot) = [1, \mathscr{R}'(t,1), ..., \mathscr{R}'(t, z)], \ t \in (-\infty, \infty),$$
(5.22)

where

$$\mathscr{R}'(t, u) \in \{\mathscr{R}'_1(t), \mathscr{R}'_2(t), \mathscr{R}'_3(t)\}, \ u = 1, 2, ..., z,$$
(5.23)

and $\mathscr{R}'_i(t)$, $i = 1,2,3$, are given by (4.26)–(4.28) with $d(t) = d(t,u)$, $u = 1,2,...z$, where $d(t,u)$ are defined by (5.19).

Motivation: For each fixed u, $u = 1,2,...,z$, co-ordinate $\mathscr{R}'(t,u)$ of the vector $\mathscr{R}'(t,\cdot)$ defined by (5.22), from Theorem 4.4, that is a consequence of Lemma 4.8, may be one of the three types of reliability functions given by (4.26)–(4.28) with $d(t) = d(t,u)$, where $d(t,u)$ are defined by (5.19). Thus the number of different multi-state limit reliability functions of the considered system is equal to the number of z-term variations of the 3-element set (5.23), i.e. 3^z, and they are of the form (5.22). $\square$

Corollary 5.6

If components of the non-homogeneous multi-state parallel system have Weibull reliability functions

$$R^{(i)}(t,\cdot) = [1,R^{(i)}(t,1),...,R^{(i)}(t,z)],\ t \in (-\infty,\infty),$$

where

$$R^{(i)}(t,u) = 1 \text{ for } t < 0,\ R^{(i)}(t,u) = \exp[-\beta_i(u)\,t^{\alpha_i(u)}]\text{ for } t \geq 0,\ \alpha_i(u) > 0,\ \beta_i(u) > 0,$$
$$i = 1,2,...,a,\ u = 1,2,...,z,$$

and

$$a_n(u) = b_n(u)/(\alpha(u)\log n),\ b_n(u) = (\log n/\beta(u))^{1/\alpha(u)},\ u = 1,2,...,z,$$

where

$$\alpha(u) = \min_{1 \leq i \leq a}\{\alpha_i(u)\},\ \beta(u) = \min_{(i:\alpha_i(u)=\alpha(u))}\{\beta_i(u)\}\text{ for } u = 1,2,...,z,$$

then

$$\mathscr{R}'_3(t,\cdot)\ = [1,\mathscr{R}'_3(t,1),...,\mathscr{R}'_3(t,z)],\ t \in (-\infty,\infty),$$

where

$$\mathscr{R}'_3(t,u)\ = 1 - \exp[-d(t,u)\exp[-t]]\text{ for } t \in (-\infty,\infty),\ u = 1,2,...,z,$$

and

$$d(t,u) = \sum_{(i:\alpha_i(u)=\alpha(u),\beta_i(u)=\beta(u))} q_i\qquad \text{for } t \in (-\infty,\infty),\ u = 1,2,...,z,$$

is its limit reliability function.

Motivation: Since for each fixed u, we have

$$a_n(u)t + b_n(u) = b_n(u)(1 + \frac{a_n(u)}{b_n(u)} t) \to \infty \text{ as } n \to \infty,$$

then defining

$$R(t,u) = 1 \text{ for } t < 0 \text{ and } R(t,u) = 1 - \exp[-\beta(u)t^{\alpha(u)}] \text{ for } t \geq 0, u = 1,2,...,z,$$

for all $i = 1,2,...,a$ and $t \in (-\infty,\infty)$, we have

$$\lim_{n \to \infty} \frac{R^{(i)}(a_n(u)t + b_n(u), u)}{R(a_n(u)t + b_n(u), u)} = \lim_{n \to \infty} \frac{\exp[-\beta_i(u)(a_n(u)t + b_n(u))^{\alpha_i(u)}]}{\exp[-\beta(u)(a_n(u)t + b_n(u))^{\alpha(u)}]}$$

$$= \lim_{n \to \infty} \exp[-\beta(u)(a_n(u)t + b_n(u))^{\alpha(u)}[\frac{\beta_i(u)}{\beta(u)}(a_n(u)t + b_n(u))^{\alpha_i(u) - \alpha(u)} - 1]] \leq 1,$$

which means that condition (vi) of Lemma 5.4 holds with $t_0(u) = \infty$. Moreover, according to (5.19), we have

$$d(t,u) = \sum_{(i:\alpha_i(u)=\alpha(u),\beta_i(u)=\beta(u))} q_i .$$

Therefore

$$V'(t,u) = \lim_{n \to \infty} nR(a_n(u)t + b_n(u), u) d(t,u)$$

$$= \lim_{n \to \infty} n \exp[-\beta(u)(a_n(u)t + b_n(u))^{\alpha(u)}]] d(t,u)$$

$$= \lim_{n \to \infty} \exp[-\log n(1 + \frac{t}{\alpha(u)\log n})^{\alpha(u)} + \log n] d(t,u)$$

$$= d(t,u)\exp[-t] \text{ for } t \in (-\infty,\infty), u = 1,2,...,z,$$

which from Lemma 5.4 completes the proof. □

Example 5.6 (*a three-stratum rope, durability*)
Let us consider the steel rope of type *M*-80-200-10 described in the Polish Norm [102]. It is a three-stratum rope composed of 36 strands: 18 outer strands, 12 inner strands and 6 more inner strands. All strands consist of seven still wires. The rope cross-section is presented in Figure 5.4.

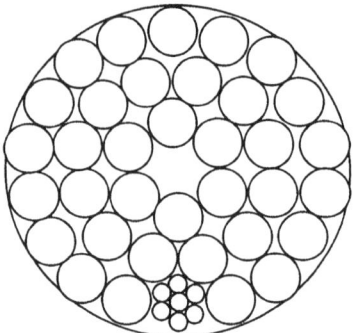

Fig. 5.4. The steel rope M-80-200-10 cross-section

Considering the strands as basic components we conclude that the rope is a parallel system composed of $n = 36$ components (strands). Due to Trade Norm [113] concerned with the evaluation of wear level, the following reliability states of the strands are distinguished:

state 3 – a strand is new, without any defects,

state 2 – the number of broken wires in the strand is greater than 0% and less than 25% of all its wires, or corrosion of wires is greater than 0% and less than 25%, abrasion is up to 25% and strain is up to 50%,

state 1 – the number of broken wires in the strand is greater than or equal to 25% and less than 50% of all its wires, or corrosion of wires is greater than or equal to 25% and less than 50%, abrasion is up to 50% and strain is up to 50%,

state 0 – otherwise (a strand is failed).

Since outer strands are more subject to damage than inner strands, then two types of strands are distinguished. Namely, it is assumed that all strands have Weibull reliability functions with two different parameters:

$$\alpha_1(u) = 2 , \ \beta_1(u) = 0.10u , \ u = 1,2,3,$$

for outer strands and

$$\alpha_2(u) = 1.5, \ \beta_2(u) = 0.25u , \ u = 1,2,3,$$

for inner strands, which means that the reliability functions of the components of the first and second type respectively are given by

$$R^{(1)}(t,u) = 1 \text{ for } t < 0 , \ R^{(1)}(t,u) = \exp[-0.10ut^2] \text{ for } t \geq 0 , \ u = 1,2,3,$$

and

$$R^{(2)}(t,u) = 1 \text{ for } t < 0 , \ R^{(2)}(t,u) = \exp[-0.25ut^{3/2}] \text{ for } t \geq 0 , \ u = 1,2,3.$$

Thus, since the considered system is non-homogeneous, according to Definition 3.18, we conclude that the rope is a non-homogeneous four-state, i.e. $z = 3$, parallel system with parameters

$n = 36$, $a = 2$, $q_1 = 1/2$, $q_2 = 1/2$.

Then, according to (3.36)–(3.37), its reliability function is given by

$$\boldsymbol{R'}_{36}(t,\cdot) = [1, \boldsymbol{R'}_{36}(t,1), \boldsymbol{R'}_{36}(t,2), \boldsymbol{R'}_{36}(t,3)],$$

where

$$\boldsymbol{R'}_{36}(t,u) = 1 \text{ for } t < 0,$$

$$\boldsymbol{R'}_{36}(t,u) = 1 - \prod_{i=1}^{2}(F^{(i)}(t,u))^{q_i n}$$

$$= [1 - \exp[-0.10ut^2]]^{18}[1 - \exp[-0.25ut^{3/2}]]^{18} \text{ for } t \geq 0, u = 1,2,3.$$

Since

$$\alpha(u) = \min\{2, 1.5\} = 1.5,$$

$$\beta(u) = \min\{0.25u\} = 0.25u, u = 1,2,3,$$

then applying Corollary 5.6 with normalising constants

$$a_n(u) = b_n(u)/(\alpha(u)\log n) = 2/(3(0.25u)^{2/3}(\log 36)^{1/3}),$$

$$b_n(u) = [\frac{1}{\beta(u)}\log n]^{1/\alpha(u)} = (\log 36/0.25u)^{2/3}, u = 1,2,3,$$

and considering that according to (5.19)

$$d(t,u) = 1/2 \text{ for } u = 1,2,3,$$

we conclude that the rope limit reliability function is

$$\mathcal{R'}_3(t,\cdot) = [1, 1 - \exp[-0.5\exp[-t]], 1 - \exp[-0.5\exp[-t]], 1 - \exp[-0.5\exp[-t]]]$$

for $t \in (-\infty, \infty)$.
Hence, using (3.49), since

$a_n(1) = 1.098$, $b_n(1) = 5.901$,

$a_n(2) = 0.692$, $b_n(2) = 3.717$,

$a_n(3) = 0.528$, $b_n(3) = 2.837$,

then the approximate formula for the rope reliability function takes the form

$$\boldsymbol{R'}_{36}(t,\cdot) \cong \boldsymbol{\mathscr{R}'}_3\left(\frac{t - b_n(u)}{a_n(u)}, \cdot\right)$$

$$= [1, 1 - \exp[-0.5\exp[-0.911t + 5.372]],$$

$$1 - \exp[-0.5\exp[-1.445t + 5.372]],$$

$$1 - \exp[-0.5\exp[-1.894t + 5.372]]], \ t \in (-\infty, \infty).$$

The approximate mean values of the rope lifetimes $T(u)$ in the state subsets and their standard deviations in years, from (3.13), calculated according to the formulae ([13], [71])

$$M(u) = E[T(u)] = Ca_n(u) + b_n(u) - \log2/a_n(u) ,$$

$$\sigma(u) = \pi a_n(u)/\sqrt{6} , \ u = 1,2,3,$$

where $C \cong 0.5772$ is Euler's constant, are:

$$M(1) \cong 5.90, \ M(2) \cong 3.11, \ M(3) \cong 1.83,$$

$$\sigma(1) \cong 1.41, \ \sigma(2) \cong 0.89, \ \sigma(3) \cong 0.68,$$

whereas the rope expected lifetimes in the particular states, from (3.17), are:

$$\overline{M}(1) \cong 2.79, \ \overline{M}(2) \cong 1.28, \ \overline{M}(3) \cong 1.83.$$

If the critical rope reliability state is $r = 2$, then according to (3.18) its risk function is given by the following approximate formula

$$r(t) \cong \exp[-0.5\exp[-1.445t + 5.371]], \ t \in (-\infty, \infty).$$

The moment when the rope risk function exceeds an admissible level $\delta = 0.05$, from (3.19), is

$$\tau \cong (5.372 - \log(-2\log \delta))/1.445 = 2.48 \text{ years.}$$

The graphs of the rope multi-state reliability function and its risk function plotted by the computer program ([71]) are given in Figure 5.5.

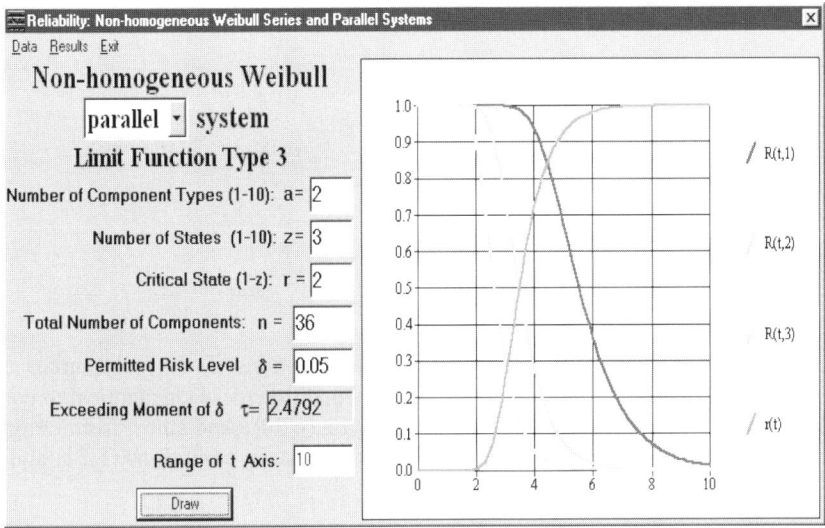

Fig. 5.5. The graphs of the rope multi-state reliability function and risk function

5.3. Reliability evaluation of multi-state "*m* out of *n*" systems

In proving facts on limit reliability functions of the homogeneous multi-state "*m* out of *n*" systems the following two extensions of Lemmas 4.9–4.11 are used.

Lemma 5.5
If

(i) $m = $ constant $(m/n \to 0$ as $n \to \infty)$,

(ii) $\mathscr{R}^{(0)}(t,u) = 1 - \sum_{i=0}^{m-1} \dfrac{[V(t,u)]^i}{i!} \exp[-V(t,u)]$, $u = 1,2,...,z$, is a non-degenerate reliability function,

(iii) $R_n^{(m)}(t,\cdot) = [1, R_n^{(m)}(t,1),...,R_n^{(m)}(t,z)]$, $t \in (-\infty,\infty)$, is the reliability function of a homogeneous multi-state "*m* out of *n*" system defined by (3.24)–(3.25),

(iv) $a_n(u) > 0$, $b_n(u) \in (-\infty,\infty)$, $u = 1,2,...,z$,

then

$$\mathfrak{R}^{(0)}(t,\cdot) = [1,\mathfrak{R}^{(0)}(t,1),...,\mathfrak{R}^{(0)}(t,z)],\ t \in (-\infty,\infty),$$

is the limit multi-state reliability function of this system, i.e.

$$\lim_{n\to\infty} \boldsymbol{R}_n^{(m)}(a_n(u)t + b_n(u),u) = \mathfrak{R}^{(0)}(t,u)\ \text{for}\ t \in C_{\mathfrak{R}^{(0)}(u)},\ u = 1,2,...,z, \tag{5.24}$$

if and only if

$$\lim_{n\to\infty} nR(a_n(u)t + b_n(u),u) = V(t,u)\ \text{for}\ t \in C_{V(u)},\ u = 1,2,...,z. \tag{5.25}$$

Motivation: For each fixed u, $u = 1,2,...,z$, assumptions (i)–(iv) of Lemma 5.5 are identical to assumptions (i)–(iv) of Lemma 4.9, condition (5.24) is identical to condition (4.29) and condition (5.25) is identical to condition (4.30). And since, from Lemma 4.9, condition (4.29) and condition (4.30) are equivalent, then conditions (5.24) and (5.25) are equivalent. □

Lemma 5.6
If

(i) $m/n \to \mu$, $0 < \mu < 1$, as $n \to \infty$,

(ii) $\mathfrak{R}^{(\mu)}(t) = 1 - \dfrac{1}{\sqrt{2\pi}} \int\limits_{-\infty}^{-v(t,u)} e^{-\frac{x^2}{2}}\,dx$, $u = 1,2,...,z$, is a non-degenerate reliability

function,

(iii) $\boldsymbol{R}_n^{(m)}(t,\cdot) = [1,\boldsymbol{R}_n^{(m)}(t,1),...,\boldsymbol{R}_n^{(m)}(t,z)]$, $t \in (-\infty,\infty)$, is the reliability function of a homogeneous multi-state "m out of n" system defined by (3.24)–(3.25),

(iv) $a_n(u) > 0$, $b_n(u) \in (-\infty,\infty)$, $u = 1,2,...,z$,

then

$$\mathfrak{R}^{(\mu)}(t,\cdot) = [1,\mathfrak{R}^{(\mu)}(t,1),...,\mathfrak{R}^{(\mu)}(t,z)],\ t \in (-\infty,\infty),$$

is the limit multi-state reliability function of this system, i.e.

$$\lim_{n\to\infty} \boldsymbol{R}_n^{(m)}(a_n(u)t + b_n(u),u) = \mathfrak{R}^{(\mu)}(t,u)\ \text{for}\ t \in C_{\mathfrak{R}^{(\mu)}(u)},\ u = 1,2,...,z, \tag{5.26}$$

if and only if

$$\lim_{n\to\infty} \frac{(n+1)R(a_n(u)t + b_n(u), u) - m}{\sqrt{\frac{m(n-m+1)}{n+1}}} = v(t,u) \text{ for } t \in C_{v(u)}. \tag{5.27}$$

Motivation: For each fixed u, $u = 1,2,...,z$, assumptions (i)–(iv) of Lemma 5.6 are identical to assumptions (i)–(iv) of Lemma 4.10, condition (5.26) is identical to condition (4.31) and condition (5.27) is identical to condition (4.32). Moreover, since, from Lemma 4.10, condition (4.31) and condition (4.32) are equivalent, then conditions (5.26) and (5.27) are equivalent. □

Lemma 5.7
If

(i) $n - m = \overline{m} = $ constant $(m/n \to 1 \text{ as } n \to \infty)$,

(ii) $\overline{\mathscr{R}}^{(1)}(t,u) = \sum_{i=0}^{\overline{m}} \frac{[\overline{V}(t)]^i}{i!} \exp[-\overline{V}(t)]$, $u = 1,2,...,z$, is a non-degenerate reliability
        function,

(iii) $\overline{R}_n^{(\overline{m})}(t,\cdot) = [1, \overline{R}_n^{(\overline{m})}(t,1),..., \overline{R}_n^{(\overline{m})}(t,z)]$, $t \in (-\infty,\infty)$, is the reliability function of
        a homogeneous multi-state "m out of n" system defined by (3.26)–(3.27),

(iv) $a_n(u) > 0$, $b_n(u) \in (-\infty,\infty)$, $u = 1,2,...,z$,

then

$$\overline{\mathscr{R}}^{(1)}(t,\cdot) = [1, \overline{\mathscr{R}}^{(1)}(t,1),..., \overline{\mathscr{R}}^{(1)}(t,z)], t \in (-\infty,\infty),$$

is the limit multi-state reliability function of this system, i.e.

$$\lim_{n\to\infty} \overline{R}_n^{(\overline{m})}(a_n(u)t + b_n(u), u) = \overline{\mathscr{R}}^{(1)}(t,u) \text{ for } t \in C_{\overline{\mathscr{R}}^{(1)}(u)}, u = 1,2,...,z, \tag{5.28}$$

if and only if

$$\lim_{n\to\infty} nF(a_n(u)t + b_n(u), u) = \overline{V}(t,u) \text{ for } t \in C_{\overline{V}(u)}, u = 1,2,...,z. \tag{5.29}$$

Motivation: For each fixed u, $u = 1,2,...,z$, assumptions (i)–(iv) of Lemma 5.7 are identical to assumptions (i)–(iv) of Lemma 4.11, condition (5.28) is identical to condition (4.33) and condition (5.29) is identical to condition (4.34). And since, from Lemma 4.9, condition (4.33) and condition (4.34) are equivalent, then conditions (5.28) and (5.29) are equivalent. □

Lemmas 5.5–5.7 and Theorem 4.5 from Chapter 4 allow us to establish the class of possible limit reliability functions of the homogeneous multi-state "*m* out of *n*" systems pointed out in the next theorem.

Theorem 5.5
The class of limit non-degenerate reliability functions of the homogeneous multi-state "*m* out of *n*" system is composed of $3^z + 4^z + 3^z$ reliability functions of the following form.

Case 1. $m = \text{constant} \ (m/n \to 0 \ as \ n \to \infty)$.

$$\mathcal{R}^{(0)}(t,\cdot) = [1, \mathcal{R}^{(0)}(t,1),...,\mathcal{R}^{(0)}(t,z)], \ t \in (-\infty,\infty), \tag{5.30}$$

where

$$\mathcal{R}^{(0)}(t,u) \in \{\mathcal{R}_1^{(0)}(t), \mathcal{R}_2^{(0)}(t), \mathcal{R}_3^{(0)}(t)\}, \ u = 1,2,...,z, \tag{5.31}$$

and $\mathcal{R}_i^{(0)}(t)$, $i = 1,2,3$, are defined by (4.35)–(4.37).

Case 2. $m/n = \mu + o(1/\sqrt{n}), \ 0 < \mu < 1, \ (m/n \to \mu \ as \ n \to \infty)$.

$$\mathcal{R}^{(\mu)}(t,\cdot) = [1, \mathcal{R}^{(\mu)}(t,1),...,\mathcal{R}^{(\mu)}(t,z)], \ t \in (-\infty,\infty), \tag{5.32}$$

where

$$\mathcal{R}^{(\mu)}(t,u) \in \{\mathcal{R}_4^{(\mu)}(t), \mathcal{R}_5^{(\mu)}(t), \mathcal{R}_6^{(\mu)}(t), \mathcal{R}_7^{(\mu)}(t)\}, \ u = 1,2,...,z, \tag{5.33}$$

and $\mathcal{R}_i^{(\mu)}(t)$, $i = 4,5,6,7$, are defined by (4.38)–(4.42).

Case 3. $n - m = \overline{m} = \text{constant} \ (m/n \to 1 \ as \ n \to \infty)$.

$$\overline{\mathcal{R}}^{(1)}(t,\cdot) = [1, \overline{\mathcal{R}}^{(1)}(t,1),...,\overline{\mathcal{R}}^{(1)}(t,z)], \ t \in (-\infty,\infty), \tag{5.34}$$

where

$$\overline{\mathcal{R}}^{(1)}(t,u) \in \{\overline{\mathcal{R}}_8^{(1)}(t), \overline{\mathcal{R}}_9^{(1)}(t), \overline{\mathcal{R}}_{10}^{(1)}(t)\}, \ u = 1,2,...,z, \tag{5.35}$$

and $\overline{\mathcal{R}}_i^{(1)}(t)$, $i = 8,9,10$, are defined by (4.43)–(4.45).

Motivation: For each fixed u, $u = 1,2,\ldots,z$, co-ordinate $\mathfrak{R}^{(0)}(t,u)$ of the vector $\mathfrak{R}^{(0)}(t,\cdot)$ defined by (5.30), co-ordinate $\mathfrak{R}^{(\mu)}(t,u)$ of the vector $\mathfrak{R}^{(\mu)}(t,\cdot)$ defined by (5.32) and co-ordinate $\overline{\mathfrak{R}}^{(1)}(t,u)$ of the vector $\overline{\mathfrak{R}}^{(1)}(t,\cdot)$ defined by (5.34), from Theorem 4.5 that is the consequence of Lemmas 4.9–4.11, may be one of the three types of reliability functions defined by (4.35)–(4.37), or one of the four types of reliability functions defined by (4.38)–(4.42), or one of the three types of reliability functions defined by (4.43)–(4.45), respectively. Thus the number of different limit multi-state reliability functions of the considered system is equal to the sum of the number of z-term variations of the 3-component set (5.31), the number of z-term variations of the 4-component set (5.33) and the number of z-term variations of the 3-component set (5.35). It means that this number is equal to $3^z + 4^z + 3^z$ and they are of the forms (5.30), (5.32) and (5.34) respectively. $\square$

Corollary 5.7
If components of the homogeneous multi-state "*m* out of *n*" system have exponential reliability functions

$$R(t,\cdot) = [1,R(t,1),\ldots,R(t,z)],\ t \in (-\infty,\infty),$$

where

$$R(t,u) = 1 \text{ for } t < 0,\ R(t,u) = \exp[-\lambda(u)t]\text{ for } t \geq 0,\ \lambda(u) > 0,\ u = 1,2,\ldots,z,$$

and

$$m = \text{constant},$$

$$a_n(u) = \frac{1}{\lambda(u)},\ b_n(u) = \frac{1}{\lambda(u)}\log n,\ u = 1,2,\ldots,z,$$

then

$$\mathfrak{R}_3^{(0)}(t,\cdot) = [1,\mathfrak{R}_3^{(0)}(t,1),\ldots,\mathfrak{R}_3^{(0)}(t,z)],\ t \in (-\infty,\infty),$$

where

$$\mathfrak{R}_3^{(0)}(t,u) = 1 - \sum_{i=0}^{m-1}\frac{\exp[-it]}{i!}\exp[-\exp[t]]\ \text{for } t \in (-\infty,\infty),\ u = 1,2,\ldots,z,$$

is its limit reliability function.
Motivation: Since for each fixed u, sufficiently large n and all $t \in (-\infty,\infty)$, we have

$$a_n(u)t + b_n(u) = \frac{t + \log n}{\lambda(u)} > 0,$$

then for sufficiently large n

$$R(a_n(u)t + b_n(u),u) = \exp[-\lambda(u)(a_n(u)t + b_n(u))] \text{ for } t \in (-\infty,\infty).$$

Hence

$$V(t,u) = \lim_{n\to\infty} n R(a_n(u)t + b_n(u),u)$$

$$= \lim_{n\to\infty} n \exp[-\lambda(u)(a_n(u)t + b_n(u))]$$

$$= \lim_{n\to\infty} n \exp[-t - \log n]$$

$$= \exp[-t] \text{ for } t \in (-\infty,\infty),$$

which from Lemma 5.5 completes the proof.

Example 5.7 (*a steel rope, durability*)
Let us consider a steel rope composed of $n = 36$ four-state, i.e. $z = 3$, identical wires having exponential reliability functions with transitions rates between the state subsets

$$\lambda(u) = 0.2u/\text{year}, \ u = 1,2,3.$$

Assuming that the rope is in the state subset $\{u, u+1,..., z\}$ if at least $m = 10$ of its wires are in this state subset, according to Definition 3.9, we conclude the rope is a homogeneous four-state "10 out of 36" system. Thus, according to (3.24)–(3.25), its reliability function is given by

$$R_{36}^{(10)}(t,\cdot) = [1, R_{36}^{(10)}(t,1), R_{36}^{(10)}(t,2), R_{36}^{(10)}(t,3)],$$

where

$$R_{36}^{(10)}(t,u) = 1 \text{ for } t < 0,$$

$$R_{36}^{(10)}(t,u) = 1 - \sum_{i=0}^{9} \binom{36}{i} \exp[-i0.2ut][1 - \exp[-0.2ut]]^{36-i}, \ u = 1,2,3.$$

Applying Corollary 5.7 with normalising constants

$$a_n(u) = \frac{5}{u}, \ b_n(u) = \frac{5}{u} \log 36, \ u = 1,2,3,$$

we conclude that the rope limit reliability function is given by

$$\mathscr{R}_3^{(0)}(t,\cdot) = [1, \mathscr{R}_3^{(0)}(t,1), \mathscr{R}_3^{(0)}(t,2), \mathscr{R}_3^{(0)}(t,3)], \ t \in (-\infty,\infty),$$

where

$$\mathscr{R}_3^{(0)}(t,u) = 1 - \sum_{i=0}^{9} \frac{\exp[-it]}{i!} \exp[-\exp[-t]].$$

Hence, considering (3.49), since

$$a_n(1) = 5.00, \ b_n(1) = 17.92,$$

$$a_n(2) = 2.5, \ b_n(2) = 8.96,$$

$$a_n(3) = 1.67, \ b_n(3) = 5.97,$$

then the approximate formula for the rope reliability function takes the form

$$R_{36}^{(10)}(t,\cdot) \cong \mathscr{R}_3^{(0)}\left(\frac{t - b_n(u)}{a_n(u)}, \cdot\right)$$

$$= [1, 1 - \sum_{i=0}^{9} \frac{\exp[-i(0.2t - 3.58)]}{i!} \exp[-\exp[-0.2t + 3.58]],$$

$$1 - \sum_{i=0}^{9} \frac{\exp[-i(0.4t - 3.58)]}{i!} \exp[-\exp[-0.4t + 3.58]],$$

$$1 - \sum_{i=0}^{9} \frac{\exp[-i(0.6t - 3.58)]}{i!} \exp[-\exp[-0.6t + 3.58]] \ ], \ t \in (-\infty,\infty).$$

The approximate mean values of the rope lifetimes $T(u)$ in the state subsets and their standard deviations in years, by (3.13), are:

$$M(1) \cong 6.66, \ M(2) \cong 3.33, \ M(3) \cong 2.22,$$

$$\sigma(1) \cong 1.62, \ \sigma(2) \cong 0.81, \ \sigma(3) \cong 0.54,$$

whereas, from (3.17), the approximate mean values of the rope sojourn times in the particular reliability states are:

$\overline{M}(1) \cong 3.33, \ \overline{M}(2) \cong 1.11, \ \overline{M}(3) \cong 2.22.$

If the critical state is $r = 2$, then from (3.18) the rope risk function is approximately given by

$$r(t) \cong \sum_{i=0}^{9} \frac{\exp[-i(0.4t - 3.58)]}{i!} \exp[-\exp[-0.4t + 3.58]], \ t \in (-\infty, \infty).$$

The moment when the risk exceeds an admissible level $\delta = 0.05$, after applying (3.19), is

$\tau \cong 2.0738$ years.

The behaviour of the rope system reliability function and its risk function are illustrated in Table 5.2 and Figure 5.6.

Table 5.2. The values of the still rope multi-state reliability function and risk function

t	$\Re_3^{(0)}(\frac{t-b_n(1)}{a_n(1)},1)$	$\Re_3^{(0)}(\frac{t-b_n(2)}{a_n(2)},2)$	$\Re_3^{(0)}(\frac{t-b_n(3)}{a_n(3)},3)$	$r(t)$
0.2	1.000000	0.999999	0.999998	0.000000
0.6	0.999998	0.999977	0.999795	0.000023
1.0	0.999990	0.999609	0.994249	0.000391
1.4	0.999950	0.996412	0.945898	0.003588
1.8	0.999795	0.980140	0.776749	0.019860
2.2	0.999283	0.927919	0.493317	0.072081
2.6	0.997831	0.815200	0.231067	0.184800
3.0	0.994249	0.642210	0.080576	0.357790
3.4	0.986488	0.444152	0.021678	0.555848
3.8	0.971569	0.267825	0.004693	0.732175
4.2	0.945898	0.141305	0.000851	0.858695
4.6	0.906025	0.065843	0.000134	0.934157
5.0	0.849686	0.027422	0.000019	0.972578
5.4	0.776749	0.010338	0.000002	0.989662
5.8	0.689654	0.003571	0.000000	0.996429
6.2	0.593139	0.001143	0.000000	0.998857
6.6	0.493317	0.000342	0.000000	0.999658
7.0	0.396455	0.000097	0.000000	0.999903
7.4	0.307841	0.000026	0.000000	0.999974
7.8	0.231067	0.000007	0.000000	0.999993

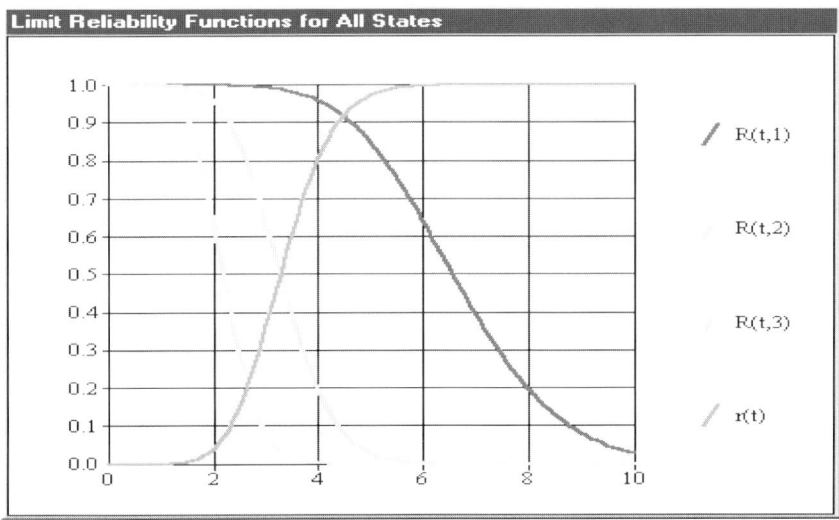

Fig. 5.6. The graphs of the still rope multi-state reliability function and risk function

Unfortunately, similarly to the case of two-state systems, there are no extensions of Lemmas 4.5–4.7 to non-homogeneous multi-state "*m* out of *n*" systems.

5.4. Reliability evaluation of multi-state series-parallel systems

In proving facts on limit reliability functions for homogeneous regular multi-state series-parallel systems the following extensions of Lemmas 4.12–4.13 are used ([74]).

Lemma 5.8
If

(i) $k_n \to \infty$,

(ii) $\mathscr{R}(t,u) = 1 - \exp[-V(t,u)]$, $u = 1,2,...,z$, is a non-degenerate reliability function,

(iii) $\boldsymbol{R}_{k_n,l_n}(t,\cdot) = [1,\boldsymbol{R}_{k_n,l_n}(t,1),...,\boldsymbol{R}_{k_n,l_n}(t,z)]$, $t \in (-\infty,\infty)$, is the reliability function of a homogeneous regular multi-state series-parallel system defined by (3.28)–(3.29),

(iv) $a_n(u) > 0$, $b_n(u) \in (-\infty,\infty)$, $u = 1,2,...,z$,

then

$$\mathfrak{R}(t,\cdot) = [1,\mathfrak{R}(t,1),...,\mathfrak{R}(t,z)],\ t \in (-\infty,\infty),$$

is the limit multi-state reliability function of this system, i.e.

$$\lim_{n\to\infty} \mathbf{R}_{k_n,l_n}(a_n(u)t + b_n(u),u) = \mathfrak{R}(t,u) \text{ for } t \in C_{\mathfrak{R}(u)},\ u = 1,2,...,z, \tag{5.36}$$

if and only if

$$\lim_{n\to\infty} k_n [R(a_n(u)t + b_n(u),u)]^{l_n} = V(t,u) \text{ for } t \in C_{V(u)},\ u = 1,2,...,z. \tag{5.37}$$

Motivation: For each fixed u, $u = 1,2,...,z$, assumptions (i)–(iv) of Lemma 5.8 are identical to assumptions (i)–(iv) of Lemma 4.12, condition (5.36) is identical to condition (4.47) and condition (5.37) is identical to condition (4.48). Moreover, since from Lemma 4.12, conditions (4.47) and (4.48) are equivalent, then conditions (5.36) and (5.37) are equivalent.

Lemma 5.9
If

(i) $k_n \to k,\ k > 0,\ l_n \to \infty,$

(ii) $\mathfrak{R}(t,u),\ u = 1,2,...,z$, is a non-degenerate reliability function,

(iii) $\mathbf{R}_{k_n,l_n}(t,\cdot) = [1,\mathbf{R}_{k_n,l_n}(t,1),...,\mathbf{R}_{k_n,l_n}(t,z)],\ t \in (-\infty,\infty)$, is the reliability function
of a homogeneous regular multi-state series-parallel system defined by (3.28)–
(3.29),

(iv) $a_n(u) > 0,\ b_n(u) \in (-\infty,\infty),\ u = 1,2,...,z,$

then

$$\mathfrak{R}(t,\cdot) = [1,\mathfrak{R}(t,1),...,\mathfrak{R}(t,z)],\ t \in (-\infty,\infty),$$

is its multi-state limit reliability function, i.e.

$$\lim_{n\to\infty} \mathbf{R}_{k_n,l_n}(a_n(u)t + b_n(u),u) = \mathfrak{R}(t,u) \text{ for } t \in C_{\mathfrak{R}(u)},\ u = 1,2,...,z, \tag{5.38}$$

if and only if

$$\lim_{n\to\infty} [R(a_n(u)t + b_n(u),u)]^{l_n} = \mathfrak{R}_0(t,u) \text{ for } t \in C_{\mathfrak{R}_0(u)},\ u = 1,2,...,z, \tag{5.39}$$

where $\mathcal{R}_0(t,u)$ is a non-degenerate reliability function and moreover

$$\mathcal{R}(t,u) = 1 - [1 - \mathcal{R}_0(t,u)]^k \quad \text{for } t \in (-\infty,\infty), u = 1,2,...,z. \tag{5.40}$$

Motivation: For each fixed u, $u = 1,2,...,z$, assumptions (i)–(iv) of Lemma 5.9 are identical to assumptions (i)–(iv) of Lemma 4.13. Moreover, condition (5.38) is equivalent to condition (4.49) and condition (5.39) is equivalent to condition (4.50). And moreover, since from Lemma 4.13, condition (4.49) and condition (4.50) are equivalent, then condition (5.38) and condition (5.39) are equivalent and moreover, considering (4.51), the equality (5.40) holds.

Lemmas 5.8–5.9 and Theorem 4.6 from Chapter 4 are the basis for formulating the next theorem ([74]).

Theorem 5.6
The class of limit non-degenerate reliability functions of the homogeneous regular multi-state series-parallel system is composed of $3^z + 4^z + 3^z$ reliability functions of the form

$$\mathcal{R}(t,\cdot) = [1,\mathcal{R}(t,1),...,\mathcal{R}(t,z)], \, t \in (-\infty,\infty), \tag{5.41}$$

where

Case 1. $k_n = n$, $|l_n - c \log n| >> s, s > 0, c > 0$ (under Assumption 4.1).

$$\mathcal{R}(t,u) \in \{\mathcal{R}_1(t),\mathcal{R}_2(t),\mathcal{R}_3(t)\}, u = 1,2,...,z, \tag{5.42}$$

and $\mathcal{R}_i(t)$, $i = 1,2,3$, are defined by (4.52)–(4.54),

Case 2. $k_n = n$, $l_n - c \log n \sim s, s \in (-\infty,\infty), c > 0$.

$$\mathcal{R}(t,u) \in \{\mathcal{R}_4(t),\mathcal{R}_5(t),\mathcal{R}_6(t),\mathcal{R}_7(t)\}, u = 1,2,...,z, \tag{5.43}$$

and $\mathcal{R}_i(t)$, $i = 4,5,6,7$, are defined by (4.55)–(4.58),

Case 3. $k_n \rightarrow k, k > 0, l_n \rightarrow \infty$.

$$\mathcal{R}(t,u) \in \{\mathcal{R}_8(t),\mathcal{R}_9(t),\mathcal{R}_{10}(t)\}, u = 1,2,...,z, \tag{5.44}$$

and $\mathcal{R}_i(t)$, $i = 8,9,10$, are defined by (4.59)–(4.61).
Motivation: For each fixed u, $u = 1,2,...,z$, co-ordinate $\mathcal{R}(t,u)$ of the vector $\mathcal{R}(t,\cdot)$ defined by (5.41), from Theorem 4.6 that is the consequence of Lemma 4.12 and Lemma 4.13, can be a reliability function that is one of three types defined by (4.52)–(4.54), or one of four types defined by (4.55)–(4.58), or one of three types defined by (4.59)–(4.61). Thus the number of different multi-state reliability functions of the

considered system is equal to the sum of the number of z-term variations of the 3-component sets defined by (5.42) and (5.44) and the 4-component set defined by (5.43), i.e. $3^z + 4^z + 3^z$, and they are of the form (5.41).

Corollary 5.8
If components of the homogeneous regular multi-state series-parallel system have Weibull reliability functions

$$R(t,\cdot) = [1,R(t,1),...,R(t,z)], \ t \in (-\infty,\infty),$$

where

$$R(t,u) = 1 \text{ for } t < 0, \ R(t,u) = \exp[-\beta(u)t^{\alpha(u)}] \text{ for } t \geq 0, \ \alpha(u) > 0, \ \beta(u) > 0,$$
$$u = 1,2,...,z,$$

and

$$k_n = n, \ l_n > 0,$$

$$a_n(u) = b_n(u)/(\alpha(u)\log n), \ b_n(u) = (\log n/(\beta(u)l_n))^{1/\alpha(u)}, \ u = 1,2,...,z,$$

then

$$\mathcal{R}_3(t,\cdot) = [1,\mathcal{R}_3(t,1),...,\mathcal{R}_3(t,z)], \ t \in (-\infty,\infty),$$

where

$$\mathcal{R}_3(t,u) = 1 - \exp[-\exp[-t]] \text{ for } t \in (-\infty,\infty), \ u = 1,2,...,z,$$

is its limit reliability function.
Motivation: Since for each fixed u, sufficiently large n and all $t \in (-\infty,\infty)$, we have

$$a_n(u)t + b_n(u) = b_n(u)(t/(\alpha(u)\log n) + 1) > 0,$$

then for $t \in (-\infty,\infty)$ we get

$$R(a_n(u)t + b_n(u),u) = \exp[-\beta(u)(a_n(u)t + b_n(u))^{\alpha(u)}]$$

$$= \exp[-(\log n)/l_n \cdot (t/(\alpha(u)\log n) + 1)^{\alpha(u)}]$$

$$= \exp[-(\log n)/l_n - t/l_n - o(1/l_n)].$$

Further, for all $t \in (-\infty,\infty)$, we have

$$V(t,u) = \lim_{n\to\infty} k_n[R(a_n(u)t + b_n(u),u)]^{l_n}$$

$$= \lim_{n \to \infty} n \exp[-t - \log n + l_n o(1/l_n)] = \exp[-t],$$

which from Lemma 5.8 completes the proof.

Example 5.8 (*a model series-parallel system*)
If the homogeneous regular multi-state series-parallel system is such that

$$k_n = 30, \ l_n = 10, \ z = 5,$$

and its components have Weibull reliability functions with parameters

$$\beta(u) = 10^{-5}, \ \alpha(u) = (11 + u)/4, \ u = 1,2,3,4,5,$$

then, according to Corollary 5.8, assuming normalising constants

$$a_n(u) = 4b_n(u)/((11 + u)\cdot\log30), \ b_n(u) = (10^4 \cdot \log30)^{4/(11 + u)}, \ u = 1,2,3,4,5,$$

considering (3.49), we get the following approximate formula

$$[1, \mathbf{R}_{30,10}(t,1), \mathbf{R}_{30,10}(t,2), \mathbf{R}_{30,10}(t,3), \mathbf{R}_{30,10}(t,4), \mathbf{R}_{30,10}(t,5)]$$

$$\cong [1, 1 - \exp[-\exp[-0.315t + 10.204]], 1 - \exp[-\exp[-0.446t + 11.054]],$$

$$1 - \exp[-\exp[-0.604t + 11.904]], 1 - \exp[-\exp[-0.789t + 12.755]],$$

$$1 - \exp[-\exp[-1.002t + 13.605]]] \text{ for } t \in (-\infty,\infty).$$

According to (3.4), the expected values of the system components' sojourn times $T_{ij}(u)$ in the state subsets are given by the formula ([71])

$$M_{ij}(u) = E[T_{ij}(u)] = (\beta(u))^{-1/\alpha(u)} \Gamma((\alpha(u) + 1)/\alpha(u))$$

$$= 10^{20/(11 + u)} \Gamma((15 + u)/(11 + u)), \ u = 1,2,...,5.$$

Hence, in particular, we have

$$M_{ij}(1) \cong 41.46, \ M_{ij}(2) \cong 31.01, \ M_{ij}(3) \cong 24.19, \ M_{ij}(4) \cong 19.40, \ M_{ij}(5) \cong 16.12,$$

and according to (3.8) the expected values of the system components' lifetimes in particular states are:

$$\overline{M}_{ij}(1) \cong 10.45, \ \overline{M}_{ij}(2) \cong 6.82, \ \overline{M}_{ij}(3) \cong 4.79, \ \overline{M}_{ij}(4) \cong 3.28, \ \overline{M}_{ij}(5) \cong 16.12.$$

The mean values of the system sojourn times $T(u)$ in the state subsets after applying the formula (3.13), are given by ([13], [71])

$$M(u) = E[T(u)] \cong 0.5772 \, a_n(u) + b_n(u), \, u = 1,2,...,5,$$

i.e.

$$M(1) \cong 34.23, \, M(2) \cong 26.09, \, M(3) \cong 20.67, \, M(4) \cong 16.89, \, M(5) \cong 14.16.$$

Hence, from (3.17), the mean values of the system lifetimes in particular states are:

$$\overline{M}(1) \cong 8.14, \, \overline{M}(2) \cong 5.42, \, \overline{M}(3) \cong 3.78, \, \overline{M}(4) \cong 2.73, \, \overline{M}(5) \cong 14.16.$$

If the critical state is $r = 2$, then from (3.18) the system risk function is

$$r(t) \cong \exp[-\exp[-0.446t + 11.054]].$$

The moment when the system risk exceeds an admissible level $\delta = 0.05$, from (3.19), is

$$\tau = r^{-1}(\delta) \cong [11.054 - \log[-\log \delta]]/0.446 = 22.32.$$

Corollary 5.9
If components of the homogeneous regular multi-state series-parallel system have Weibull reliability functions

$$R(t,\cdot) = [1, R(t,1),...,R(t,z)], \, t \in (-\infty,\infty),$$

where

$$R(t,u) = 1 \text{ for } t < 0, \, R(t,u) = \exp[-\beta(u) \, t^{\alpha(u)}] \text{ for } t \ge 0, \, \alpha(u) > 0, \, \beta(u) > 0,$$
$$u = 1,2,...,z,$$

and

$$k_n \to k, \, l_n \to \infty,$$

$$a_n(u) = (\beta(u)l_n)^{-1/\alpha(u)}, \, b_n(u) = 0, \tag{5.45}$$

then

$$\mathcal{R}_9(t,\cdot) = [1, \mathcal{R}_9(t,1),...,\mathcal{R}_9(t,z)]$$

where

$$\mathcal{R}_9(t,u) = 1 \text{ for } t < 0, \, \mathcal{R}_9(t,u) = 1 - [1 - \exp[-t^{\alpha(u)}]]^3 \text{ for } t \ge 0,$$

is its limit reliability function.

Motivation: Corollary 5.9 is a particular case of more general Corollary 5.10, which will be proved in a later part of this chapter. Therefore, we omit its proof.

Example 5.9 (*a pipeline system*)
Let us consider the pipeline system composed of $k_n = 3$ lines of pipe segments linked in parallel, each of them composed of $l_n = 100$ five-state identical segments linked in series. Considering pipe segments as basic components of the pipeline system, according to Definitions 3.12–3.13, we conclude that it is a homogeneous regular five-state series-parallel system. Therefore, from (3.28)–(3.29), the pipeline system reliability function is given by

$$R_{3,100}(t,\cdot) = [1, R_{3.1000}(t,1), R_{3,100}(t,2), R_{3,100}(t,3), R_{3,100}(t,4)], \qquad (5.46)$$

where

$$R_{3,100}(t,u) = 1 - [1 - [R(t,u)]^{100}]^3, \ t \in (-\infty,\infty), \ u = 1,2,3,4. \qquad (5.47)$$

Taking into account pipe segment reliability data given in their technical certificates and expert opinions we assume that they have Weibull reliability functions

$$R(t,u) = 1 \text{ for } t < 0, \ R(t,u) = \exp[-\beta(u)t^{\alpha(u)}] \text{ for } t \geq 0, \ u = 1,2,3, 4,$$

with the following parameters:

$$\alpha(1) = 3, \ \beta(1) = 0.00001, \ \alpha(2) = 2.5, \ \beta(2) = 0.0001,$$

$$\alpha(3) = 2, \ \beta(3) = 0.0016, \ \alpha(4) = 1, \ \beta(4) = 0.05.$$

Hence and from (5.46)–(5.47) it follows that the pipeline system exact reliability function is given by

$$R_{3,100}(t,\cdot) = [1, 1 - [1 - \exp[-0.001t^3]]^3, 1 - [1 - \exp[-0.01t^{5/2}]]^3,$$

$$1 - [1 - \exp[-0.16t^2]]^3, 1 - [1 - \exp[-5t]]^3] \text{ for } t \geq 0.$$

From (3.4), the mean values $M_{ij}(u)$, $u = 1,2,3,4$, of the pipe segments in the state subsets in years are:

$$M_{ij}(1) = \Gamma(4/3)(0.00001)^{-1/3} \cong 41.45, \ M_{ij}(2) = \Gamma(7/5)(0.0001)^{-2/5} \cong 35.32,$$

$$M_{ij}(3) = \Gamma(3/2)(0.0016)^{-1/2} \cong 22.16, \ M_{ij}(4) = \Gamma(2)(0.5)^{-2} \cong 20.00,$$

while from (3.8), the mean values $\overline{M}_{ij}(u)$, $u = 1,2,3,4$, of the pipe segments in particular states are:

$$\overline{M}_{ij}(1) \cong 6.13,\ \overline{M}_{ij}(2) \cong 13.16,\ \overline{M}_{ij}(3) \cong 2.16,\ \overline{M}_{ij}(4) \cong 20.00.$$

Assuming, according to (5.45), normalising constants

$$a_n(u) = (\beta(u)l_n)^{-1/\alpha(u)},\ b_n(u) = 0,\ u = 1,2,3,4,$$

and applying Corollary 5.9, we conclude that the limit reliability function of the pipeline system is

$$\mathcal{R}_9(t,\cdot) = [1, \mathcal{R}_9(t,1), \mathcal{R}_9(t,2), \mathcal{R}_9(t,3), \mathcal{R}_9(t,4)],\ t \in (-\infty,\infty),$$

where

$$\mathcal{R}_9(t,1) = 1 - [1 - \exp[-t^3]]^3,\ \mathcal{R}_9(t,2) = 1 - [1 - \exp[-t^{5/2}]]^3,$$

$$\mathcal{R}_9(t,3) = 1 - [1 - \exp[-t^2]]^3,\ \mathcal{R}_9(t,4) = 1 - [1 - \exp[-t]]^3 \text{ for } t \geq 0.$$

Since, in particular from (5.45), we have

$$a_n(1) = 10,\ a_n(2) = 6.31,\ a_n(3) = 2.5,\ a_n(4) = 0.2,\ b_n(u) = 0,\ u = 1,2,3,4.$$

then applying the approximate formula (3.49) for $t \geq 0$, we get

$$R_{3,100}(t,\cdot) \cong \mathcal{R}_9((t - b_n(u))/a_n(u),\cdot)$$

$$= [1, 1 - [1 - \exp[-0.001t^3]]^3, 1 - [1 - \exp[-0.01t^{5/2}]]^3,$$

$$1 - [1 - \exp[-0.16t^2]]^3, 1 - [1 - \exp[-5t]]^3]. \tag{5.48}$$

The expected values $M(u)$, $u = 1,2,3,4$, of the system sojourn times in the state subsets in years, calculated on the basis of the approximate formula (5.48), according to (3.13), are:

$$M(1) = \Gamma(4/3)[3(0.001)^{-1/3} - 3(0.002)^{-1/3} + (0.003)^{-1/3}] \cong 11.72,$$

$$M(2) = \Gamma(7/5)[3(0.01)^{-2/5} - 3(0.02)^{-2/5} + (0.03)^{-2/5}] \cong 7.67,$$

$$M(3) = \Gamma(3/2)[3(0.16)^{-1/2} - 3(0.32)^{-1/2} + (0.48)^{-1/2}] \cong 3.23,$$

$$M(4) = \Gamma(2)[3(5)^{-1} - 3(10)^{-1} + (15)^{-1}] \cong 0.37.$$

Hence, from (3.17), the system mean lifetimes $\overline{M}(u)$ in particular states are:

$$\overline{M}(1) \cong 4.05, \ \overline{M}(2) \cong 4.44, \ \overline{M}(3) \cong 2.86, \ \overline{M}(4) \cong 0.37.$$

If the critical state is $r = 2$, then the system risk function, according (3.18), is given by

$$r(t) = [1 - \exp[-0.01t^{5/2}]]^3.$$

The moment when the system risk exceeds an admissible level $\delta = 0.05$, from (3.19), is

$$\tau = r^{-1}(\delta) = [-100 \log(1 - \sqrt[3]{\delta})]^{2/5} \cong 4.62.$$

The behaviour of the exact and approximate multi-state system reliability function co-ordinate $u = 2$ and the risk function are presented in Tables 5.3–5.4 and Figures 5.7–5.8.

Table 5.3. The values of the component $u = 2$ of the exact and approximate piping system reliability function

t	$R_{3,100}(t,2) = \mathcal{R}_9((t - b_n(2))/a_n(2))$
0.0	1.000
1.5	1.000
3.0	0.997
4.5	0.957
6.0	0.799
7.5	0.515
9.0	0.242
10.5	0.082
12.0	0.020
13.5	0.004
15.0	0.000

Table 5.4. The values of the piping system risk function

t	$r(t)$
0.0	0.000
1.5	0.000
3.0	0.003
4.5	0.043
6.0	0.201
7.5	0.485
9.0	0.758
10.5	0.918
12.0	0.980
13.5	0.996
15.0	1.000

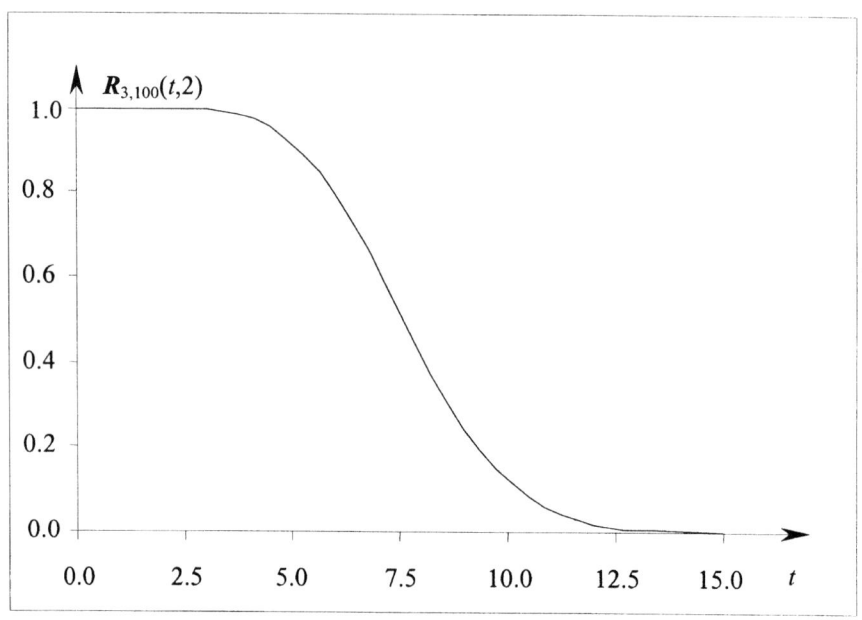

Fig. 5.7. The graph of the component $u = 2$ of the exact and approximate piping system reliability function

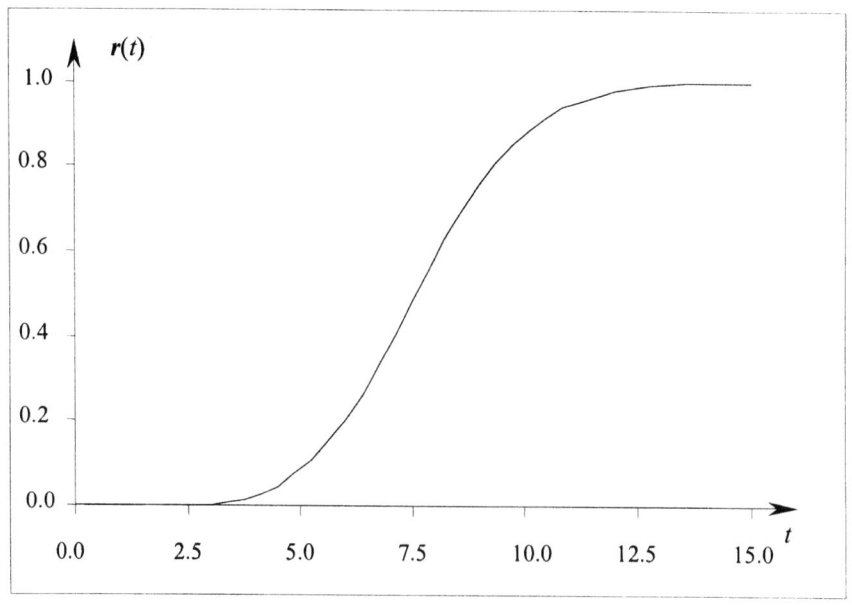

Fig. 5.8. The graph of the piping system risk function

The following extensions of Lemmas 4.15-4.16 are necessary tools in determination of limit reliability functions of non-homogeneous regular multi-state series-parallel systems ([74]).

Lemma 5.10
If

(i) $k_n \to \infty$,

(ii) $\mathcal{R}'(t,u) = 1 - \exp[-V'(t,u)]$, $u = 1,2,...,z$, is a non-degenerate reliability function,

(iii) $\mathbf{R}'_{k_n,l_n}(t,\cdot) = [1, \mathbf{R}'_{k_n,l_n}(t,1) ,..., \mathbf{R}'_{k_n,l_n}(t,z)]$, $t \in (-\infty,\infty)$, is the reliability function of a non-homogeneous regular multi-state series-parallel system defined by (3.43)–(3.45),

(iv) $a_n(u) > 0$, $b_n(u) \in (-\infty,\infty)$, $u = 1,2,...,z$,

(v) $R(t,u)$ for each fixed u, is one of reliability functions $R^{(1)}(t,u)$, $R^{(2)}(t,u)$, ..., $R^{(a)}(t,u)$ defined by (3.45) such that

(vi) $\exists N(u)$ $\forall n > N(u)$ $R(a_n(u)t + b_n(u),u) \neq 0$ for $t < t_0(u)$, $R(a_n(u)t + b_n(u),u) = 0$ for $t \geq t_0(u)$, where $t_0(u) \in (-\infty,\infty>$,

(vii) $\lim\limits_{n\to\infty} \dfrac{R^{(i)}(a_n(u)t + b_n(u),u)}{R(a_n(u)t + b_n(u),u)} \leq 1$ for $t < t_0(u)$, $i = 1,2,...,a$, $u = 1,2,...,z$,

and moreover there exist non-increasing functions

(viii) $d(t,u) = \begin{cases} \lim\limits_{n\to\infty} \sum\limits_{i=1}^{a} q_i d_i(a_n(u)t + b_n(u),u) & \text{for } t < t_0(u) \\ 0 & \text{for } t \geq t_0(u), \end{cases}$ (5.49)

where

$$d_i(a_n(u)t + b_n(u),u) = [\frac{R^{(i)}(a_n(u)t + b_n(u),u)}{R(a_n(u)t + b_n(u),u)}]^{l_n},$$

then

$\mathcal{R}'(t,\cdot) = [1, \mathcal{R}'(t,1) ,..., \mathcal{R}'(t,z)]$, $t \in (-\infty,\infty)$,

is its multi-state limit reliability function, i.e.

$$\lim_{n\to\infty} \boldsymbol{R'}_{k_n,l_n}(a_n(u)t+b_n(u),u) = \boldsymbol{\mathfrak{R}'}(t,u) \text{ for } t\in C_{\mathfrak{R'}(u)}, \ u=1,2,...,z, \tag{5.50}$$

if and only if

$$\lim_{n\to\infty} k_n[R(a_n(u)t+b_n(u),u)]^{l_n}d(t,u) = V'(t,u) \text{ for } t\in C_{V'(u)}, \ u=1,2,...,z. \tag{5.51}$$

Motivation: Since for each fixed u, $u=1,2,...,z$, assumptions (i)–(viii) of Lemma 5.10 are identical to assumptions (i)–(viii) of Lemma 4.15, condition (5.50) is identical to condition (4.66) and condition (5.51) is identical to condition (4.67), which from Lemma 4.15 are equivalent, then Lemma 5.10 is valid.

Lemma 5.11
If

(i) $k_n\to k$, $k>0$, $l_n\to\infty$,

(ii) $\boldsymbol{\mathfrak{R}'}(t,u)$, $u=1,2,...,z$, is a non-degenerate reliability function,

(iii) $\boldsymbol{R'}_{k_n,l_n}(t,\cdot) = [1, \boldsymbol{R'}_{k_n,l_n}(t,1),..., \boldsymbol{R'}_{k_n,l_n}(t,z)]$, $t\in(-\infty,\infty)$, is the reliability function of a non-homogeneous regular multi-state series-parallel system defined by (3.43)–(3.45),

(iv) $a_n(u)>0$, $b_n(u)\in(-\infty,\infty)$, $u=1,2,...,z$,

(v) $R(t,u)$ for each fixed u, is one of reliability functions $R^{(1)}(t,u)$, $R^{(2)}(t,u)$, ..., $R^{(a)}(t,u)$ defined by (3.45) such that

(vi) $\exists\, N(u)\ \forall\, n>N(u)\ R(a_n(u)t+b_n(u),u)\neq 0$ for $t<t_0(u)$, $R(a_n(u)t+b_n(u),u)=0$ for $t\geq t_0(u)$, where $t_0(u)\in(-\infty,\infty>$,

(vii) $\displaystyle\lim_{n\to\infty}\frac{R^{(i)}(a_n(u)t+b_n(u),u)}{R(a_n(u)t+b_n(u),u)}\leq 1$ for $t<t_0(u)$, $i=1,2,...,a$, $u=1,2,...,z$,

and moreover there exist non-increasing functions

$$(viii)\ \ d_i(t,u) = \begin{cases} \displaystyle\lim_{n\to\infty} d_i(a_n(u)t+b_n(u),u) & \text{for } t<t_0(u) \\[2mm] 0 & \text{for } t\geq t_0(u), \end{cases} \tag{5.52}$$

where

$$d_i(a_n(u)t + b_n(u), u) = [\frac{R^{(i)}(a_n(u)t + b_n(u), u)}{R(a_n(u)t + b_n(u), u)}]^{l_n},$$ (5.53)

then

$$\mathcal{R}'(t, \cdot) = [1, \mathcal{R}'(t,1), ..., \mathcal{R}'(t, z)], t \in (-\infty, \infty),$$

is its multi-state limit reliability function, i.e.

$$\lim_{n \to \infty} \mathbf{R'}_{k_n, l_n}(a_n(u)t + b_n(u), u) = \mathcal{R}'(t, u) \text{ for } t \in C_{\mathcal{R}'(u)}, u = 1, 2, ..., z,$$ (5.54)

if and only if

$$\lim_{n \to \infty} [R(a_n(u)t + b_n(u), u)]^{l_n} = \mathcal{R}_0(t, u) \text{ for } t \in C_{\mathcal{R}_0(u)}, u = 1, 2, ..., z,$$ (5.55)

where $\mathcal{R}_0(t, u)$ is a non-degenerate reliability function and moreover

$$\mathcal{R}'(t, u) = 1 - \prod_{i=1}^{a} [1 - d_i(t, u) \mathcal{R}_0(t, u)]^{q_i k}, \; t \in (-\infty, \infty), u = 1, 2, ..., z.$$ (5.56)

Motivation: For each fixed u, $u = 1, 2, ..., z$, assumptions (i)–(viii) of Lemma 5.11 and assumptions (i)–(viii) of Lemma 4.16 are equivalent. Condition (5.54) is identical to condition (4.69) and condition (5.55) is identical to condition (4.70). Moreover, from Lemma 4.16 conditions (4.69) and (4.70) are equivalent, which means that conditions (5.54) and (5.55) are also equivalent and from (4.71) the equality (5.56) holds.

Lemma 5.10, Lemma 5.11 and Theorem 4.7 from Chapter 4 establish the class of limit reliability functions for non-homogeneous regular multi-state series-parallel systems given in the following theorem ([74]).

Theorem 5.7
The class of limit non-degenerate reliability functions of the non-homogeneous regular multi-state series-parallel system is composed of $3^z + 4^z + 3^z$ reliability functions of the form

$$\mathcal{R}'(t, \cdot) = [1, \mathcal{R}'(t,1), ..., \mathcal{R}'(t, z)], t \in (-\infty, \infty),$$ (5.57)

where

Case 1. $k_n = n$, $|l_n - c \log n| \gg s$, $s > 0$, $c > 0$ (under Assumption 4.1 and the assumptions of Lemma 5.10).

$$\mathcal{R}'(t, u) \in \{ \mathcal{R}'_1(t), \mathcal{R}'_2(t), \mathcal{R}'_3(t) \}, u = 1, 2, ..., z,$$ (5.58)

and $\mathcal{R}'_i(t)$, $i = 1,2,3$, are defined by (4.72)–(4.74) with $d(t) = d(t,u)$, $u = 1,2,...z$, where $d(t,u)$ are defined by (5.49),

Case 2. $k_n = n$, $l_n - c \log n \sim s$, $s \in (-\infty,\infty)$, $c > 0$ (under the assumptions of Lemma 5.10).

$$\mathcal{R}'(t,u) \in \{\mathcal{R}'_4(t), \mathcal{R}'_5(t), \mathcal{R}'_6(t), \mathcal{R}'_7(t)\}, \quad u = 1,2,...,z, \tag{5.59}$$

and $\mathcal{R}'_i(t)$, $i = 4,5,6,7$, are defined by (4.75)–(4.78) with $d(t) = d(t,u)$, $u = 1,2,...z$, where $d(t,u)$ are defined by (5.49),

Case 3. $k_n \to k$, $k > 0$, $l_n \to \infty$ (under the assumptions of Lemma 5.11).

$$\mathcal{R}'(t,u) \in \{\mathcal{R}'_8(t), \mathcal{R}'_9(t), \mathcal{R}'_{10}(t)\}, \quad u = 1,2,...,z, \tag{5.60}$$

and $\mathcal{R}'_i(t)$, $i = 8,9,10$, are defined by (4.79)–(4.81) with $d_i(t) = d_i(t,u)$, $u = 1,2,...z$, where $d_i(t,u)$ are defined by (5.52).

Motivation: For each fixed u, $u = 1,2,...,z$, co-ordinate $\mathcal{R}'(t,u)$ of the vector $\mathcal{R}'(t,\cdot)$ defined by (5.57), according to Theorem 4.7, that is a consequence of Lemma 4.15 and Lemma 4.16, can be one of the three types of reliability functions given by (4.72)–(4.74), or one of the four types of reliability function given by (4.75)–(4.78) with $d(t) = d(t,u)$, where $d(t,u)$ are defined by (5.49), or one of the three types of reliability functions given by (4.79)–(4.81) with $d_i(t) = d_i(t,u)$, where $d_i(t,u)$ are defined by (5.52). Thus the number of different limit reliability functions of the considered system is equal to the sum of the number of z-term variations of the 3-element sets defined by (5.58) and (5.60) and the 4-element set defined by (5.59), i.e. $3^z + 4^z + 3^z$, and they are of the form (5.57).

Corollary 5.10
If components of the non-homogeneous regular multi-state series-parallel system have Weibull reliability functions

$$R^{(i,j)}(t,\cdot) = [1, R^{(i,j)}(t,1),..., R^{(i,j)}(t,z)], \quad t \in (-\infty,\infty),$$

where

$R^{(i,j)}(t,u) = 1$ for $t < 0$, $R^{(i,j)}(t,u) = \exp[-\beta_{ij}(u)\, t^{\alpha_{ij}(u)}]$ for $t \geq 0$, $\alpha_{ij}(u) > 0$, $\beta_{ij}(u) > 0$, $u = 1,2,...,z$, $i = 1,2,...,a$, $j = 1,2,...,e_i$,

and

$$k_n \to k, \, l_n \to \infty,$$

$$a_n(u) = (\beta(u)l_n)^{-1/\alpha(u)}, \ b_n(u) = 0, \tag{5.61}$$

where

$$\alpha_i(u) = \min_{1 \le j \le e_i} \{\alpha_{ij}(u)\}, \ \beta_i(u) = \sum_{(j:\alpha_{ij}(u)=\alpha_i(u))} p_{ij}\beta_{ij}(u), \tag{5.62}$$

$$\alpha(u) = \max_{1 \le i \le a} \{\alpha_i(u)\}, \ \beta(u) = \min\{\beta_i(u) : \alpha_i(u) = \alpha(u)\}, \tag{5.63}$$

then

$$\mathfrak{R}'_9(t, \cdot) = [1, \mathfrak{R}'_9(t,1), \dots, \mathfrak{R}'_9(t,z)]$$

where

$$\mathfrak{R}'_9(t,u) = 1 \text{ for } t < 0,$$

$$\mathfrak{R}'_9(t,u) = 1 - \prod_{(i:\alpha_i(u)=\alpha(u))}[1-\exp[-(\beta_i(u)/\beta(u))t^{\alpha(u)}]]^{q_ik} \text{ for } t \ge 0,$$

is its limit reliability function.

Motivation: Since for each fixed u, according to (5.61), we have

$$a_n(u)t + b_n(u) \to 0^- \text{ for } t < 0$$

and

$$a_n(u)t + b_n(u) \to 0^+ \text{ for } t \ge 0 \text{ as } n \to \infty,$$

then for all $i = 1,2,\dots,a$, we get

$$R^{(i)}(a_n(u)t + b_n(u),u) = 1 \text{ for } t < 0$$

and considering (5.62)

$$R^{(i)}(a_n(u)t + b_n(u),u) = \exp[-\sum_{j=1}^{e_i} p_{ij}\beta_{ij}(u)(a_n(u)t)^{\alpha_{ij}(u)}]$$

$$= \exp[-(a_n(u)t)^{\alpha_i(u)}\sum_{j=1}^{e_i} p_{ij}\beta_{ij}(u)(a_n(u)t)^{\alpha_{ij}(u)-\alpha_i(u)}]$$

$$= \exp[-\beta_i(a_n(u)t)^{\alpha_i(u)} + o(1)] \text{ for } t \ge 0.$$

Defining

$R(t,u) = 1$ for $t < 0$ and $R(t,u) = \exp[-\beta(u)t^{\alpha(u)}]$ for $t \geq 0$, $u = 1,2,\ldots,z$,

for all $i = 1,2,\ldots,a$, by (5.63), we get

$$\lim_{n \to \infty} \frac{R^{(i)}(a_n(u)t + b_n(u), u)}{R(a_n(u)t + b_n(u), u)} = 1 \text{ for } t < 0$$

and for $t \geq 0$

$$\lim_{n \to \infty} \frac{R^{(i)}(a_n(u)t + b_n(u), u)}{R(a_n(u)t + b_n(u), u)} = \lim_{n \to \infty} \frac{\exp[-\beta_i(u)(a_n(u)t + b_n(u))^{\alpha_i(u)}]}{\exp[-\beta(u)(a_n(u)t + b_n(u))^{\alpha}]}$$

$$= \lim_{n \to \infty} \exp[-\beta(u)(a_n(u)t)^{\alpha(u)}[\frac{\beta_i(u)}{\beta(u)}(a_n(u)t)^{\alpha_i(u)-\alpha(u)} - 1]] \leq 1.$$

The above means that condition (vii) of Lemma 5.11 is satisfied with $t_0(u) = \infty$ and moreover, from (5.53) it follows that

$$d_i(t,u) = \begin{cases} 1 & \text{for } t < 0 \\ \exp[-(\beta_i(u)/\beta(u)-1)t^{\alpha(u)}] & \text{for } t \geq 0 \end{cases}$$

for i such that

$$\alpha_i(u) = \alpha(u)$$

and

$$d_i(t,u) = \begin{cases} 1 \text{ for } t < 0 \\ 0 \text{ for } t \geq 0 \end{cases}$$

otherwise. Further, we have

$$\mathscr{R}_0(t,u) = \lim_{n \to \infty} (R(a_n t + b_n))^{l_n} = 1 \text{ for } t < 0$$

and

$$\mathscr{R}_0(t,u) = \lim_{n \to \infty} [R(a_n(u)t + b_n(u))]^{l_n}$$

$$= \lim_{n \to \infty} \exp[-l_n \beta(u)(a_n(u)t)^{\alpha(u)}] = \exp[-t^{\alpha(u)}] \text{ for } t \geq 0,$$

which from Lemma 5.11 completes the proof.

Example 5.10 (*a piping system*)
The piping system is composed of $k_n = 3$ pipeline lines linked in parallel, each of them composed of $l_n = 100$ five-state pipe segments. In two of the pipelines there are 40 pipe segments with exponential reliability functions

$$R^{(1,1)}(t,1) = \exp[-0.025t], R^{(1,1)}(t,2) = \exp[-0.026t],$$

$$R^{(1,1)}(t,3) = \exp[-0.028t], R^{(1,1)}(t,4) = \exp[-0.30t] \text{ for } t \geq 0,$$

and 60 pipe segments with Weibull reliability functions

$$R^{(1,2)}(t,1) = \exp[-0.0015t^2], R^{(1,2)}(t,2) = \exp[-0.0016t^2],$$

$$R^{(1,2)}(t,3) = \exp[-0.002t^2], R^{(1,2)}(t,4) = \exp[-0.0025t^2] \text{ for } t \geq 0.$$

The third pipeline is composed of 50 pipe segments with Weibull reliability functions

$$R^{(2,1)}(t,1) = \exp[-0.0007t^3], R^{(2,1)}(t,2) = \exp[-0.0008t^3],$$

$$R^{(2,1)}(t,3) = \exp[-0.0010t^3], R^{(2,1)}(t,4) = \exp[-0.0016t^3] \text{ for } t \geq 0,$$

and 50 pipe segments with Weibull reliability functions

$$R^{(2,2)}(t,1) = \exp[-0.15 \sqrt{t}\,], R^{(2,2)}(t,2) = \exp[-0.16 \sqrt{t}\,],$$

$$R^{(2,2)}(t,3) = \exp[-0.18 \sqrt{t}\,], R^{(2,2)}(t,4) = \exp[-0.2 \sqrt{t}\,] \text{ for } t \geq 0.$$

Thus the piping system is a non-homogeneous regular multi-state series-parallel system in which, according to Definition 3.20, we have

$$k_n = k = 3, l_n = 100, a = 2, q_1 = 2/3, q_2 = 1/3.$$

Therefore, from (3.44), we get

$$\boldsymbol{R'}_{3,100}(t,u) = 1 - \prod_{i=1}^{2}[1 - (R^{(i)}(t,u))^{100}]^{q_i 3}$$

$$= 1 - [1 - (R^{(1)}(t,u))^{100}]^2 \, [1 - (R^{(2)}(t,u))^{100}],$$

where substituting

$$e_1 = 2, p_{11} = 0.4, p_{12} = 0.6,$$

according to (3.45), we get

$$R^{(1)}(t,u) = \prod_{j=1}^{e_1} (R^{(1,j)}(t,u))^{p_{1j}} = (R^{(1,1)}(t,u))^{0.4}(R^{(1,2)}(t,u))^{0.6}$$

and substituting

$$e_2 = 2,\ p_{21} = 0.5,\ p_{22} = 0.5,$$

we get

$$R^{(2)}(t,u) = \prod_{j=1}^{e_2} (R^{(2,j)}(t,u))^{p_{2j}} = (R^{(2,1)}(t,u))^{0.5}(R^{(2,2)}(t,u))^{0.5}.$$

Hence, from (3.43), the exact reliability function of the piping system takes the form

$$\boldsymbol{R}'_{3,100}\,(t,\cdot) = [1, 1 - [1 - \exp[-t - 0.09t^2]]^2 \cdot [1 - \exp[-0.035t^3 - 7.5\sqrt{t}\,],$$

$$1 - [1 - \exp[-1.04t - 0.096t^2]]^2 \cdot [1 - \exp[-0.04t^3 - 8\sqrt{t}\,],$$

$$1 - [1 - \exp[-1.12t - 0.12t^2]]^2 \cdot [1 - \exp[-0.05t^3 - 10\sqrt{t}\,],$$

$$1 - [1 - \exp[-1.2t - 0.15t^2]]^2 \cdot [1 - \exp[-0.08t^3 - 10\sqrt{t}\,]]\ \text{for } t \geq 0.$$

Further, applying Corollary 5.10 and considering (5.62), (5.63) and (5.61), we have

$\alpha_1(u) = \min\{1,2\} = 1$ for $u = 1,2,3,4,$

$\beta_1(1) = 0.4\cdot 0.025 = 0.01,\ \beta_1(2) = 0.4\cdot 0.026 = 0.0104,$

$\beta_1(3) = 0.4\cdot 0.028 = 0.0112,\ \beta_1(4) = 0.4\cdot 0.03 = 0.012,$

$\alpha_2(u) = \min\{3, 0.5\} = 0.5$ for $u = 1,2,3,4,$

$\beta_2(1) = 0.5\cdot 0.15 = 0.075,\ \beta_2(2) = 0.5\cdot 0.16 = 0.08,$

$\beta_2(3) = 0.5\cdot 0.18 = 0.09,\ \beta_2(1) = 0.5\cdot 0.2 = 0.1,$

$\alpha(u) = \max\{1,0.5\} = 1$ for $u = 1,2,3,4,$

$\beta(1) = \min\{0.01\} = 0.01,\ \beta(2) = \min\{0.0104\} = 0.0104,$

$\beta(3) = \min\{0.0112\} = 0.0112,\ \beta(1) = \min\{0.012\} = 0.012,$

$a_n(1) = 1/(0.01 \cdot 100) = 1$, $a_n(2) = 1/(0.0104 \cdot 100) = 0.962$,

$a_n(3) = 1/(0.0112 \cdot 100) = 0.893$, $a_n(4) = 1/(0.012 \cdot 100) = 0.833$,

$b_n(u) = 0$ for $u = 1,2,3,4$,

and we conclude that the system limit reliability function is

$$\mathfrak{R}'_9 (t,\cdot) = [1, 1 - [1 - \exp[-t]]^2, 1 - [1 - \exp[-t]]^2, 1 - [1 - \exp[-t]]^2,$$

$$1 - [1 - \exp[-t]]^2] \text{ for } t \geq 0.$$

Thus, from (3.49), the approximate formula for the piping system reliability function takes the form

$$\boldsymbol{R'}_{3,100} (t,\cdot) \cong \mathfrak{R}'_9 ((t - b_n(u))/a_n(u),\cdot)$$

$$= [1, 1 - [1 - \exp[-t]]^2, 1 - [1 - \exp[-1.04t]]^2, 1 - [1 - \exp[-1.12t]]^2,$$

$$1 - [1 - \exp[-1.2t]]^2] \text{ for } t \geq 0.$$

For instance, the expected values of the first type pipe segments' lifetimes in the state subsets, according to (3.4), are:

$M_{11}(1) = 1/0.025 = 40$, $M_{11}(2) = 1/0.026 \cong 38.46$,

$M_{11}(3) = 1/0.028 \cong 35.71$, $M_{11}(4) = 1/0.030 \cong 33.33$,

and their lifetimes in particular states, according to (3.8), are:

$\overline{M}_{11}(1) \cong 1.54$, $\overline{M}_{11}(2) \cong 2.75$, $\overline{M}_{11}(3) \cong 2.38$, $\overline{M}_{11}(4) \cong 33.33$.

The approximate mean values of the piping system sojourn times in the state subsets, according to (3.13), are:

$M(1) \cong 1.5$, $M(2) \cong 1.44$, $M(3) \cong 1.34$, $M(4) \cong 1.25$.

Hence, from (3.17), the mean values of the piping system lifetime in particular states are:

$\overline{M}(1) \cong 0.06$, $\overline{M}(2) \cong 0.10$, $\overline{M}(3) \cong 0.09$, $\overline{M}(4) \cong 1.25$.

If the critical reliability state of the system is $r = 2$, then according to (3.18) its risk function is given by

$r(t) \cong [1 - \exp[-1.04t]]^2$.

Further, from (3.19), the moment when the system risk exceeds an admissible level $\delta = 0.05$ is

$$\tau = r^{-1}(\delta) = -(1/1.04)\log(1 - \sqrt{\delta}) \cong 0.24 \text{ years.}$$

5.5. Reliability evaluation of multi-state parallel-series systems

In proving facts on limit reliability functions for homogeneous regular multi-state parallel-series systems the following slight extensions of Lemmas 4.18–4.19 are used ([74]).

Lemma 5.12
If

    (i) $k_n \to \infty$,

    (ii) $\overline{\mathcal{R}}(t,u) = \exp[-\overline{V}(t,u)]$, $u = 1,2,...,z$, is a non-degenerate reliability function,

    (iii) $\overline{\boldsymbol{R}}_{k_n,l_n}(t,\cdot) = [1, \overline{\boldsymbol{R}}_{k_n,l_n}(t,1),..., \overline{\boldsymbol{R}}_{k_n,l_n}(t,z))]$, $t \in (-\infty,\infty)$, is the reliability function of a homogeneous regular multi-state parallel-series system defined by (3.30)–(3.31),

    (iv) $a_n(u) > 0$, $b_n(u) \in (-\infty,\infty)$, $u = 1,2,...,z$,

then

$$\overline{\mathcal{R}}(t,\cdot) = [1, \overline{\mathcal{R}}(t,1),..., \overline{\mathcal{R}}(t,z)], t \in (-\infty,\infty),$$

is its limit multi-state reliability function, i.e.

$$\lim_{n\to\infty} \overline{\boldsymbol{R}}_{k_n,l_n}(a_n(u)t + b_n(u),u) = \overline{\mathcal{R}}(t,u) \text{ for } t \in C_{\overline{\mathcal{R}}(u)}, u = 1,2,...,z, \tag{5.64}$$

if and only if

$$\lim_{n\to\infty} k_n[F(a_n(u)t + b_n(u),u)]^{l_n} = \overline{V}(t,u) \text{ for } t \in C_{\overline{V}(u)}, u = 1,2,...,z. \tag{5.65}$$

Motivation: For each fixed u, $u = 1,2,...,z$, assumptions (i)–(iv) of Lemma 5.12 are identical to assumptions (i)–(iv) of Lemma 4.18. Moreover, condition (5.64) is identical to condition (4.88) and condition (5.65) is identical to condition (4.89). Therefore, since from Lemma 4.18, condition (4.88) and condition (4.89) are equivalent, then conditions (5.64) and (5.65) are also equivalent.

Lemma 5.13
If

(i) $k_n \to k$, $k > 0$, $l_n \to \infty$,

(ii) $\overline{\mathcal{R}}(t,u)$, $u = 1,2,...,z$, is a non-degenerate reliability function,

(iii) $\overline{R}_{k_n,l_n}(t,\cdot) = [1, \overline{R}_{k_n,l_n}(t,1), ..., \overline{R}_{k_n,l_n}(t,z)]$, $t \in (-\infty,\infty)$, is the reliability function of a homogeneous regular multi-state parallel-series system defined by (3.30)–(3.31),

(iv) $a_n(u) > 0$, $b_n(u) \in (-\infty,\infty)$, $u = 1,2,...,z$,

then

$$\overline{\mathcal{R}}(t,\cdot) = [1, \overline{\mathcal{R}}(t,1), ..., \overline{\mathcal{R}}(t,z)], \ t \in (-\infty,\infty),$$

is its limit multi-state reliability function, i.e.

$$\lim_{n\to\infty} \overline{R}_{k_n,l_n}(a_n(u)t + b_n(u),u) = \overline{\mathcal{R}}(t,u) \ \text{for} \ t \in C_{\overline{\mathcal{R}}(u)}, u = 1,2,...,z, \tag{5.66}$$

if and only if

$$\lim_{n\to\infty}[F(a_n(u)t + b_n(u),u)]^{l_n} = \mathcal{F}_0(t,u) \ \text{for} \ t \in C_{\mathcal{F}_0(u)}, u = 1,2,...,z, \tag{5.67}$$

where $\mathcal{F}_0(t,u)$ is a non-degenerate distribution function and moreover

$$\overline{\mathcal{R}}(t,u) = [1 - \mathcal{F}_0(t,u)]^k \ \text{for} \ t \in (-\infty,\infty), u = 1,2,...,z. \tag{5.68}$$

Motivation: For each fixed u, $u = 1,2,...,z$, assumptions (i)–(iv) of Lemma 5.13 are identical to assumptions (i)–(iv) of Lemma 4.19. Moreover, condition (5.66) is identical to condition (4.90) and condition (5.67) is identical to condition (4.91). Since from Lemma 4.19, condition (4.90) and condition (4.91) are equivalent, then condition (5.66) and condition (5.67) are also equivalent and moreover due to (4.92) the equality (5.68) holds.

Lemma 5.12, Lemma 5.13 and Theorem 4.8 from Chapter 4 yield the next theorem ([74]).

Theorem 5.8
The class of limit non-degenerate reliability functions of the homogeneous regular multi-state parallel-series system is composed of $3^z + 4^z + 3^z$ reliability functions of the form

$$\overline{\mathcal{R}}(t,\cdot) = [1, \overline{\mathcal{R}}(t,1), ..., \overline{\mathcal{R}}(t,z)], \; t \in (-\infty,\infty), \tag{5.69}$$

where

Case 1. $k_n = n$, $|l_n - c \log n| \gg s$, $s > 0$, $c > 0$ (under Assumption 4.1).

$$\overline{\mathcal{R}}(t,u) \in \{\overline{\mathcal{R}}_1(t), \overline{\mathcal{R}}_2(t), \overline{\mathcal{R}}_3(t)\}, \; u = 1,2,...,z, \tag{5.70}$$

and $\overline{\mathcal{R}}_i(t)$, $i = 1,2,3$, are defined by (4.93)–(4.95),

Case 2. $k_n = n$, $l_n - c \log n \sim s$, $s \in (-\infty,\infty)$, $c > 0$.

$$\overline{\mathcal{R}}(t,u) \in \{\overline{\mathcal{R}}_4(t), \overline{\mathcal{R}}_5(t), \overline{\mathcal{R}}_6(t), \overline{\mathcal{R}}_7(t)\}, \; u = 1,2,...,z, \tag{5.71}$$

and $\overline{\mathcal{R}}_i(t)$, $i = 4,5,6,7$, are defined by (4.96)–(4.99),

Case 3. $k_n \to k$, $k > 0$, $l_n \to \infty$.

$$\overline{\mathcal{R}}(t,u) \in \{\overline{\mathcal{R}}_8(t), \overline{\mathcal{R}}_9(t), \overline{\mathcal{R}}_{10}(t)\}, \; u = 1,2,...,z, \tag{5.72}$$

and $\overline{\mathcal{R}}_i(t)$, $i = 8,9,10$, are defined by (4.100)–(4.102).

Motivation: For each fixed u, $u = 1,2,...,z$, co-ordinate $\overline{\mathcal{R}}(t,u)$ of the vector $\overline{\mathcal{R}}(t,\cdot)$ defined by (5.69), according to Theorem 4.8 that is the consequence of Lemma 4.18 and Lemma 4.19, can be one of the three types of reliability function given by (4.93)–(4.95), or one of the four types of reliability function given by (4.96)–(4.99), or one of the three types of reliability function given by (4.100)–(4.102). Thus the number of different limit multi-state reliability functions of the considered system is equal to the sum of the numbers of z-term variations of the 3-component sets defined by (5.70) and (5.72) and the number of the 4-component set defined by (5.71). It means that this number is equal to $3^z + 4^z + 3^z$ and they are of the form (5.69).

Corollary 5.11
If components of the homogeneous regular multi-state parallel-series system have Weibull reliability functions

$$R(t,\cdot) = [1, R(t,1), ..., R(t,z)], \; t \in (-\infty,\infty),$$

where

$$R(t,u) = 1 \text{ for } t < 0, R(t,u) = \exp[-\beta(u) t^{\alpha(u)}] \text{ for } t \geq 0, \alpha(u) > 0, \beta(u) > 0, \quad (5.73)$$
$$u = 1,2,...,z,$$

and

$$k_n \to k, \; k > 0, \; l_n \to \infty, \quad (5.74)$$

$$a_n(u) = b_n(u)/(\alpha(u)\beta(u)(b_n(u))^{\alpha(u)}), \; b_n(u) = [(\log l_n)/\beta(u)]^{1/\alpha(u)}, \; u = 1,2,...,z, \quad (5.75)$$

then

$$\overline{\mathcal{R}}_{10}(t,\cdot) = [1, \overline{\mathcal{R}}_{10}(t,1), ..., \overline{\mathcal{R}}_{10}(t,z)], \; t \in (-\infty,\infty),$$

where

$$\overline{\mathcal{R}}_{10}(t,u) = [1 - \exp[-\exp[-t]]^k \text{ for } t \in (-\infty,\infty), u = 1,2,...,z,$$

is its limit reliability function.

Motivation: Since for each fixed u, sufficiently large n and all $t \in (-\infty,\infty)$, according to (5.74) and (5.75), we have

$$a_n(u)t + b_n(u) > 0,$$

and from (5.73)

$$F(a_n(u)t + b_n(u),u) = 1 - \exp[-\beta(u)(a_n(u)t + b_n(u))^{\alpha(u)}] \text{ for } t \in (-\infty,\infty),$$

and further from (5.67)

$$\mathfrak{I}_0(t,u) = \lim_{n \to \infty} [F(a_n(u)t + b_n(u),u)]^{l_n}$$

$$= \lim_{n \to \infty} [1 - \exp[-\beta(u)(b_n(u))^{\alpha(u)}((1 + ((a_n(u)/b_n(u))t)^{\alpha(u)}]]^{l_n}$$

$$= \lim_{n \to \infty} [1 - \exp[-(\log l_n)(1 + t/(\alpha(u)\log l_n))^{\alpha(u)}]]^{l_n}$$

$$= \lim_{n \to \infty} [1 - (1/l_n)\exp[-t + o(1)]]^{l_n}$$

$$= \exp[-\exp[-t]] \text{ for } t \in (-\infty,\infty).$$

Thus, after considering (5.68), from Lemma 5.13, $\overline{\mathscr{R}}_{10}(t,\cdot)$ is the limit reliability function of the considered system.

Example 5.11 (*an electrical energy distribution system*)

Let us consider a model energetic network stretched between two poles and composed of three energetic cables, six insulators and two bearers and analyse the reliability of all cables only. Each cable consists of 36 identical wires. Assuming that the cable is able to conduct the current if at least one of its wires is not failed we conclude that it is a homogeneous parallel-series system composed of $k_n = 3$ parallel subsystems linked in series, each of them consisting of $l_n = 36$ basic components. Further, assuming that the wires are four-state components, i.e. $z = 3$, having Weibull reliability functions with parameters

$$\alpha(u) = 2,\ \beta(u) = (7.07)^{2u-8},\ u = 1,2,3,$$

according to Corollary 5.11, assuming normalising constants

$$a_n(u) = (7.07)^{4-u}/(2\sqrt{\log 36}),\ b_n(u) = (7.07)^{4-u}\sqrt{\log 36},\ u = 1,2,3,$$

and applying (3.49) we obtain the following approximate form of the system multi-state reliability function

$$\overline{R}_{3,36}(t,\cdot) = [1,\overline{R}_{3,36}(t,1),\overline{R}_{3,36}(t,2),\overline{R}_{3,36}(t,3)]$$

$$\cong [1,[1-\exp[-\exp[-0.01071t+7.167]]]^3,$$

$$[1-\exp[-\exp[-0.07572t+7.167]]]^3,$$

$$[1-\exp[-\exp[-0.53543t+7.167]]]^3]\ \text{for}\ t \in (-\infty,\infty).$$

The values of the system sojourn times $T(u)$ in the state subsystems in months, after applying (3.13), are given by

$$E[T(u)] \cong \int_0^\infty [1-\exp[-\exp[-(7.07)^{u-4}2\sqrt{\log 36}t+2\log 36]]]^3\,dt,\ u = 1,2,3,$$

and particularly

$$M(1) \cong 650,\ M(2) \cong 100,\ M(3) \cong 15.$$

Hence, from (3.17), the system mean lifetimes in particular states are:

$$\overline{M}(1) \cong 550,\ \overline{M}(2) \cong 85,\ \overline{M}(3) \cong 15.$$

If the critical reliability state of the system is $r = 2$, then its risk function, according to (3.18), is given by

$$r(t) \cong 1 - [1 - \exp[-\exp[-0.07572t + 7.167]]]^3.$$

The moment when the system risk exceeds an admissible level $\delta = 0.05$, calculated due to (3.19), is

$$\tau = r^{-1}(\delta) \cong [7.167 - \log[-\log[1 - (1 - \delta)^{1/3}]]]/0.07572 = 76 \text{ months}.$$

The extensions of Lemmas 4.22–4.23 are essential tools for finding limit reliability functions for non-homogeneous regular multi-state parallel-series systems. They may be formulated as follows ([74]).

Lemma 5.14
If

(i) $k_n \to \infty$,

(ii) $\overline{\mathscr{R}}'(t,u) = \exp[-\overline{V}'(t,u)]$, $u = 1,2,...,z$, is a non-degenerate reliability function,

(iii) $\overline{R}'_{k_n,l_n}(t,\cdot) = [1, \overline{R}'_{k_n,l_n}(t,1),..., \overline{R}'_{k_n,l_n}(t,z)]$, $t \in (-\infty,\infty)$, is the reliability function of a non-homogeneous regular multi-state parallel-series system defined by (3.46)–(3.48),

(iv) $a_n(u) > 0$, $b_n(u) \in (-\infty,\infty)$, $u = 1,2,...,z$,

(v) $F(t,u)$ for each fixed u, is one of the distribution functions $F^{(1)}(t,u), F^{(2)}(t,u),..., F^{(a)}(t,u)$ defined by (3.48) such that

(vi) $\exists\, N(u)\ \forall\, n > N(u)\ F(a_n(u)t + b_n(u),u) = 0$ for $t < t_0(u)$, $F(a_n(u)t + b_n(u),u) \neq 0$ for $t \geq t_0(u)$, where $t_0(u) \in <-\infty,\infty)$,

(vii) $\displaystyle\lim_{n\to\infty} \frac{F^{(i)}(a_n(u)t + b_n(u),u)}{F(a_n(u)t + b_n(u),u)} \leq 1$ for $t \geq t_0(u)$, $i = 1,2,...,a$, $u = 1,2,...,z$,

and moreover there exist non-decreasing functions

(viii) $\overline{d}(t,u) = \begin{cases} 0 & \text{for } t < t_o(u) \\ \displaystyle\lim_{n\to\infty} \sum_{i=1}^{a} q_i \overline{d}_i(a_n(u)t + b_n(u),u) & \text{for } t \geq t_o(u), \end{cases}$ (5.76)

where

$$\bar{d}_i(a_n(u)t+b_n(u),u) = [\frac{F^{(i)}(a_n(u)t+b_n(u),u)}{F(a_n(u)t+b_n(u),u)}]^{l_n},$$

then

$$\overline{\mathscr{R}}'(t,\cdot) = [1, \overline{\mathscr{R}}'(t,1),..., \overline{\mathscr{R}}'(t,z)], t \in (-\infty,\infty),$$

is its limit multi-state reliability function, i.e.

$$\lim_{n\to\infty} \overline{R}'_{k_n,l_n}(a_n(u)t + b_n(u),u) = \overline{\mathscr{R}}'(t,u) \text{ for } t \in C_{\overline{\mathscr{R}}'(u)}, u = 1,2,...,z, \qquad (5.77)$$

if and only if

$$\lim_{n\to\infty} k_n(F(a_n(u)t+b_n(u),u))^{l_n} \bar{d}(t,u) = \overline{V}'(t,u) \text{ for } t \in C_{\overline{V}'(u)}, u = 1,2,...,z. \quad (5.78)$$

Motivation: Since for each fixed u, $u = 1,2,...,z$, assumptions (i)–(viii) of Lemma 5.14 are identical with assumptions (i)–(viii) of Lemma 4.22, condition (5.77) is identical to condition (4.105) and condition (5.78) is identical to condition (4.106), which from Lemma 4.22 are equivalent, then Lemma 5.14 is valid.

Lemma 5.15
If

(i) $k_n \to k$, $k > 0$, $l_n \to \infty$,

(ii) $\overline{\mathscr{R}}'(t,\cdot)$, $u = 1,2,...,z$, is a non-degenerate reliability function,

(iii) $\overline{R}'_{k_n,l_n}(t,\cdot) = [1, \overline{R}'_{k_n,l_n}(t,1),..., \overline{R}'_{k_n,l_n}(t,z)]$, $t \in (-\infty,\infty)$, is the reliability function of a non-homogeneous regular multi-state parallel-series system defined by (3.46)–(3.48),

(iv) $a_n(u) > 0$, $b_n(u) \in (-\infty,\infty)$, $u = 1,2,...,z$,

(v) $F(t,u)$ for each fixed u, is one of the distribution functions $F^{(1)}(t,u),F^{(2)}(t,u),...,$ $F^{(a)}(t,u)$ defined by (3.48) such that

(vi) $\exists N(u)$ $\forall n > N(u)$ $F(a_n(u)t + b_n(u),u) = 0$ for $t < t_0(u)$, $F(a_n(u)t + b_n(u),u) \neq 0$ for $t \geq t_0(u)$, where $t_0(u) \in <-\infty,\infty)$,

(vii) $\lim_{n\to\infty} \dfrac{F^{(i)}(a_n(u)t+b_n(u),u)}{F(a_n(u)t+b_n(u),u)} \leq 1$ for $t \geq t_0(u)$, $i = 1,2,...,a$, $u = 1,2,...,z$,

and moreover there exist non-decreasing functions

$$\text{(viii)} \quad \bar{d}_i(t,u) = \begin{cases} 0 & \text{for } t < t_0(u) \\ \lim\limits_{n \to \infty} \bar{d}_i(a_n(u)t + b_n(u), u) & \text{for } t \geq t_0(u), \end{cases} \tag{5.79}$$

where

$$\bar{d}_i(a_n(u)t + b_n(u), u) = \left[\frac{F^{(i)}(a_n(u)t + b_n(u), u)}{F(a_n(u)t + b_n(u), u)}\right]^{l_n},$$

then

$$\overline{\mathscr{R}}'(t,\cdot) = [1, \overline{\mathscr{R}}'(t,1), ..., \overline{\mathscr{R}}'(t,z)], \, t \in (-\infty, \infty),$$

is its limit multi-state reliability function, i.e.

$$\lim_{n \to \infty} \overline{R}'_{k_n, l_n}(a_n(u)t + b_n(u), u) = \overline{\mathscr{R}}'(t,u) \text{ for } t \in C_{\overline{\mathscr{R}}'(u)}, \, u = 1, 2, ..., z, \tag{5.80}$$

if and only if

$$\lim_{n \to \infty}[F(a_n(u)t + b_n(u), u)]^{l_n} = \mathfrak{I}_0(t,u) \text{ for } t \in C_{\mathfrak{I}_0(u)}, \, u = 1, 2, ..., z, \tag{5.81}$$

where $\mathfrak{I}_0(t,u)$ is a non-degenerate distribution function and moreover

$$\overline{\mathscr{R}}'(t,u) = \prod_{i=1}^{a}[1 - \bar{d}_i(t,u) \, \mathfrak{I}_0(t,u)]^{q_i k}, \, t \in (-\infty, \infty), \, u = 1, 2, ..., z. \tag{5.82}$$

Motivation: For each fixed u, $u = 1, 2, ..., z$, assumptions (i)–(viii) of Lemma 5.15 are identical to assumptions (i)–(viii) of Lemma 4.23, condition (5.80) is identical to condition (4.108) and condition (5.81) is identical to condition (4.109). Since from Lemma 4.23, condition (4.108) and condition (4.109) are equivalent, then condition (5.80) and condition (5.81) are equivalent and moreover from (4.110), the equality (5.82) holds.

The application of Lemmas 5.14–5.15 and Theorem 4.9 from Chapter 4, results in the following theorem ([74]).

Theorem 5.9
The class of limit non-degenerate reliability functions of the non-homogeneous regular multi-state parallel-series system is composed of $3^z + 4^z + 3^z$ reliability functions of the form

$$\overline{\mathscr{R}}'(t,\cdot) = [1, \overline{\mathscr{R}}'(t,1), ..., \overline{\mathscr{R}}'(t,z)], \, t \in (-\infty, \infty), \tag{5.83}$$

where

Case 1. $k_n = n$, $|l_n - c \log n| \gg s$, $s > 0$, $c > 0$ (under Assumption 4.1 and the assumptions of Lemma 5.14).

$$\overline{\mathscr{R}}'(t,u) \in \{ \overline{\mathscr{R}}'_1(t), \overline{\mathscr{R}}'_2(t), \overline{\mathscr{R}}'_3(t) \}, \, u = 1,2,...,z, \tag{5.84}$$

and $\overline{\mathscr{R}}'_i(t)$, $i = 1,2,3$, are given by (4.111)–(4.113) with $\overline{d}_i(t) = \overline{d}_i(t,u)$, $u = 1,2,...z$, where $\overline{d}_i(t,u)$ are defined by (5.76),

Case 2. $k_n = n$, $l_n - c \log n \sim s$, $s \in (-\infty, \infty)$, $c > 0$ (under the assumptions of Lemma 5.14).

$$\overline{\mathscr{R}}'(t,u) \in \{ \overline{\mathscr{R}}'_4(t), \overline{\mathscr{R}}'_5(t), \overline{\mathscr{R}}'_6(t), \overline{\mathscr{R}}'_7(t) \}, \, u = 1,2,...,z, \tag{5.85}$$

and $\overline{\mathscr{R}}'_i(t)$, $i = 4,5,6,7$, are given by (4.114)–(4.117) with $\overline{d}(t) = \overline{d}(t,u)$, $u = 1,2,...z$, where $\overline{d}(t,u)$ are defined by (5.76),

Case 3. $k_n \rightarrow k$, $k > 0$, $l_n \rightarrow \infty$ (under the assumptions of Lemma 5.15).

$$\overline{\mathscr{R}}'(t,u) \in \{ \overline{\mathscr{R}}'_8(t), \overline{\mathscr{R}}'_9(t), \overline{\mathscr{R}}'_{10}(t) \}, \, u = 1,2,...,z, \tag{5.86}$$

and $\overline{\mathscr{R}}'_i(t)$, $i = 8,9,10$, are given by (4.118)–(4.120) with $\overline{d}_i(t) = \overline{d}_i(t,u)$, $u = 1,2,...z$, where $\overline{d}_i(t,u)$ are defined by (5.79).

Motivation: For each fixed u, $u = 1,2,...,z$, co-ordinate $\overline{\mathscr{R}}'(t,u)$ of the vector $\overline{\mathscr{R}}'(t,\cdot)$ defined by (5.83), from Theorem 4.9 being the consequence of Lemma 4.22 and Lemma 4.23, may be one of the three types of reliability function defined by (4.111)–(4.113), or one of the four types of reliability function defined by (4.114)–(4.117) with $\overline{d}(t) = \overline{d}(t,u)$, where $\overline{d}(t,u)$ are given by (5.76), or one of the three types of reliability function given by (4.118)–(4.120) with $\overline{d}_i(t) = \overline{d}_i(t,u)$, where $\overline{d}_i(t,u)$ are defined by (5.79). Thus the number of different limit multi-state reliability functions of the system is equal to the sum of the number of z-term variations of the 3-element sets given by (5.84) and (5.86) and the number of the 4-element set given by (5.85). It means this number is equal to $3^z + 4^z + 3^z$ and they are of the form (5.83).

CHAPTER 6

RELIABILITY EVALUATION OF PORT AND SHIPYARD TRANSPORTATION SYSTEMS

The multi-state asymptotic approach is applied to the reliability and risk characteristics evaluation of selected large transportation systems used in ports and shipyards. Reliability analysis of multi-state series, series-parallel and parallel-series transportation systems is based on corollaries formulated and proved in this chapter and corollaries given in the previous chapter. Corollaries are used to evaluate reliability characteristics of three transportation systems used at the Port of Gdynia and one operating at the Naval Shipyard of Gdynia. The port grain transportation system built of three-state non-homogeneous series-parallel subsystems, the port oil piping transportation system composed of three-state non-homogeneous series-parallel subsystems, the port bulk transportation system built of four-state non-homogeneous series-parallel and series subsystems and the shipyard rope transportation system that is a four-state homogeneous parallel-series system are considered. Multi-state reliability functions, mean values of sojourn times in the state subsets and their standard deviations, mean values of lifetimes in the particular states, risk functions, and exceeding moments of a permitted risk level are determined for these systems. The accuracy of the asymptotic approach to the reliability evaluation of these systems is also illustrated. System components reliability data and system operation processes data come from experts operating these systems, component technical certificates and obligatory norms. Reliability data by necessity are approximate and concerned only with the mean values of the system components' sojourn times in the state subsets and hypothetical distributions of these lifetimes.

6.1. Auxiliary results

Assuming in Corollary 5.4

$$a_n(u) = [\bar{d}(t,u)\beta(u)n]^{-1/\alpha(u)} = [n \sum_{(i:\alpha_i(u)=\alpha(u))} q_i \beta_i(u)]^{-1/\alpha(u)},$$

$b_n(u) = 0$ for $u = 1,2,\ldots,z$,

after considering the justification given in Chapter 1, prior to Definition 1.6, we obtain the following result.

Corollary 6.1
If components of the non-homogeneous multi-state series system have Weibull reliability functions

$$R^{(i)}(t,\cdot) = [1, R^{(i)}(t,1),\ldots, R^{(i)}(t,z)], t \in (-\infty,\infty),$$

where

$$R^{(i)}(t,u) = 1 \text{ for } t < 0, R^{(i)}(t,u) = \exp[-\beta_i(u)\, t^{\alpha_i(u)}] \text{ for } t \geq 0, \ \alpha_i(u) > 0, \ \beta_i(u) > 0, \quad (6.1)$$
$$i = 1,2,\ldots,a, \ u = 1,2,\ldots,z,$$

and

$$a_n(u) = [\beta(u)n]^{-1/\alpha(u)}, \ b_n(u) = 0 \text{ for } u = 1,2,\ldots,z, \quad (6.2)$$

where

$$\alpha(u) = \min_{1 \leq i \leq a}\{\alpha_i(u)\}, \ \beta(u) = \sum_{(i:\alpha_i(u)=\alpha(u))} q_i\beta_i(u) \text{ for } u = 1,2,\ldots,z, \quad (6.3)$$

then

$$\overline{\mathfrak{R}}'_2(t,\cdot) = [1, \overline{\mathfrak{R}}'_2(t,1),\ldots, \overline{\mathfrak{R}}'_2(t,z)],$$

where

$$\overline{\mathfrak{R}}'_2(t,u) = 1 \text{ for } t < 0, \ \overline{\mathfrak{R}}'_2(t,u) = \exp[-t^{\alpha(u)}] \text{ for } t \geq 0, \ u = 1,2,\ldots,z,$$

is its limit reliability function.

Corollary 6.2
If components of the homogeneous regular multi-state parallel-series system have Weibull reliability functions

$$R(t,\cdot) = [1, R(t,1),\ldots, R(t, z)], t \in (-\infty,\infty),$$

where

$R(t,u) = 1$ for $t < 0$, $R(t,u) = \exp[-\beta(u) t^{\alpha(u)}]$ for $t \geq 0$, $\alpha(u) > 0$, $\beta(u) > 0$, $\quad$ (6.4)
$u = 1,2,...,z,$

and

$$k_n = n, \quad l_n - c \log n >> s, \; c > 0, \; s > 0, \qquad (6.5)$$

$$a_n(u) = b_n(u)/(\alpha(u)\beta(u)(b_n(u))^{\alpha(u)}\log n), \; b_n(u) = [(1/\beta(u))\log(l_n/\log n)]^{1/\alpha(u)}, \qquad (6.6)$$
$u = 1,2,...,z,$

then

$$\overline{\mathscr{R}}_3(t,\cdot) = [1, \overline{\mathscr{R}}_3(t,1), ..., \overline{\mathscr{R}}_3(t,z)], \; t \in (-\infty,\infty),$$

where

$$\overline{\mathscr{R}}_3(t,u) = \exp[-\exp[t]] \text{ for } t \in (-\infty,\infty), \; u = 1,2,...,z,$$

is its limit reliability function.
Motivation: Since for each fixed u, sufficiently large n and all $t \in (-\infty,\infty)$, from (6.5) and (6.6), we have

$$a_n(u)t + b_n(u) > 0 \text{ and } a_n(u)/b_n(u) \to 0 \text{ as } n \to \infty,$$

then according to (6.4), we get

$$F(a_n(u)t + b_n(u),u) = 1 - \exp[-\beta(u)(a_n(u)t + b_n(u))^{\alpha(u)}]$$

$$= 1 - \exp[-\beta(u)(b_n(u))^{\alpha(u)}(1 + (a_n(u)/b_n(u))t)^{\alpha(u)}]$$

$$= 1 - \exp[-\beta(u)(b_n(u))^{\alpha(u)}(1 + \alpha(u)(a_n(u)/b_n(u))t)$$

$$+ o(a_n(u)/b_n(u))].$$

Moreover, from the condition

$$\alpha(u)\beta(u)(b_n(u))^{\alpha(u)} a_n(u)/b_n(u) = 1/\log n \to 0 \text{ as } n \to \infty,$$

it follows that

$$F(a_n(u)t + b_n(u),u) = 1 - \exp[-\log(l_n/\log n) - t/\log n + o(1/\log n)]$$

$$= 1 - (\log n)/l_n(1 - t/\log n + o(1/\log n)$$

$$= 1 - (\log n)/l_n + t/l_n + o(1/l_n) \text{ for } t \in (-\infty,\infty),$$

and according to (5.65)

$$\bar{V}(t,u) = \lim_{n\to\infty} k_n[F(a_n(u)t + b_n(u),u)]^{l_n}$$

$$= \lim_{n\to\infty} n[1 - (\log n)/l_n + t/l_n + o(1/l_n)]^{l_n}$$

$$= \lim_{n\to\infty} \exp[\log n - \log n + t + l_n o(1/l_n)]$$

$$= \exp[t] \text{ for } t \in (-\infty,\infty),$$

which from Lemma 5.12 completes the proof.

6.2. Reliability of a port grain transportation system

The grain elevator, presented in Figure 6.1, is the basic structure in the Baltic Grain Terminal of the Port of Gdynia assigned to handle the clearing of exported and imported grain. Elevator technological potentialities allow us to join different loading and unloading relations of ships, cars and railway trucks. Its output in the grain unloading process is 400 ton/hour and in grain loading is 360 ton/hour. The whole technological process is controlled electronically. A computer station delivers full visual information about the grain stream (flow) and its balance and the elevator's working state.

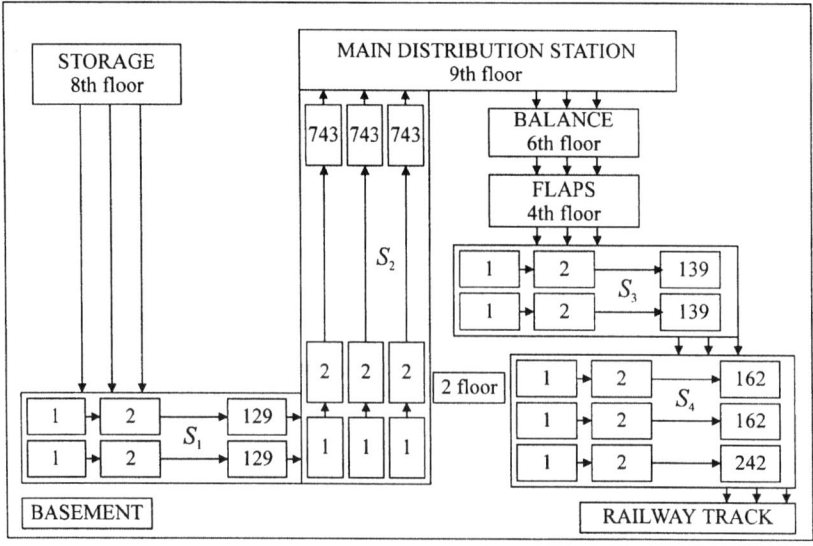

Fig. 6.1. The scheme of the grain transportation system

One of the basic elevator functions is loading railway trucks with grain. The railway truck loading is performed in the following successive elevator operation steps ([70]):

– gravitational passing of grain from the storage placed on the 8th elevator floor through 45 hall to horizontal conveyors placed in the elevator basement,

– transport of grain through horizontal conveyors to vertical bucket elevators transporting grain to the main distribution station placed on the 9th floor,

– gravitational dumping of grain through the main distribution station to the balance placed on the 6th floor,

– dumping weighed grain through the complex of flaps placed on the 4th floor to horizontal conveyors placed on the 2nd floor,

– dumping of grain from horizontal conveyors to worm conveyors,

– dumping of grain from worm conveyors to railway trucks.

In loading the railway trucks with grain the following elevator transportation subsystems take part:

S_1 – horizontal conveyors of the first type,

S_2 – vertical bucket elevators,

S_3 – horizontal conveyors of the second type,

S_4 – worm conveyors,

the main distribution station and the balance.

The main distribution station is the system of dumping channels in the form of a steel box composed of dividing walls, which direct the grain from bucket conveyors to the balance. Its executive elements are composed of three steel sleeves and pneumatic elements in the form of three servomotors. The electronic balance weighs the dumped grain with electronic indicators. Its executive elements during loading and unloading with grain are flaps, which are opened and closed by five pneumatic servomotors.

The transporting subsystems have steel covers and they are provided with drives in the form of electrical engines with gears. In their reliability analysis we omit their drives as they are different types mechanisms. We also omit their covers as they have a high reliability and, practically, do not fail.

Taking into account the efficiency of the considered transportation system we distinguish the following three reliability states of its components ([70]):

state 2 – the state ensuring the largest efficiency of the conveyor,

state 1 – the state ensuring less efficiency of the conveyor caused by throwing grain off the belt,

state 0 – the state involving failure of the conveyor.

Subsystem S_1 is composed of two identical belt conveyors of the first type each composed of a ribbon belt, a drum driving the belt, a reversible drum, 117 channelled rollers and nine rollers supporting the belt. Thus subsystem S_1 consists of $k_n = 2$ conveyors each composed of $l_n = 129$ components. In each conveyor there is one belt with reliability functions

$$R^{(1,1)}(t,1) = \exp[-0.0125t^2], \ R^{(1,1)}(t,2) = \exp[-0.022t^2] \text{ for } t \geq 0,$$

two drums with reliability functions

$$R^{(1,2)}(t,1) = \exp[-0.0015t^2], \ R^{(1,2)}(t,2) = \exp[-0.0018t^2] \text{ for } t \geq 0,$$

117 channelled rollers with reliability functions

$$R^{(1,3)}(t,1) = \exp[-0.005t^2], \ R^{(1,3)}(t,2) = \exp[-0.0075t^2] \text{ for } t \geq 0,$$

and nine supporting rollers with reliability functions

$$R^{(1,4)}(t,1) = \exp[-0.004t^2], \ R^{(1,4)}(t,2) = \exp[-0.005t^2] \text{ for } t \geq 0.$$

Thus it is a non-homogeneous regular multi-state series-parallel system where, according to Definition 3.20, we have

$$k_n = k = 2, \ l_n = 129, \ a = 1, \ q_1 = 1,$$

$$e_1 = 4, \ p_{11} = 1/129, \ p_{12} = 2/129, \ p_{13} = 117/129, \ p_{14} = 9/129$$

and according to (3.43)–(3.45) its exact reliability function is

$$\boldsymbol{R}'_{2,129}(t,\cdot) = [1,1 - [1 - \exp[-0.6365t^2]]^2, 1 - [1 - \exp[-0.9481t^2]]^2] \text{ for } t \geq 0.$$

Further, applying Corollary 5.10, according to (5.62), we have

$$\alpha_1(1) = \min\{2, 2, 2, 2\} = 2,$$

$$\beta_1(1) = \frac{1}{129}0.0125 + \frac{2}{129}0.0015 + \frac{117}{129}0.005 + \frac{9}{129}0.004 = 0.004934108,$$

$$\alpha_1(2) = \min\{2, 2, 2, 2\} = 2,$$

$$\beta_1(2) = \frac{1}{129}0.022 + \frac{2}{129}0.0018 + \frac{117}{129}0.0075 + \frac{9}{129}0.005 = 0.007349612,$$

and according to (5.63)

$$\alpha(1) = \max\{2\} = 2, \ \beta(1) = \min\{0.004934108\} = 0.004934108,$$

$$\alpha(2) = \max\{2\} = 2, \ \beta(2) = \min\{0.007349612\} = 0.007349612,$$

and according to (5.61)

$$a_n(1) = (0.004934108 \cdot 129)^{-1/2} = 1.253432119, \ b_n(1) = 0,$$

$$a_n(2) = (0.007349612 \cdot 129)^{-1/2} = 1.027005872, \ b_n(2) = 0,$$

and we conclude that the limit reliability function of subsystem S_1 is

$\mathcal{R}'_9 (t, \cdot) = [1, 1 - [1 - \exp[-t^2]]^2, 1 - [1 - \exp[-t^2]]^2]$ for $t \geq 0$.

Hence, from (3.49), we get the approximate subsystem reliability function (the formula is exact in this case) given by

$$R'_{2,129} (t, \cdot) \cong \mathcal{R}'_9 ((t - b_n(u))/a_n(u), \cdot)$$

$$= [1, 1 - [1 - \exp[-0.6365t^2]]^2, 1 - [1 - \exp[-0.9481t^2]]^2] \text{ for } t \geq 0.$$

The expected values of the subsystem lifetimes in the state subsets and their standard deviations, from (3.13)–(3.15), are:

$M(1) \cong 1.44$ years, $M(2) \cong 1.18$ years,

$\sigma(1) \cong 0.53$ years, $\sigma(2) \cong 0.44$ years.

Hence and from (3.17), the subsystem mean lifetimes in particular states are:

$\overline{M}(1) \cong 0.26$ years, $\overline{M}(2) \cong 1.18$ years.

If the critical reliability state of the subsystem is $r = 2$, then according to (3.18) its risk function is given by

$r(t) \cong [1 - \exp[-0.9481t^2]]^2.$

Hence and from (3.19), the moment when the subsystem risk function exceeds an admissible level $\delta = 0.05$ is

$$\tau = r^{-1}(\delta) = [-(1/0.9481)\log(1 - \sqrt{\delta}\,)]^{1/2} \cong 0.52 \text{ years.}$$

Subsystem S_2 is composed of three identical bucket elevators ([103]) each composed of a buttress belt, a drum driving the belt, a reversible drum and 740 buckets. Thus subsystem S_2 consists of $k_n = 3$ elevators, each composed of $l_n = 743$ components. In each elevator there is one belt with reliability functions

$R^{(1,1)}(t,1) = \exp[-0.025t^2], R^{(1,1)}(t,2) = \exp[-0.026t^2]$ for $t \geq 0$,

two drums with reliability functions

$R^{(1,2)}(t,1) = \exp[-0.0015t^2], R^{(1,2)}(t,2) = \exp[-0.0018t^2]$ for $t \geq 0$,

740 buckets with reliability functions

$R^{(1,3)}(t,1) = \exp[-0.03t^2], R^{(1,3)}(t,2) = \exp[-0.06t^2]$ for $t \geq 0$.

Thus it is a non-homogeneous regular multi-state series-parallel system where, according to Definition 3.20, we have

$k_n = k = 3$, $l_n = 743$, $a = 1$, $q_1 = 1$,

$e_1 = 3$, $p_{11} = 1/743$, $p_{12} = 2/743$, $p_{13} = 740/743$

and according to (3.43)–(3.45) its exact reliability function is given by

$$\boldsymbol{R'}_{3,743} (t,\cdot) = [1,1 - [1 - \exp[-22.228t^2]]^3, 1 - [1 - \exp[-44.4296t^2]]^3] \text{ for } t \geq 0.$$

Further, applying Corollary 5.10, according to (5.62), we have

$\alpha_1(1) = \min\{2, 2, 2\} = 2,$

$$\beta_1(1) = \frac{1}{743}0.025 + \frac{2}{743}0.0015 + \frac{740}{743}0.03 = 0.029916554,$$

$\alpha_1(2) = \min\{2, 2, 2\} = 2,$

$$\beta_1(2) = \frac{1}{743}0.026 + \frac{2}{743}0.0018 + \frac{740}{743}0.06 = 0.059797577,$$

and according to (5.63)

$\alpha(1) = \max\{2\} = 2,$ $\beta(1) = \min\{0.029916554\} = 0.029916554,$

$\alpha(2) = \max\{2\} = 2,$ $\beta(2) = \min\{0.059797577\} = 0.059797577,$

and according to (5.61)

$a_n(1) = (0.029916554 \cdot 743)^{-1/2} = 0.212104464$, $b_n(1) = 0,$

$a_n(2) = (0.059797577 \cdot 743)^{-1/2} = 0.0150025056$, $b_n(2) = 0,$

and we conclude that the limit reliability function of subsystem S_2 is

$$\boldsymbol{\mathfrak{R}'}_9 (t,\cdot) = [1,1 - [1 - \exp[-t^2]]^3, 1 - [1 - \exp[-t^2]]^3] \text{ for } t \geq 0.$$

Thus, from (3.49), the approximate formula (it is exact in this case) takes the form

$$\boldsymbol{R'}_{3,743} (t,\cdot) \cong \boldsymbol{\mathfrak{R}'}_9 ((t - b_n(u))/a_n(u), \cdot)$$

$$= [1,1 - [1 - \exp[-22.228t^2]]^3, 1 - [1 - \exp[-44.4296t^2]]^3] \text{ for } t \geq 0.$$

The expected values and standard deviations of the subsystem lifetimes in the state subsets, according to (3.13)–(3.15), are:

$M(1) \cong 0.27$ years, $M(2) \cong 0.19$ years,

$\sigma(1) \cong 0.10$ years, $\sigma(2) \cong 0.07$ years.

Hence and from (3.17), the subsystem mean lifetimes in particular states are:

$\overline{M}(1) \cong 0.08$ years, $\overline{M}(2) \cong 0.19$ years.

If the critical reliability state of the subsystem is $r = 2$, then from (3.18) its risk function has the form

$r(t) \cong [1 - \exp[-44.2296t^2]]^3.$

Hence, and from (3.19), the moment when the subsystem risk function exceeds an admissible level $\delta = 0.05$ is

$\tau = r^{-1}(\delta) = [-(1/44.2296)\log(1 - \sqrt[3]{\delta})]^{1/2} \cong 0.10$ years.

Subsystem S_3 is composed of two identical belt conveyors of the second type each composed of a buttress belt, a drum driving the belt, a reversible drum, 117 channelled rollers and 19 rollers supporting the belt. Thus subsystem S_3 consists of $k_n = 2$ conveyors, each composed of $l_n = 139$ components. In each conveyor there is one belt with reliability functions

$R^{(1,1)}(t,1) = \exp[-0.0125t^2]$, $R^{(1,1)}(t,2) = \exp[-0.022t^2]$ for $t \geq 0$,

two drums with reliability functions

$R^{(1,2)}(t,1) = \exp[-0.0015t^2]$, $R^{(1,2)}(t,2) = \exp[-0.0018t^2]$ for $t \geq 0$,

117 channelled rollers with reliability functions

$R^{(1,3)}(t,1) = \exp[-0.005t^2]$, $R^{(1,3)}(t,2) = \exp[-0.0075t^2]$ for $t \geq 0$,

and 19 supporting rollers with reliability functions

$R^{(1,4)}(t,1) = \exp[-0.004t^2]$, $R^{(1,4)}(t,2) = \exp[-0.005t^2]$ for $t \geq 0$.

Thus it is a non-homogeneous regular multi-state series-parallel system where, according to Definition 3.20, we have

$k_n = k = 2, \ l_n = 139, \ a = 1, \ q_1 = 1,$

$e_1 = 4$, $p_{11} = 1/139$, $p_{12} = 2/139$, $p_{13} = 117/139$, $p_{14} = 19/139$

and according to (3.43)–(3.45) its exact reliability function is

$$\boldsymbol{R'}_{2,139}(t,\cdot) = [1, 1 - [1 - \exp[-0.6765t^2]]^2, 1 - [1 - \exp[-0.9981t^2]]^2] \text{ for } t \geq 0.$$

Further, applying Corollary 5.10, according to (5.62), we have

$$\alpha_1(1) = \min\{2, 2, 2, 2\} = 2,$$

$$\beta_1(1) = \frac{1}{139}0.0125 + \frac{2}{139}0.0015 + \frac{117}{139}0.005 + \frac{19}{139}0.004 = 0.004866906,$$

$$\alpha_1(2) = \min\{2, 2, 2, 2\} = 2,$$

$$\beta_1(2) = \frac{1}{139}0.022 + \frac{2}{139}0.0018 + \frac{117}{139}0.0075 + \frac{19}{139}0.005 = 0.007180575,$$

and according to (5.63)

$$\alpha(1) = \max\{2\} = 2,\ \beta(1) = \min\{0.004866906\} = 0.004866906,$$

$$\alpha(2) = \max\{2\} = 2,\ \beta(2) = \min\{0.007180575\} = 0.007180575,$$

and according to (5.61)

$$a_n(1) = (0.004866906 \cdot 139)^{-1/2} = 1.215811147,\ b_n(1) = 0,$$

$$a_n(2) = (0.007180575 \cdot 139)^{-1/2} = 1.000951394,\ b_n(2) = 0,$$

and we conclude that the limit reliability function of subsystem S_3 is

$$\boldsymbol{\mathscr{R}'}_9(t,\cdot) = [1, 1 - [1 - \exp[-t^2]]^2, 1 - [1 - \exp[-t^2]]^2] \text{ for } t \geq 0.$$

Thus, from (3.49), the approximate formula (it is exact in this case) for the subsystem reliability function takes the form

$$\boldsymbol{R'}_{2,139}(t,\cdot) \cong \boldsymbol{\mathscr{R}'}_9((t - b_n(u))/a_n(u), \cdot)$$

$$= [1, 1 - [1 - \exp[-0.6765t^2]]^2, 1 - [1 - \exp[-0.9981t^2]]^2] \text{ for } t \geq 0.$$

The expected values and standard deviations of the subsystem sojourn times in the state subsets, according to (3.13)–(3.15), are:

$M(1) \cong 1.39$ years, $M(2) \cong 1.15$ years,

$\sigma(1) \cong 0.53$ years, $\sigma(2) \cong 0.42$ years.

Hence and from (3.17), the mean lifetimes of the subsystem in particular states are:

$\overline{M}(1) \cong 0.24$ years, $\overline{M}(2) \cong 1.15$ years.

If the critical reliability state of the subsystem is $r = 2$, then according to (3.18) its risk function takes the form

$r(t) \cong [1 - \exp[-0.9981t^2]]^2.$

Hence, and from (3.19), the moment when the subsystem risk exceeds an admissible level $\delta = 0.05$ is

$\tau = r^{-1}(\delta) = [-(1/0.9981)\log(1 - \sqrt{\delta})]^{1/2} \cong 0.50$ year.

Subsystem S_4 is composed of three chain conveyors, each composed of a wheel driving the belt, a reversible wheel and 160, 160 and 240 links respectively. The subsystem consists of three conveyors. Two of these are composed of 162 components and the remaining one is composed of 242 components. Thus it is a non-homogeneous non-regular multi-state series-parallel system. In order to make it a regular system we conventionally complete two first conveyors having 162 components with 80 components that do not fail. After this supplement subsystem S_4 consists of $k_n = 3$ conveyors, each composed of $l_n = 242$ components. In two of them there are two driving wheels with reliability functions

$R^{(1,1)}(t,1) = \exp[-0.005t^2]$, $R^{(1,1)}(t,2) = \exp[-0.008t^2]$ for $t \geq 0$,

160 links with reliability functions

$R^{(1,2)}(t,1) = \exp[-0.012t^2]$, $R^{(1,2)}(t,2) = \exp[-0.018t^2]$ for $t \geq 0$,

and 80 components with "reliability functions"

$R^{(1,3)}(t,1) = \exp[-\beta_1 t^2]$, $R^{(1,3)}(t,2) = \exp[-\beta_2 t^2]$ for $t \geq 0$, where $\beta_1 = \beta_2 = 0$.

The third conveyer is composed of two driving wheels with reliability functions

$R^{(2,1)}(t,1) = \exp[-0.022t^{3/2}]$, $R^{(2,1)}(t,2) = \exp[-0.026t^{3/2}]$ for $t \geq 0$,

and 240 links with reliability functions

$R^{(2,2)}(t,1) = \exp[-0.034t^{3/2}]$, $R^{(2,2)}(t,2) = \exp[-0.042t^{3/2}]$ for $t \geq 0$.

Thus the subsystem is a non-homogeneous regular multi-state series-parallel system where, according to Definition 3.20, we have

$k_n = k = 3, l_n = 242, a = 2, q_1 = 2/3, q_2 = 1/3,$

$e_1 = 3, p_{11} = 2/242, p_{12} = 160/242, p_{13} = 80/242,$

$e_2 = 2, p_{21} = 2/242, p_{22} = 240/242,$

and according to (3.43)–(3.45) its exact reliability function is given by

$$\boldsymbol{R'}_{3,242}(t,\cdot) = [1, 1 - [1 - \exp[-1.93t^2]]^2 \cdot [1 - \exp[-8.204t^{3/2}]],$$

$$1 - [1 - \exp[-2.896t^2]]^2 \cdot [1 - \exp[-10.132t^{3/2}]]] \text{ for } t \geq 0.$$

Further, applying Corollary 5.10, from (5.62), we have

$\alpha_1(1) = \min\{2, 2, 2\} = 2,$

$\beta_1(1) = \dfrac{2}{242} 0.005 + \dfrac{160}{242} 0.012 + \dfrac{80}{242} 0 = 0.007975206,$

$\alpha_2(1) = \min\{3/2, 3/2\} = 3/2,$

$\beta_2(1) = \dfrac{2}{242} 0.022 + \dfrac{240}{242} 0.034 = 0.033900826,$

$\alpha_1(2) = \min\{2, 2, 2\} = 2,$

$\beta_1(2) = \dfrac{2}{242} 0.008 + \dfrac{160}{242} 0.018 + \dfrac{80}{242} 0 = 0.011966942,$

$\alpha_2(2) = \min\{3/2, 3/2\} = 3/2,$

$\beta_2(2) = \dfrac{2}{242} 0.026 + \dfrac{240}{242} 0.042 = 0.041867768,$

and from (5.63)

$\alpha(1) = \max\{2, 3/2\} = 2, \beta(1) = \min\{0.007975206\} = 0.007975206,$

$\alpha(2) = \max\{2, 3/2\} = 2, \beta(2) = \min\{0.011966942\} = 0.011966942,$

and from (5.61)

$a_n(1) = (0.007975206 \cdot 242)^{-1/2} = 0.719815778, b_n(1) = 0,$

$a_n(2) = (0.011966942 \cdot 242)^{-1/2} = 0.587625622, b_n(2) = 0,$

and we conclude that the limit reliability function of subsystem S_4 is

$\mathscr{R}'_9 (t,\cdot) = [1, 1 - [1 - \exp[-t]]^2, 1 - [1 - \exp[-t]]^2]$ for $t \geq 0$.

Thus, from (3.49), we obtain the following approximate formula

$R'_{3,242} (t,\cdot) \cong \mathscr{R}'_9 ((t - b_n(u))/a_n(u), \cdot)$

$= [1,1 - [1 - \exp[-1.93t^2]]^2, 1 - [1 - \exp[-2.896t^2]]^2]$ for $t \geq 0$.

The approximate mean values and standard deviations of the subsystem lifetimes in the state subsets, according to (3.13)–(3.15), are:

$M(1) \cong 0.82$ years, $M(2) \cong 0.67$ years, $\sigma(1) \cong 0.32$ years, $\sigma(2) \cong 0.26$ years.

Hence, and from (3.17), the subsystem mean lifetimes in particular states are:

$\overline{M}(1) \cong 0.15$ years, $\overline{M}(2) \cong 0.67$ years.

If the critical reliability state of the subsystem is $r = 2$, then due to (3.18) its risk function is given by

$r(t) \cong [1 - \exp[-2.896t^2]]^2.$

Hence, and from (3.19), it follows that the moment when the subsystem risk exceeds an admissible level $\delta = 0.05$ is

$\tau = r^{-1}(\delta) = [-(1/2.896)\log(1 - \sqrt{\delta})]^{1/2} \cong 0.30$ years.

Since the considered subsystems create a series structure in a reliability sense, then according to Definition 3.17 and formulae (3.33)–(3.34), the reliability function of the whole transportation system is given by

$\overline{R}'(t,\cdot) \cong [1, \overline{R}'(t,1), \overline{R}'(t,2)], t \geq 0,$

where

$\overline{R}'(t,1) \cong 24c\exp[-25.471t^2] - 12\exp[-27.401t^2] - 12\exp[-26.1475t^2]$

$+ 6\exp[-28.0775t^2] - 24\exp[-47.699t^2] + 12\exp[-49.629t^2]$

$$+ 12\exp[-48.3755t^2] - 6\exp[-50.3055t^2] + 8\exp[-69.927t^2]$$

$$- 4\exp[-71.857t^2] - 4\exp[-70.6035t^2] + 2\exp[-72.5335t^2]$$

$$- 12\exp[-26.1075t^2] + 6\exp[-28.0375t^2] + 6\exp[-26.784t^2]$$

$$- 3\exp[-28.714t^2] + 12\exp[-48.3355t^2] - 6\exp[-50.2655t^2]$$

$$- 6\exp[-49.012t^2] + 3\exp[-50.942t^2] - 4\exp[-70.5635t^2]$$

$$+ 2\exp[-72.4935t^2] + 2\exp[-71.24t^2] - \exp[-73.17t^2],$$

$$\overline{R}'(t,2) \cong 24\exp[-49.2718t^2] - 12\exp[-52.1678t^2] - 12\exp[-50.2699t^2]$$

$$+ 6\exp[-53.1659t^2] - 24\exp[-93.7014t^2] + 12\exp[-96.5974t^2]$$

$$+ 12\exp[-94.6995t^2] - 6\exp[-97.5955t^2] + 8\exp[-138.131t^2]$$

$$- 4\exp[-141.027t^2] - 4\exp[-139.1291t^2] + 2\exp[-142.0251t^2]$$

$$- 12\exp[-50.2199t^2] + 6\exp[-53.1159t^2] + 6\exp[-51.218t^2]$$

$$- 3\exp[-54.114t^2] + 12\exp[-94.6495t^2] - 6\exp[-97.5455t^2]$$

$$- 6\exp[-95.6476t^2] + 3\exp[-98.5436t^2] - 4\exp[-139.0791t^2]$$

$$+ 2\exp[-141.9751t^2] + 2\exp[-140.0772t^2] - \exp[-142.9732t^2].$$

The approximate mean values and standard deviations of the system lifetimes in the state subsets, from (3.13)–(3.15), are:

$$M(1) \cong 0.27 \text{ years}, \ M(2) \cong 0.19 \text{ years},$$

$$\sigma(1) \cong 0.09 \text{ years}, \ \sigma(2) \cong 0.07 \text{ years}.$$

Hence, from (3.17), the mean values of the system lifetime in particular states are:

$$\overline{M}(1) \cong 0.08 \text{ years}, \ \overline{M}(2) \cong 0.19 \text{ years}.$$

If a critical reliability state of the system is $r = 2$, then from (3.18) its risk function takes the form

$$r(t) \cong 1 - \overline{R}'(t,2).$$

Hence, from (3.19), the moment when the risk exceeds the critical level $\delta = 0.05$ is

$\tau = r^{-1}(\delta) \cong 0.10$ years.

The behaviour of the exact and approximate reliability functions and the risk function of the considered system is illustrated in Table 6.1 and Figure 6.2. Moreover, in Table 6.1 the differences between the exact and approximate values of the components of system multi-state reliability functions marked by $\Delta(1)$ and $\Delta(2)$ are given. These differences show that replacement of the system's exact reliability function by its approximate form does not result in large mistakes in this evaluation.

Table 6.1. The values of the components of multi-state reliability functions and the risk function of the port grain transportation system

t	$\overline{R}'(t,1)$		$\Delta(1)$	$\overline{R}'(t,2)$		$\Delta(2)$	$r(t)$
	exact reliability function	approximate reliability function		exact reliability function	approximate reliability function		
0.00	1.00000	1.00000	0.0000	1.00000	1.00000	0.0000	0.0000
0.05	0.99984	0.99981	0.0000	0.99882	0.99877	0.0001	0.0012
0.10	0.99191	0.99164	0.0003	0.95345	0.95288	0.0006	0.0471
0.15	0.93800	0.93695	0.0011	0.74554	0.74389	0.0017	0.2561
0.20	0.79233	0.79023	0.0021	0.42211	0.42005	**0.0021**	0.5779
0.25	0.57025	0.56759	**0.0027**	0.17068	0.16933	0.0014	0.8307
0.30	0.34445	0.34213	0.0023	0.05099	0.05046	0.0005	0.9495
0.35	0.17559	0.17411	0.0015	0.01162	0.01148	0.0001	0.9885
0.40	0.07652	0.07579	0.0007	0.00205	0.00203	0.0000	0.9980
0.45	0.02885	0.02857	0.0003	0.00028	0.00028	0.0000	0.9997
0.50	0.00948	0.00940	0.0001	0.00003	0.00003	0.0000	1.0000
0.55	0.00273	0.00271	0.0000	0.00000	0.00000	0.0000	1.0000
0.60	0.00069	0.00069	0.0000	0.00000	0.00000	0.0000	1.0000

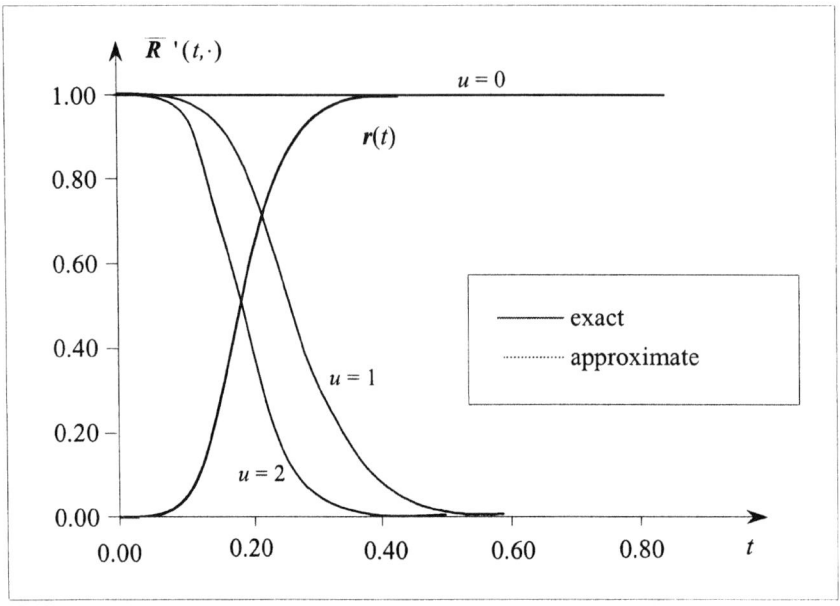

Fig. 6.2. The graphs of the components of multi-state reliability functions and the risk function of the port grain transportation system

6.3. Reliability of a port oil transportation system

Oil Terminal No. 21 in Dębogórze, presented in Figure 6.3, is set up to receive from ships, store and send by carriages or cars oil products such as petrol, driving oil and fuel oil. Three terminal parts A, B and C fulfil these purposes. They are linked by the piping transportation systems. The unloading of tankers is performed at the pier placed in the Port of Gdynia. The pier is connected to terminal part A through the transportation subsystem S_1 built of two piping lines composed of steel pipe segments with diameter of 600 mm. In part A there is a supporting station fortifying tankers' pumps and making possible further transport of oil by means of subsystem S_2 to terminal part B. Subsystem S_2 is built of two piping lines composed of steel pipe segments of diameter 600 mm. Terminal part B is connected to terminal part C by subsystem S_3. Subsystem S_3 is built of one piping line composed of steel pipe segments of diameter 500 mm and two piping lines composed of steel pipe segments of diameter 350 mm. Terminal part C is set up for loading the rail cisterns with oil products and for the wagon carrying these to the railway station of the Port of Gdynia and further into the country.

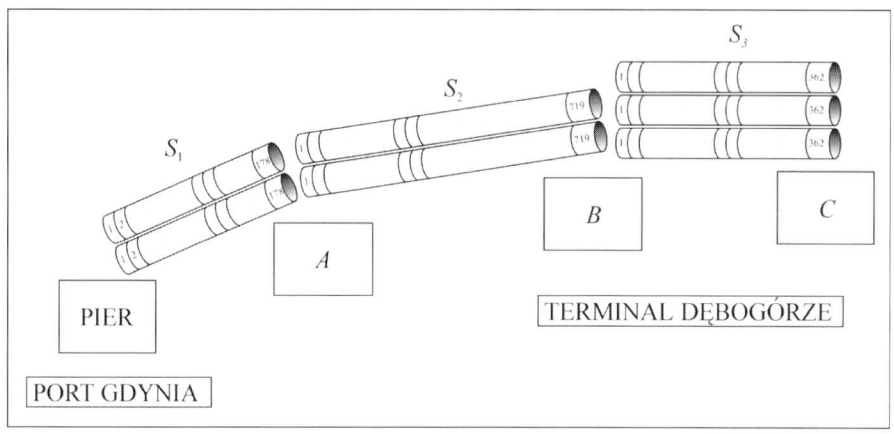

Fig. 6.3. The scheme of the oil transportation system

Considering the safety of the operation of the piping in the system, the following three reliability states of the piping components are distinguished:

state 2 – the piping operation is fully safe,

state 1 – the piping operation is less safe and dangerous because of the possibility of environment pollution,

state 0 – the piping is destroyed.

Subsystem S_1 consists of two identical piping lines, each composed of 176 pipe segments of length 12 m and two valves. It means that it is built of $k_n = 2$ piping lines, each composed of $l_n = 178$ components. In each of the lines there are 176 pipe segments with reliability functions

$$R^{(1,1)}(t,1) = \exp[-0.000000001t^4], \quad R^{(1,1)}(t,2) = \exp[-0.000000004t^4] \text{ for } t \geq 0,$$

and two valves with reliability functions

$$R^{(1,2)}(t,1) = \exp[-0.000000052t^4], \quad R^{(1,2)}(t,2) = \exp[-0.000000107t^4] \text{ for } t \geq 0.$$

Thus subsystem S_1 is a non-homogeneous regular multi-state series-parallel system, where according to Definition 3.20

$$k_n = k = 2, \ l_n = 178, \ a = 1, \ q_1 = 1,$$

$$e_1 = 2, \ p_{11} = 176/178, \ p_{12} = 2/178.$$

The exact subsystem reliability function for $t \geq 0$, according to (3.43)–(3.45) is given by

$$\boldsymbol{R'}_{2,178}(t,\cdot) = [1, 1 - [1 - \exp[-0.00000028t^4]]^2, 1 - [1 - \exp[-0.000000918t^4]]^2].$$

Applying Corollary 5.10, according to (5.62), we have

$$\alpha_1(1) = \min\{4, 4\} = 4,$$

$$\beta_1(1) = \frac{176}{178}\,0.000000001 + \frac{2}{178}\,0.000000052 = 0.000000001573,$$

$$\alpha_1(2) = \min\{4, 4\} = 4,$$

$$\beta_1(2) = \frac{176}{178}\,0.000000004 + \frac{2}{178}\,0.000000107 = 0.0000000051573,$$

and from (5.63)

$$\alpha(1) = \max\{4\} = 4,$$

$$\beta(1) = \min\{0.000000001573\} = 0.000000001573,$$

$$\alpha(2) = \max\{4\} = 4,$$

$$\beta(2) = \min\{0.000000005157\} = 0.0000000051573,$$

and from (5.61)

$$a_n(1) = (0.000000001573 \cdot 178)^{-1/4} = 43.47208719,\ b_n(1) = 0,$$

$$a_n(2) = (0.000000005157 \cdot 178)^{-1/4} = 32.30645681,\ b_n(2) = 0,$$

and we conclude that the subsystem limit reliability function is given by

$$\mathscr{R}'_9(t,\cdot) = [1, 1 - [1 - \exp[-t^4]]^2, 1 - [1 - \exp[-t^4]]^2]\ \text{for}\ t \geq 0.$$

Thus, from (3.49), for $t \geq 0$, we have

$$\boldsymbol{R}'_{2,178}(t,\cdot) \cong [1, 1 - [1 - \exp[-0.00000028t^4]]^2, 1 - [1 - \exp[-0.000000918t^4]]^2].$$

The mean values and standard deviations of subsystem S_1 sojourn times in the state subsets, from (3.13)–(3.15), are:

$$M(1) \cong 45.67\ \text{years},\ M(2) \cong 33.94\ \text{years},$$

$$\sigma(1) \cong 8.92\ \text{years},\ \sigma(2) \cong 6.63\ \text{years}.$$

Hence, from (3.17), the mean lifetimes of the subsystem in the particular states are:

$\overline{M}(1) \cong 11.73$ years, $\overline{M}(2) \cong 33.94$ years.

If a critical state is $r = 2$, then according to (3.18) the system risk function takes the form

$$r(t) \cong [1 - \exp[-0.000000918t^4]]^2$$

and from (3.19) the moment when the risk exceeds the critical level $\delta = 0.05$ is

$$\tau = r^{-1}(\delta) = [-(1/0.000000918)\log(1 - \sqrt{\delta})]^{1/4} \cong 22.91 \text{ years.}$$

Subsystem S_2 consists of two identical piping lines, each composed of 717 pipe segments of length 12 m and two valves. It means that it is built of $k_n = 2$ piping lines, each composed of $l_n = 719$ components. In each of the lines there are 717 pipe segments with reliability functions

$$R^{(1,1)}(t,1) = \exp[-0.000000001t^4], R^{(1,1)}(t,2) = \exp[-0.000000004t^4] \text{ for } t \geq 0,$$

and two valves with reliability function

$$R^{(1,2)}(t,1) = \exp[-0.000000052t^4], R^{(1,2)}(t,2) = \exp[-0.000000107t^4] \text{ for } t \geq 0.$$

Thus subsystem S_2 is a non-homogeneous regular multi-state series-parallel system, where according to Definition 3.20

$$k_n = k = 2, l_n = 719, a = 1, q_1 = 1,$$

$$e_1 = 2, p_{11} = 717/719, p_{12} = 2/719.$$

The exact subsystem reliability function for $t \geq 0$, according to (3.43)–(3.45), is given by

$$\boldsymbol{R'}_{2,719}(t,\cdot) = [1,1 - [1 - \exp[-0.000000821t^4]]^2, 1 - [1 - \exp[-0.000003082t^4]]^2].$$

Applying Corollary 5.10, according to (5.62), we have

$$\alpha_1(1) = \min\{4, 4\} = 4,$$

$$\beta_1(1) = \frac{717}{719}0.000000001 + \frac{2}{719}0.000000052 = 0.0000000011418,$$

$$\alpha_1(2) = \min\{4, 4\} = 4,$$

$$\beta_1(2) = \frac{717}{719} 0.000000001 + \frac{2}{719} 0.000000052 = 0.0000000042865,$$

from (5.63)

$$\alpha(1) = \max\{4\} = 4,$$

$$\beta(1) = \min\{0.0000000011418\} = 0.0000000011418,$$

$$\alpha(2) = \max\{4\} = 4,$$

$$\beta(2) = \min\{0.0000000042865\} = 0.0000000042865,$$

and from (5.61)

$$a_n(1) = (0.0000000011418 \cdot 719)^{-1/4} = 33.22111547, \ b_n(1) = 0,$$

$$a_n(2) = (0.0000000042865 \cdot 719)^{-1/4} = 23.86667072, \ b_n(2) = 0,$$

and we conclude that the subsystem S_2 limit reliability is given by

$$\mathscr{R}'_9(t,\cdot) = [1, 1 - [1 - \exp[-t^4]]^2, 1 - [1 - \exp[-t^4]]^2] \text{ for } t \geq 0.$$

Hence, from (3.49), for $t \geq 0$, we get

$$R'_{2,719}(t,\cdot) \cong [1, 1 - [1 - \exp[-0.000000821t^4]]^2, 1 - [1 - \exp[-0.000003082t^4]]^2].$$

The mean values and standard deviations of the subsystem sojourn times in the state subsets, according to (3.13)–(3.15), are:

$$M(1) \cong 34.90 \text{ years}, \ M(2) \cong 25.07 \text{ years},$$

$$\sigma(1) \cong 6.82 \text{ years}, \ \sigma(2) \cong 4.92 \text{ years},$$

and from (3.17), the mean values of the subsystem lifetimes in the particular states are:

$$\overline{M}(1) \cong 9.83 \text{ years}, \ \overline{M}(2) \cong 25.07 \text{ years}.$$

If the critical state is $r = 2$, then from (3.18), the subsystem risk function takes the form

$$r(t) \cong [1 - \exp[-0.000003082t^4]]^2.$$

Hence, from (3.19), the moment when the subsystem risk exceeds the critical level $\delta = 0.05$ is

$\tau = r^{-1}(\delta) = [-(1/0.000003082)\log(1 - \sqrt{\delta}\,)]^{1/4} \cong 16.93$ years.

Subsystem S_3 consists of three different piping lines, each composed of 360 pipe segments of either 10 m or 7.5 m length and two valves. It means that it is built of $k_n = 3$ piping lines, each composed of $l_n = 362$ components. In two lines there are 60 pipe segments with reliability functions

$$R^{(1,1)}(t,1) = \exp[-0.0000000008t^4], \ R^{(1,1)}(t,2) = \exp[-0.000000002t^4] \text{ for } t \geq 0$$

and two valves with reliability functions

$$R^{(1,2)}(t,1) = \exp[-0.000000052t^4], \ R^{(1,2)}(t,2) = \exp[-0.000000107t^4] \text{ for } t \geq 0.$$

In the third line there are 360 pipe segments with reliability functions

$$R^{(2,1)}(t,1) = \exp[-0.00000022t^3], \ R^{(2,1)}(t,2) = \exp[-0.0000003t^3] \text{ for } t \geq 0$$

and two valves with reliability functions

$$R^{(2,2)}(t,1) = \exp[-0.000000052t^4], \ R^{(2,2)}(t,2) = \exp[-0.000000107t^4] \text{ for } t \geq 0.$$

Thus subsystem S_3 is a non-homogeneous regular multi-state series-parallel system, where according to Definition 3.20

$$k_n = k = 3, \ l_n = 362, \ a = 2, \ q_1 = 2/3, \ q_2 = 1/3,$$

$$e_1 = 2, \ p_{11} = 360/362, \ p_{12} = 2/362,$$

$$e_2 = 2, \ p_{21} = 360/362, \ p_{22} = 2/362,$$

The exact subsystem reliability function for $t \geq 0$, according to (3.43)–(3.45), is given by

$$\boldsymbol{R'}_{3,362}\,(t,\cdot)$$

$$= [1,1 - [1 - \exp[-0.000000392t^4]]^2 \cdot [1 - \exp[-0.0000792t^3 - 0.000000104t^4]],$$

$$1 - [1 - \exp[-0.000000934t^4]]^2 \cdot [1 - \exp[-0.000108t^3 - 0.000001294t^4]]].$$

Applying Corollary 5.10, according to (5.62), we have

$$\alpha_1(1) = \min\{4, 4\} = 4,$$

$$\beta_1(1) = \frac{360}{362}\,0.0000000008 + \frac{2}{362}\,0.000000052 = 0.0000000010828,$$

$$\alpha_2(1) = \min\{3, 4\} = 3,$$

$$\beta_2(1) = \frac{360}{362} 0.00000022 + \frac{2}{362} 0.000000052 = 0.000000218,$$

$$\alpha_1(2) = \min\{4, 4\} = 4,$$

$$\beta_1(2) = \frac{360}{362} 0.000000002 + \frac{2}{362} 0.000000107 = 0.0000000025801,$$

$$\alpha_2(2) = \min\{3, 4\} = 3,$$

$$\beta_2(2) = \frac{360}{362} 0.0000003 + \frac{2}{362} 0.000000107 = 0.000000298,$$

next from (5.63)

$$\alpha(1) = \max\{4, 3\} = 4, \ \beta(1) = \min\{0.0000000010828\} = 0.0000000010828,$$

$$\alpha(2) = \max\{4, 3\} = 4, \ \beta(2) = \min\{0.0000000025801\} = 0.0000000025801,$$

and from (5.61)

$$a_n(1) = (0.0000000010828 \cdot 362)^{-1/4} = 39.96487724, \ b_n(1) = 0,$$

$$a_n(2) = (0.0000000025801 \cdot 362)^{-1/4} = 32.1672016, \ b_n(2) = 0,$$

and we conclude that the subsystem S_3 limit reliability function is given by

$$\mathscr{R}'_9(t,\cdot) = [1, 1 - [1 - \exp[-t^4]]^2, 1 - [1 - \exp[-t^4]]^2] \text{ for } t \geq 0.$$

Hence, applying (3.49), for $t \geq 0$, we get

$$\boldsymbol{R}'_{3,362}(t,\cdot) \cong [1, 1 - [1 - \exp[-0.000000392t^4]]^2, 1 - [1 - \exp[-0.000000934t^4]]^2].$$

The mean values and standard deviations of the subsystem sojourn times in the state subsets, according to (3.13)–(3.15), are:

$$M(1) \cong 41.99 \text{ years}, \ M(2) \cong 33.80 \text{ years},$$

$$\sigma(1) \cong 8.17 \text{ years}, \ \sigma(2) \cong 6.57 \text{ years}.$$

Hence, from (3.17), the mean values of the subsystem lifetimes in the particular states are:

$$\overline{M}(1) \cong 8.19 \text{ years}, \ \overline{M}(2) \cong 33.80 \text{ years}.$$

If the critical state is $r = 2$, then from (3.18), the subsystem risk function takes the form

$$r(t) \cong [1 - \exp[-0.000000934t^4]]^2$$

and from (3.19) the moment when the risk exceeds the critical level $\delta = 0.05$ is

$$\tau = r^{-1}(\delta) = [-(1/0.000000934)\log(1 - \sqrt{\delta})]^{1/4} \cong 22.82 \text{ years}.$$

Since the subsystems create a series reliability structure, then by (3.33)–(3.34) the multi-state reliability function of the whole oil transportation system is given by

$$\overline{\boldsymbol{R}}'(t,\cdot) \cong [1, \overline{\boldsymbol{R}}'(t,1), \overline{\boldsymbol{R}}'(t,2)],$$

where

$$\overline{\boldsymbol{R}}'(t,1) \cong 8\exp[-0.000001493t^4] - 4\exp[-0.000001885t^4] - 4\exp[-0.000002314t^4]$$

$$+ 2\exp[-0.000002706t^4] - 4\exp[-0.000001773t^4] + 2\exp[-0.000002165t^4]$$

$$+ 2\exp[-0.000002594t^4] - \exp[-0.000002986t^4],$$

$$\overline{\boldsymbol{R}}'(t,2) \cong 8\exp[-0.000004934t^4] - 4\exp[-0.000005868t^4] - 4\exp[-0.000008016t^4]$$

$$+ 2\exp[-0.00000895t^4] - 4\exp[-0.000005852t^4] + 2\exp[-0.000006786t^4]$$

$$+ 2\exp[-0.000008934t^4] - \exp[-0.000009868t^4].$$

From (3.13)–(3.15), the approximate mean values and standard deviations of the system sojourn times in the state subsets are:

$$M(1) \cong 32.60 \text{ years}, M(2) \cong 23.98 \text{ years},$$

$$\sigma(1) \cong 5.90 \text{ years}, \sigma(2) \cong 4.43 \text{ years}.$$

While, according to (3.17), the mean values of the system lifetimes in particular states are:

$$\overline{M}(1) \cong 8.62 \text{ years}, \overline{M}(2) \cong 23.98 \text{ years}.$$

If the critical state is $r = 2$, then from (3.18) the system risk function is given by

$r(t) \cong 1 - \overline{R}'(t,2).$

Hence, from (3.19), the moment when the risk exceeds the admissible level $\delta = 0.05$ is

$\tau = r^{-1}(\delta) \cong 16.50$ years.

The behaviour of the system's exact and approximate reliability functions, the differences between them and its risk function are illustrated in Table 6.2 and Figure 6.4.

Table 6.2. The values of the multi-state reliability functions components and the risk function of the port oil transportation system

t	$\overline{R}'(t,1)$		$\Delta(1)$	$\overline{R}'(t,2)$		$\Delta(2)$	$r(t)$
	exact reliability function	approximate reliability function		exact reliability function	approximate reliability function		
0	1.00000	1.00000	0.0000	1.00000	1.00000	0.0000	0.0000
4	1.00000	1.00000	0.0000	1.00000	1.00000	0.0000	0.0000
8	0.99999	0.99998	0.0000	0.99983	0.99981	0.0000	0.0002
12	0.99967	0.99962	0.0001	0.99574	0.99545	0.0003	0.0046
16	0.99674	0.99628	0.0005	0.96187	0.95986	0.0020	0.0401
20	0.98122	0.97932	0.0019	0.82208	0.81658	0.0055	0.1834
24	0.92630	0.92181	0.0045	0.51602	0.51032	**0.0057**	0.4897
28	0.79042	0.78421	**0.0062**	0.18509	0.18326	0.0018	0.8167
32	0.55756	0.55282	0.0047	0.02933	0.02919	0.0001	0.9708
36	0.29212	0.29028	0.0018	0.00161	0.00161	0.0000	0.9984
40	0.10179	0.10146	0.0003	0.00002	0.00002	0.0000	1.0000
44	0.02121	0.02118	0.0000	0.00000	0.00000	0.0000	1.0000
48	0.00239	0.00239	0.0000	0.00000	0.00000	0.0000	1.0000
52	0.00013	0.00013	0.0000	0.00000	0.00000	0.0000	1.0000
56	0.00000	0.00000	0.0000	0.00000	0.00000	0.0000	1.0000
60	0.00000	0.00000	0.0000	0.00000	0.00000	0.0000	1.0000

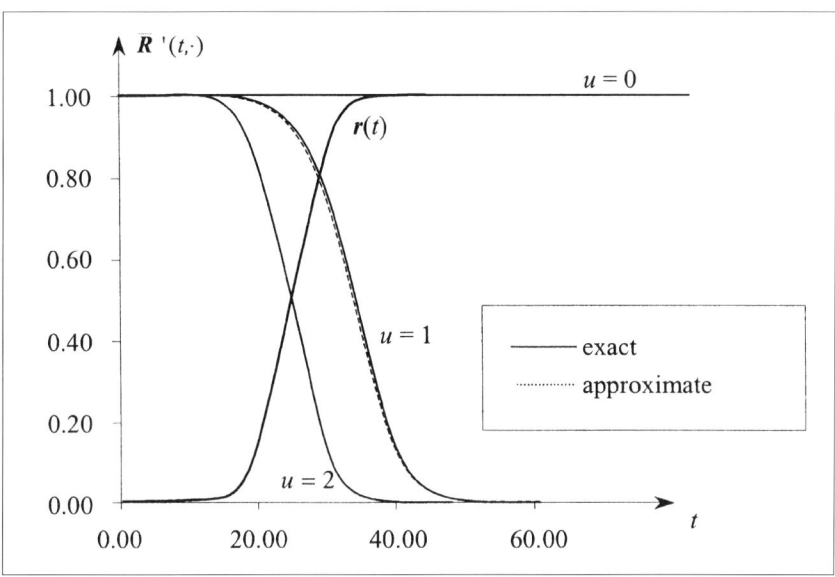

Fig. 6.4. The graphs of the multi-state reliability functions and the risk function of the port oil transportation system

The considered reliability structure of the oil transportation system is adequate in the case when each of the oil products may be transported by each of three system pipelines. However in such transportation some parts of different products are mixed and they have to be refined. If mixing of different oil products is not tolerated, the system structure should be considered as a non-homogeneous series system composed of $n = 2880$ components. In this case the system is composed of 1786 pipe segments with reliability functions

$$R^{(1)}(t,1) = \exp[-0.000000001t^4], \ R^{(1)}(t,2) = \exp[-0.000000004t^4] \text{ for } t \geq 0,$$

720 pipe segments with reliability functions

$$R^{(2)}(t,1) = \exp[-0.0000000008t^4], \ R^{(2)}(t,2) = \exp[-0.000000002t^4] \text{ for } t \geq 0,$$

360 pipe segments with reliability functions

$$R^{(3)}(t,1) = \exp[-0.00000022t^3], \ R^{(3)}(t,2) = \exp[-0.0000003t^3] \text{ for } t \geq 0,$$

and 14 valves with reliability functions

$$R^{(4)}(t,1) = \exp[-0.000000052t^4], \ R^{(4)}(t,2) = \exp[-0.000000107t^4] \text{ for } t \geq 0.$$

According to Definition 3.17, the considered system is a non-homogeneous multi-state series system with parameters

$n = 2880$, $a = 4$,

$q_1 = 1786/2880$, $q_2 = 720/2880$, $q_3 = 360/2880$, $q_4 = 14/2880$,

and according to (3.33)–(3.34) its exact reliability function is given by

$$\overline{R}'_{2880}\ (t,\cdot) = [1, \exp[-0.00000309t^4 - 0.0000792t^3],$$

$$\exp[-0.000010082t^4 - 0.000108t^3]]\ \text{for } t \geq 0.$$

The expected values of the system sojourn times in the state subset and their standard deviations, according to (3.13)–(3.15), are:

$M(1) \cong 17.30$ years, $M(2) \cong 14.00$ years,

$\sigma(1) \cong 5.60$ years, $\sigma(2) \cong 4.40$ years.

Hence, from (3.17), the system mean lifetimes in the states are:

$\overline{M}(1) \cong 3.30$ years, $\overline{M}(2) \cong 14.00$ years.

If a critical system reliability state is $r = 2$, then according to (3.18) its risk function takes the form

$$r(t) \cong 1 - \exp[-0.000010082t^4 - 0.000108t^3].$$

Further, from (3.19), the moment when the system risk exceeds a permitted level $\delta = 0.05$ is

$$\tau \cong r^{-1}(\delta) \cong 6.20 \text{ years.}$$

6.4. Reliability of a port bulk transportation system

The bulk conveyor system is the part of the Baltic Bulk Terminal of the Port of Gdynia assigned to load ships with bulk cargo from Terminal Storage. Its scheme is given in Figure 6.5. Three self-acting loading machines, the transportation system composed of belt conveyors and the coastal loading system carry out the loading of the ships.
In the conveyor loading system we distinguish the following transportation subsystems:
　　S_1 – the dosage conveyor,
　　S_2 – the horizontal conveyor,
　　S_3 – the horizontal conveyor,
　　S_4 – the sloping conveyors,

S_5 – the dosage conveyor with buffer,

S_6 – the loading system.

The transporting subsystems have steel covers and they are provided with drives in the form of electrical engines with gears. In their reliability analysis we omit their drives, as they are mechanisms of different types. We also omit their covers as they have a high reliability and, practically, do not fail.

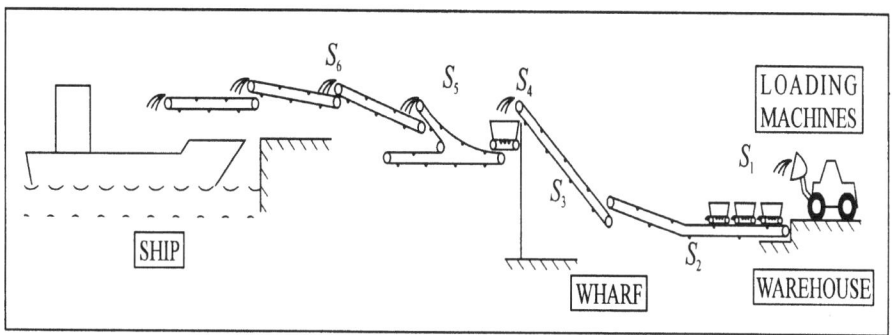

Fig. 6.5. The scheme of the bulk cargo transportation system

Taking into account the efficiency of the considered transportation system we distinguish the following three reliability states of its components:

    state 3 – the state ensuring the largest efficiency of the conveyor,

    state 2 – the state ensuring less efficiency of the conveyor caused by throwing material off the belt,

    state 1 – the state ensuring least efficiency of the conveyor caused by throwing material off the belt and needing human assistance,

    state 0 – the state involving failure of the conveyer.

Subsystem S_1 is composed of three identical belt conveyers, which consist of a ribbon belt, a drum driving the belt, a reversible drum, 12channelled rollers and three rollers supporting the belt. It means that subsystem S_1 consists of $k_n = 3$ conveyors, each composed of $l_n = 18$ components. In each conveyer there is one belt with reliability functions

$$R^{(1,1)}(t,1) = \exp[-0.012t^2],\ R^{(1,1)}(t,2) = \exp[-0.022t^2],\ R^{(1,1)}(t,3) = \exp[-0.049t^2]$$

for $t \geq 0$, two drums driving the belt with reliability functions

$$R^{(1,2)}(t,1) = \exp[-0.0019t^2],\ R^{(1,2)}(t,2) = \exp[-0.0024t^2],\ R^{(1,2)}(t,3) = \exp[-0.0029t^2]$$

for $t \geq 0$, 12 channelled rollers with reliability functions

$$R^{(1,3)}(t,1) = \exp[-0.028t^2],\ R^{(1,3)}(t,2) = \exp[-0.03t^2],\ R^{(1,3)}(t,3) = \exp[-0.032t^2]$$

for $t \geq 0$, and three supporting rollers with reliability functions

$$R^{(1,4)}(t,1) = \exp[-0.0075t^2],\ R^{(1,4)}(t,2) = \exp[-0.01t^2],\ R^{(1,4)}(t,3) = \exp[-0.02t^2]$$

for $t \geq 0$.

Thus, according to Definition 3.20, it is a non-homogeneous regular multi-state series-parallel system with parameters:

$k_n = k = 3, l_n = 18, a = 1, q_1 = 1.$

$e_1 = 4, p_{11} = 1/18, p_{12} = 2/18, p_{13} = 12/18, p_{14} = 3/18,$

and from (3.43)–(3.45) its exact reliability function is given by

$$\boldsymbol{R'}_{3,18}(t,\cdot) = [1,1 - [1 - \exp[-0.3743t^2]]^3, 1 - [1 - \exp[-0.4168t^2]]^3,$$

$$1 - [1 - \exp[-0.4988t^2]]^3] \text{ for } t \geq 0.$$

Applying Corollary 5.10, according to (5.62), we have

$\alpha_1(1) = \min\{2, 2, 2, 2\} = 2,$

$\beta_1(1) = \dfrac{1}{18} \cdot 0.012 + \dfrac{2}{18} \cdot 0.0019 + \dfrac{12}{18} \cdot 0.028 + \dfrac{3}{18} \cdot 0.0075 = 0.020794444,$

$\alpha_1(2) = \min\{2, 2, 2, 2\} = 2,$

$\beta_1(2) = \dfrac{1}{18} \cdot 0.022 + \dfrac{2}{18} \cdot 0.0024 + \dfrac{12}{18} \cdot 0.03 + \dfrac{3}{18} \cdot 0.01 = 0.023155555,$

$\alpha_1(3) = \min\{2, 2, 2, 2\} = 2,$

$\beta_1(3) = \dfrac{1}{18} \cdot 0.049 + \dfrac{2}{18} \cdot 0.0029 + \dfrac{12}{18} \cdot 0.032 + \dfrac{3}{18} \cdot 0.02 = 0.027711111,$

and according to (5.63)

$\alpha(1) = \max\{2\} = 2, \beta(1) = \min\{0.020794444\} = 0.020794444,$

$\alpha(2) = \max\{2\} = 2, \beta(2) = \min\{0.023155555\} = 0.023155555,$

$\alpha(3) = \max\{2\} = 2, \beta(3) = \min\{0.027711111\} = 0.027711111,$

and according to (5.61)

$a_n(1) = (0.020794444 \cdot 18)^{-1/2} = 1.634519443, b_n(1) = 0,$

$a_n(2) = (0.023155555 \cdot 18)^{-1/2} = 1.548945546, b_n(2) = 0,$

$a_n(3) = (0.027711111 \cdot 18)^{-1/2} = 1.415913682, \ b_n(3) = 0,$

and we conclude that the limit reliability function of the subsystem S_1 is given by

$\mathscr{R}'_9 (t, \cdot) = [1, 1 - [1 - \exp[-t^2]]^3, 1 - [1 - \exp[-t^2]]^3, 1 - [1 - \exp[-t^2]]^3]$ for $t \geq 0$.

Therefore, from (3.49), the approximate reliability function (the formula is exact in this case) takes the form

$R'_{3,18} (t, \cdot) \cong \mathscr{R}'_9 ((t - b_n(u))/a_n(u))$

$= [1, 1 - [1 - \exp[-0.3743 t^2]]^3, 1 - [1 - \exp[-0.4168 t^2]]^3,$

$1 - [1 - \exp[-0.4988 t^2]]^3]$ for $t \geq 0$.

The expected values of the subsystem lifetimes in the state subsets and their standard deviations, according to (3.13)–(3.15), are:

$M(1) \cong 2.11$ years, $M(2) \cong 2.00$ years, $M(3) \cong 1.83$ years,

$\sigma(1) \cong 0.67$ years, $\sigma(2) \cong 0.63$ years, $\sigma(3) \cong 0.57$ years.

Hence, from (3.17), the subsystem mean lifetimes in the states are:

$\overline{M}(1) \cong 0.11$ years, $\overline{M}(2) \cong 0.17$ years, $\overline{M}(3) \cong 1.83$ years.

If a critical state is $r = 2$, then from (3.18) the subsystem risk function is given by

$r(t) \cong [1 - \exp[-0.4168 t^2]]^3$.

Hence, from (3.19), the moment when the subsystem risk exceeds the permitted level $\delta = 0.05$ is

$\tau = r^{-1}(\delta) = [-(1/0.4168)\log(1 - \sqrt[3]{\delta}\ )]^{1/2} \cong 1.05$ year.

Subsystem S_2 is composed of one belt conveyor that consists of a ribbon belt, a drum driving the belt, a reversible drum, 125 channelled rollers and 45 rollers supporting the belt. It means that subsystem S_2 consists of one conveyor composed of $n = 173$ components. In the conveyor there is one belt with reliability functions

$R^{(1)}(t,1) = \exp[-0.012 t^2], \ R^{(1)}(t,2) = \exp[-0.022 t^2], \ R^{(1)}(t,3) = \exp[-0.049 t^2]$ for $t \geq 0$,

two drums driving the belt with reliability functions

$R^{(2)}(t,1) = \exp[-0.0019t^2]$, $R^{(2)}(t,2) = \exp[-0.0024t^2]$, $R^{(2)}(t,3) = \exp[-0.0029t^2]$

for $t \geq 0$, 125 channelled rollers with reliability functions

$R^{(3)}(t,1) = \exp[-0.0074t^2]$, $R^{(3)}(t,2) = \exp[-0.012t^2]$, $R^{(3)}(t,3) = \exp[-0.021t^2]$ for $t \geq 0$

and 45 supporting rollers with reliability functions

$R^{(4)}(t,1) = \exp[-0.002t^2]$, $R^{(4)}(t,2) = \exp[-0.0025t^2]$, $R^{(4)}(t,3) = \exp[-0.003t^2]$

for $t \geq 0$.
Thus, according to Definition 3.17, it is a non-homogeneous multi-state series system with parameters

$n = 173$, $a = 4$,

$q_1 = 1/173$, $q_2 = 2/173$, $q_3 = 125/173$, $q_4 = 45/173$.

and according to (3.33)–(3.34) its exact multi-state reliability function is given by

$$\overline{R}'_{173}(t,\cdot) = [1, \exp[-1.0308t^2], \exp[-1.6393t^2], \exp[-2.8148t^2]] \text{ for } t \geq 0.$$

Next, applying Corollary 5.4, we have

$\alpha(1) = \min\{2, 2, 2, 2\} = 2$,

$\beta(1) = \max\{0.012, 0.0019, 0.0074, 0.002\} = 0.012$,

$\alpha(2) = \min\{2, 2, 2, 2\} = 2$,

$\beta(2) = \max\{0.022, 0.0024, 0.012, 0.0025\} = 0.022$,

$\alpha(3) = \min\{2, 2, 2, 2\} = 2$,

$\beta(3) = \max\{0.049, 0.0029, 0.021, 0.003\} = 0.049$,

$a_n(1) = (0.012 \cdot 173)^{-1/2} = 0.694042915$, $b_n(1) = 0$,

$a_n(2) = (0.022 \cdot 173)^{-1/2} = 0.512584663$, $b_n(2) = 0$,

$a_n(3) = (0.049 \cdot 173)^{-1/2} = 0.343462169$, $b_n(3) = 0$,

and moreover

$$\bar{d}(t,1) = \frac{1}{173} \cdot \frac{0.012}{0.012} + \frac{2}{173} \cdot \frac{0.0019}{0.012} + \frac{125}{173} \cdot \frac{0.0074}{0.012} + \frac{45}{173} \cdot \frac{0.002}{0.012} = 0.49653175,$$

$$\bar{d}(t,2) = \frac{1}{173} \cdot \frac{0.022}{0.022} + \frac{2}{173} \cdot \frac{0.0024}{0.022} + \frac{125}{173} \cdot \frac{0.012}{0.022} + \frac{45}{173} \cdot \frac{0.0025}{0.022} = 0.430714636,$$

$$\bar{d}(t,3) = \frac{1}{173} \cdot \frac{0.049}{0.049} + \frac{2}{173} \cdot \frac{0.0029}{0.049} + \frac{125}{173} \cdot \frac{0.021}{0.049} + \frac{45}{173} \cdot \frac{0.003}{0.049} = 0.332051428,$$

and we conclude that the subsystem limit reliability function is given by

$$\boldsymbol{\mathfrak{R}}'_2(t,\cdot) = [1, \exp[-0.49653175t^2], \exp[-0.430714636t^2], \exp[-0.332051428t^2]], \ t \geq 0.$$

Therefore, from (3.49), the approximate reliability function (the formula is exact in this case) takes the form

$$\boldsymbol{\overline{R}}'_{173}(t,\cdot) \cong \boldsymbol{\overline{\mathfrak{R}}}'_2((t - b_n(u))/a_n(u))$$

$$= [1, \exp[-1.0308t^2], \exp[-1.6393t^2], \exp[-2.8148t^2]] \text{ for } t \geq 0.$$

The expected values of the subsystem lifetimes in the state subsets and their standard deviations, according to (3.13)–(3.15), are:

$M(1) \cong 0.87$ years, $M(2) \cong 0.69$ years, $M(2) \cong 0.53$ years.

$\sigma(1) \cong 0.46$ years, $\sigma(2) \cong 0.36$ years, $\sigma(3) \cong 0.28$ years.

Hence, from (3.17), the subsystem mean lifetimes in the states are:

$\overline{M}(1) \cong 0.18$ years, $\overline{M}(2) \cong 0.16$ years, $\overline{M}(3) \cong 0.53$ years.

If a critical state is $r = 2$, then from (3.18) the subsystem risk function is given by

$$r(t) \cong 1 - \exp[-1.6393t^2].$$

Hence, from (3.19), the moment when the subsystem risk exceeds the permitted level $\delta = 0.05$ is

$$\tau = r^{-1}(\delta) = [-(1/1.6393)\log(1 - \delta)]^{1/2} \cong 0.18 \text{ years.}$$

We will perform the reliability evaluation of the remaining subsystems using Corollary 6.1, which is a simplified modification of Corollary 5.4 and is much easier in use. Subsystem S_3 is composed of one belt conveyor that consists of a ribbon belt, a drum driving the belt, a reversible drum, 65 channelled rollers and 20 rollers supporting the belt. It means that subsystem S_3 consists of one conveyor composed of $n = 88$ components. In the conveyor there is one belt with reliability functions

$$R^{(1)}(t,1) = \exp[-0.012t^2],\ R^{(1)}(t,2) = \exp[-0.022t^2],\ R^{(1)}(t,3) = \exp[-0.049t^2]\ \text{for}\ t \geq 0,$$

two drums driving the belt with reliability functions

$$R^{(2)}(t,1) = \exp[-0.0019t^2],\ R^{(2)}(t,2) = \exp[-0.0024t^2],\ R^{(2)}(t,3) = \exp[-0.0029t^2]$$

for $t \geq 0$, 65 channelled rollers with reliability functions

$$R^{(3)}(t,1) = \exp[-0.0074t^2],\ R^{(3)}(t,2) = \exp[-0.012t^2],\ R^{(3)}(t,3) = \exp[-0.021t^2]\ \text{for}\ t \geq 0$$

and 20 supporting rollers with reliability functions

$$R^{(4)}(t,1) = \exp[-0.002t^2],\ R^{(4)}(t,2) = \exp[-0.0025t^2],\ R^{(4)}(t,3) = \exp[-0.003t^2]$$

for $t \geq 0$.
Thus, according to Definition 3.17, it is a non-homogeneous multi-state series system with parameters

$$n = 88,\ a = 4,$$

$$q_1 = 1/88,\ q_2 = 2/88,\ q_3 = 65/88,\ q_4 = 20/88.$$

and according to (3.33)–(3.34) its exact multi-state reliability function is given by

$$\overline{R}'_{88}(t,\cdot) = [1, \exp[-0.5368t^2], \exp[-0.8568t^2], \exp[-1.4798t^2]]\ \text{for}\ t \geq 0.$$

Next, applying Corollary 6.1, according to (6.3), we have

$$\alpha(1) = \min\{2, 2, 2, 2\} = 2,$$

$$\beta(1) = \frac{1}{88} \cdot 0.012 + \frac{2}{88} \cdot 0.0019 + \frac{65}{88} \cdot 0.0074 + \frac{20}{88} \cdot 0.002 = 0.0061,$$

$$\alpha(2) = \min\{2, 2, 2, 2\} = 2,$$

$$\beta(2) = \frac{1}{88} \cdot 0.022 + \frac{2}{88} \cdot 0.0024 + \frac{65}{88} \cdot 0.012 + \frac{20}{88} \cdot 0.0025 = 0.009736363,$$

$$\alpha(3) = \min\{2, 2, 2, 2\} = 2,$$

$$\beta(3) = \frac{1}{88} \cdot 0.049 + \frac{2}{88} \cdot 0.0029 + \frac{65}{88} \cdot 0.021 + \frac{20}{88} \cdot 0.003 = 0.016815909,$$

and according to (6.2)

$$a_n(1) = (0.0061 \cdot 88)^{-1/2} = 1.364877726, \ b_n(1) = 0,$$

$$a_n(2) = (0.009736363 \cdot 88)^{-1/2} = 1.080339575, \ b_n(2) = 0,$$

$$a_n(3) = (0.016815909 \cdot 88)^{-1/2} = 0.822050484, \ b_n(3) = 0,$$

and we conclude that the subsystem limit reliability function is given by

$$\mathcal{\overline{R}}'_2 (t,\cdot) = [1, \exp[-t^2], \exp[-t^2], \exp[-t^2]] \text{ for } t \geq 0.$$

Therefore, from (3.49), the approximate reliability function (the formula is exact in this case) takes the form

$$\overline{R}'_{88} (t,\cdot) \cong \mathcal{\overline{R}}'_2 ((t - b_n(u))/a_n(u))$$

$$= [1, \exp[-0.5368t^2], \exp[-0.8568t^2], \exp[-1.4798t^2]] \text{ for } t \geq 0.$$

The expected values of the subsystem lifetimes in the state subsets and their standard deviations, according to (3.13)–(3.15), are:

$$M(1) \cong 1.21 \text{ years, } M(2) \cong 0.96 \text{ years, } M(2) \cong 0.73 \text{ years.}$$

$$\sigma(1) \cong 0.63 \text{ years, } \sigma(2) \cong 0.50 \text{ years, } \sigma(3) \cong 0.38 \text{ years.}$$

Hence, from (3.17), the subsystem mean lifetimes in the states are:

$$\overline{M}(1) \cong 0.25 \text{ years, } \overline{M}(2) \cong 0.23 \text{ years, } \overline{M}(3) \cong 0.73 \text{ years.}$$

If a critical state is $r = 2$, then from (3.18) the subsystem risk function is given by

$$r(t) \cong 1 - \exp[-0.8568t^2].$$

Hence, from (3.19), the moment when the subsystem risk exceeds the permitted level $\delta = 0.05$ is

$$\tau = r^{-1}(\delta) = [-(1/0.8568)\log(1-\delta)]^{1/2} \cong 0.24 \text{ years.}$$

Subsystem S_4 is composed of one belt conveyor that consists of a ribbon belt, a drum driving the belt, a reversible drum, 12 channelled rollers and three rollers supporting the belt. It means that subsystem S_4 consists of one conveyor composed of $n = 18$ components. In the conveyor there is one belt with reliability functions

$R^{(1)}(t,1) = \exp[-0.012t^2]$, $R^{(1)}(t,2) = \exp[-0.022t^2]$, $R^{(1)}(t,3) = \exp[-0.049t^2]$ for $t \geq 0$,

two drums driving the belt with reliability functions

$R^{(2)}(t,1) = \exp[-0.0019t^2]$, $R^{(2)}(t,2) = \exp[-0.0024t^2]$, $R^{(2)}(t,3) = \exp[-0.0029t^2]$

for $t \geq 0$, 12 channelled rollers with reliability functions

$R^{(3)}(t,1) = \exp[-0.028t^2]$, $R^{(3)}(t,2) = \exp[-0.03t^2]$, $R^{(3)}(t,3) = \exp[-0.032t^2]$ for $t \geq 0$,

and three supporting rollers with reliability functions

$R^{(4)}(t,1) = \exp[-0.0075t^2]$, $R^{(4)}(t,2) = \exp[-0.01t^2]$, $R^{(4)}(t,3) = \exp[-0.02t^2]$ for $t \geq 0$.

Thus, according to Definition 3.17, it is a non-homogeneous multi-state series system with parameters

$n = 18$, $a = 1$,

$q_1 = 1/18$, $q_2 = 2/18$, $q_3 = 12/18$, $q_4 = 3/18$,

and according to (3.33)–(3.34) its exact multi-state reliability function is given by

$$\overline{R}'_{18}(t,\cdot) = [1, \exp[-0.3743t^2], \exp[-0.4168t^2], \exp[-0.4988t^2]] \text{ for } t \geq 0.$$

Next, applying Corollary 6.1, according to (6.3), we have

$\alpha(1) = \min \{2, 2, 2, 2\} = 2$,

$\beta(1) = \dfrac{1}{18} \cdot 0.012 + \dfrac{2}{18} \cdot 0.0019 + \dfrac{12}{18} \cdot 0.028 + \dfrac{3}{18} \cdot 0.0075 = 0.020794444$,

$\alpha(2) = \min \{2, 2, 2, 2\} = 2$,

$\beta(2) = \dfrac{1}{18} \cdot 0.022 + \dfrac{2}{18} \cdot 0.0024 + \dfrac{12}{18} \cdot 0.03 + \dfrac{3}{18} \cdot 0.01 = 0.023155555$,

$\alpha(3) = \min\{2, 2, 2, 2\} = 2$,

$\beta(3) = \dfrac{1}{18} \cdot 0.049 + \dfrac{2}{18} \cdot 0.0029 + \dfrac{12}{18} \cdot 0.032 + \dfrac{3}{18} \cdot 0.02 = 0.027711111$,

and according to (6.2)

$a_n(1) = (0.020794444 \cdot 18)^{-1/2} = 1.634519443, \; b_n(1) = 0,$

$a_n(2) = (0.023155555 \cdot 18)^{-1/2} = 1.548945546, \; b_n(2) = 0,$

$a_n(3) = (0.027711111 \cdot 18)^{-1/2} = 1.415913682, \; b_n(3) = 0,$

and we conclude that the subsystem limit reliability function is given by

$$\overline{\mathfrak{R}}'_2 (t,\cdot) = [1,\exp[-t^2],\exp[-t^2],\exp[-t^2]] \text{ for } t \geq 0.$$

Therefore, from (3.49), the approximate reliability function (the formula is exact in this case) takes the form

$$\overline{R}'_{18} (t,\cdot) \cong \overline{\mathfrak{R}}'_2 ((t - b_n(u))/a_n(u))$$

$$= [1,\exp[-0.3743t^2],\exp[-0.4168t^2],\exp[-0.4988t^2]] \text{ for } t \geq 0.$$

The expected values of the subsystem lifetimes in the state subsets and their standard deviations, according to (3.13)–(3.15), are:

$M(1) \cong 1.45$ years, $M(2) \cong 1.37$ years, $M(3) \cong 1.25$ years.

$\sigma(1) \cong 0.76$ years, $\sigma(2) \cong 0.72$ years, $\sigma(3) \cong 0.66$ years.

Hence, from (3.17), the subsystem mean lifetimes in the states are:

$\overline{M}(1) \cong 0.08$ years, $\overline{M}(2) \cong 0.12$ years, $\overline{M}(3) \cong 1.25$ years.

If a critical state is $r = 2$, then from (3.18) the subsystem risk function is given by

$$r(t) \cong 1 - \exp[-0.4168t^2].$$

Hence, from (3.19), the moment when the subsystem risk exceeds the permitted level $\delta = 0.05$ is

$$\tau = r^{-1}(\delta) = [-(1/0.4168)\log(1 - \delta)]^{1/2} \cong 0.35 \text{ years.}$$

Subsystem S_5 is composed of one belt conveyor, which consists of a ribbon belt, a drum driving the belt, a reversible drum, 162 channelled rollers and 53 rollers supporting the belt. It means that subsystem S_5 consists of one conveyor composed of $n = 218$ components. In the conveyor there is one belt with reliability functions

$$R^{(1)}(t,1) = \exp[-0.012t^2], \; R^{(1)}(t,2) = \exp[-0.022t^2], \; R^{(1)}(t,3) = \exp[-0.049t^2] \text{ for } t \geq 0,$$

two drums driving the belt with reliability functions

$$R^{(2)}(t,1) = \exp[-0.0019t^2], \ R^{(2)}(t,2) = \exp[-0.0024t^2], \ R^{(2)}(t,3) = \exp[-0.0029t^2]$$

for $t \geq 0$, 162 channelled rollers with reliability functions

$$R^{(3)}(t,1) = \exp[-0.0074t^2], \ R^{(3)}(t,2) = \exp[-0.012t^2], \ R^{(3)}(t,3) = \exp[-0.021t^2] \text{ for } t \geq 0$$

and 53 supporting rollers with reliability functions

$$R^{(4)}(t,1) = \exp[-0.002t^2], \ R^{(4)}(t,2) = \exp[-0.0025t^2], \ R^{(4)}(t,3) = \exp[-0.003t^2]$$

for $t \geq 0$.

Thus, according to Definition 3.17, it is a non-homogeneous multi-state series system with parameters

$$n = 218, \ a = 4,$$

$$q_1 = 1/218, \ q_2 = 2/218, \ q_3 = 162/218, \ q_4 = 53/218.$$

and according to (3.33)–(3.34) its exact multi-state reliability function is given by

$$\boldsymbol{R'}_{218} (t,\cdot) = [1,\exp[-1.3206t^2],\exp[-2.1033t^2],\exp[-3.6158t^2]] \text{ for } t \geq 0.$$

Next, applying Corollary 6.1, according to (6.3), we have

$$\alpha(1) = \min\{2, 2, 2, 2\} = 2,$$

$$\beta(1) = \frac{1}{218} \cdot 0.012 + \frac{2}{218} \cdot 0.0019 + \frac{162}{218} \cdot 0.0074 + \frac{53}{218} \cdot 0.002 = 0.006057798,$$

$$\alpha(2) = \min\{2, 2, 2, 2\} = 2,$$

$$\beta(2) = \frac{1}{218} \cdot 0.022 + \frac{2}{218} \cdot 0.0024 + \frac{162}{218} \cdot 0.012 + \frac{53}{218} \cdot 0.0025 = 0.009648165,$$

$$\alpha(3) = \min\{2, 2, 2, 2\} = 2,$$

$$\beta(3) = \frac{1}{218} \cdot 0.049 + \frac{2}{218} \cdot 0.0029 + \frac{162}{218} \cdot 0.021 + \frac{53}{218} \cdot 0.003 = 0.016586238,$$

and according to (6.2)

$$a_n(1) = (0.006057798 \cdot 218)^{-1/2} = 0.870190543, \ b_n(1) = 0,$$

$a_n(2) = (0.009648165 \cdot 218)^{-1/2} = 0.689524008$, $b_n(2) = 0$,

$a_n(3) = (0.016586238 \cdot 218)^{-1/2} = 0.525893504$, $b_n(3) = 0$,

and we conclude that the subsystem limit reliability function is given by

$$\mathscr{R}'_2(t,\cdot) = [1, \exp[-t^2], \exp[-t^2], \exp[-t^2]] \text{ for } t \geq 0.$$

Therefore, from (3.49), the approximate reliability function (the formula is exact in this case) takes the form

$$\overline{R}'_{218}(t,\cdot) \cong \overline{\mathscr{R}}'_2((t - b_n(u))/a_n(u))$$

$$= [1, \exp[-1.3206t^2], \exp[-2.1033t^2], \exp[-3.6158t^2]] \text{ for } t \geq 0.$$

The expected values of the subsystem lifetimes in the state subsets and their standard deviations, according to (3.13)–(3.15), are:

$M(1) \cong 0.77$ years, $M(2) \cong 0.61$ years, $M(2) \cong 0.47$ years.

$\sigma(1) \cong 0.40$ years, $\sigma(2) \cong 0.32$ years, $\sigma(3) \cong 0.24$ years.

Hence, from (3.17), the subsystem mean lifetimes in the states are:

$\overline{M}(1) \cong 0.16$ years, $\overline{M}(2) \cong 0.14$ years, $\overline{M}(3) \cong 0.47$ years.

If a critical state is $r = 2$, then from (3.18) the subsystem risk function is given by

$$r(t) \cong 1 - \exp[-2.1033t^2].$$

Hence, from (3.19), the moment when the subsystem risk exceeds permitted level $\delta = 0.05$ is

$$\tau = r^{-1}(\delta) = [-(1/2.1033)\log(1 - \delta)]^{1/2} \cong 0.16 \text{ years.}$$

Subsystem S_6 is a set of belt conveyors placed on the moving tracks and having the possibility of moving in relation to each other. It is composed of a rotary mechanism, two stable belt conveyors and one moving belt conveyer. One of the stable conveyors consists of one ribbon belt, a drum driving the belt, a reversible drum, 34channelled rollers and 10 rollers supporting the belt. The second stable conveyor consists of one ribbon belt, a drum driving the belt, a reversible drum, 15channelled rollers and five rollers supporting the belt. The moving conveyor consists of one ribbon belt, a drum driving the belt, a reversible drum, 15channelled rollers and five rollers supporting the belt. Thus subsystem S_6 is a non-homogeneous multi-state series system composed of n

= 93 components of four types. In the subsystem there are three belts with reliability functions

$$R^{(1)}(t,1) = \exp[-0.012t^2], \ R^{(1)}(t,2) = \exp[-0.022t^2], \ R^{(1)}(t,3) = \exp[-0.049t^2] \text{ for } t \geq 0,$$

six drums driving the belt with reliability functions

$$R^{(2)}(t,1) = \exp[-0.0019t^2], \ R^{(2)}(t,2) = \exp[-0.0024t^2], \ R^{(2)}(t,3) = \exp[-0.0029t^2]$$

for $t \geq 0$, 64 channelled rollers with reliability functions

$$R^{(3)}(t,1) = \exp[-0.0046t^2], \ R^{(3)}(t,2) = \exp[-0.0075t^2], \ R^{(3)}(t,3) = \exp[-0.012t^2]$$

for $t \geq 0$ and 20 supporting rollers with reliability functions

$$R^{(4)}(t,1) = \exp[-0.0012t^2], \ R^{(4)}(t,2) = \exp[-0.0018t^2], \ R^{(4)}(t,3) = \exp[-0.0024t^2]$$

for $t \geq 0$.

Thus, according to Definition 3.17, it is a non-homogeneous multi-state series system with parameters

$$n = 93, \ a = 4,$$

$$q_1 = 3/93, \ q_2 = 6/93, \ q_3 = 64/93 \ q_4 = 20/93.$$

and according to (3.33)–(3.34) its exact multi-state reliability function is given by

$$\overline{R}'_{93} (t,\cdot) = [1, \exp[-0.3658t^2], \exp[-0.5964t^2], \exp[-0.9804t^2]] \text{ for } t \geq 0.$$

Next, applying Corollary 6.1, according to (6.3), we have

$$\alpha(1) = \min\{2, 2, 2, 2\} = 2,$$

$$\beta(1) = \frac{3}{93} \cdot 0.012 + \frac{6}{93} \cdot 0.0019 + \frac{64}{93} \cdot 0.0046 + \frac{20}{93} \cdot 0.0012 = 0.003933333,$$

$$\alpha(2) = \min\{2, 2, 2, 2\} = 2,$$

$$\beta(2) = \frac{3}{93} \cdot 0.022 + \frac{6}{93} \cdot 0.0024 + \frac{64}{93} \cdot 0.0075 + \frac{20}{93} \cdot 0.0018 = 0.006412903,$$

$$\alpha(3) = \min\{2, 2, 2, 2\} = 2,$$

$$\beta(3) = \frac{3}{93} \cdot 0.049 + \frac{6}{93} \cdot 0.0029 + \frac{64}{93} \cdot 0.012 + \frac{20}{93} \cdot 0.0024 = 0.010541935,$$

and according to (6.2)

$$a_n(1) = (0.003933333 \cdot 93)^{-1/2} = 1.6534000893, \; b_n(1) = 0,$$

$$a_n(2) = (0.006412903 \cdot 93)^{-1/2} = 1.294884971, \; b_n(2) = 0,$$

$$a_n(3) = (0.010541935 \cdot 93)^{-1/2} = 1.009946454, \; b_n(3) = 0,$$

and we conclude that the subsystem limit reliability function is given by

$$\mathscr{R}'_2 (t,\cdot) = [1, \exp[-t^2], \exp[-t^2], \exp[-t^2]] \text{ for } t \geq 0.$$

Therefore, from (3.49), the approximate reliability function (the formula is exact in this case) takes the form

$$\overline{R}'_{93} (t,\cdot) \cong \mathscr{R}'_2 ((t - b_n(u))/a_n(u))$$

$$= [1, \exp[-0.3658t^2], \exp[-0.5964t^2], \exp[-0.9804t^2]] \text{ for } t \geq 0.$$

The expected values of the subsystem lifetimes in the state subsets and their standard deviations, according to (3.13)–(3.15), are:

$$M(1) \cong 1.47 \text{ years}, \; M(2) \cong 1.15 \text{ years}, \; M(3) \cong 0.90 \text{ years}.$$

$$\sigma(1) \cong 0.77 \text{ years}, \; \sigma(2) \cong 0.60 \text{ years}, \; \sigma(3) \cong 0.47 \text{ years}.$$

Hence, from (3.17), the subsystem mean lifetimes in the states are:

$$\overline{M}(1) \cong 0.32 \text{ years}, \; \overline{M}(2) \cong 0.25 \text{ years}, \; \overline{M}(3) \cong 0.90 \text{ years}.$$

If a critical state is $r = 2$, then from (3.18) the subsystem risk function is given by

$$r(t) \cong 1 - \exp[-0.5964t^2].$$

Hence, from (3.19), the moment when the subsystem risk exceeds the permitted level $\delta = 0.05$ is

$$\tau = r^{-1}(\delta) = [-(1/0.5964)\log(1 - \delta)]^{1/2} \cong 0.29 \text{ years}.$$

Since, according to Definition 3.17, all subsystems create a series reliability structure, then from (3.33)–(3.34) the reliability function of the whole transportation system is given by

$$\overline{R}'(t,\cdot) \cong [1, \overline{R}'(t,1), \overline{R}'(t,2), \overline{R}'(t,3)],$$

where

$$\overline{R}'(t,1) = 3\exp[-4.0026t^2] - 3\exp[-4.3769t^2] + \exp[-4.7512t^2],$$

$$\overline{R}'(t,2) = 3\exp[-6.0294t^2] - 3\exp[-6.4462t^2] + \exp[-6.8630t^2],$$

$$\overline{R}'(t,3) = 3\exp[-9.8884t^2] - 3\exp[-10.3872t^2] + \exp[-10.8860t^2].$$

The expected values of the subsystem lifetimes in the state subsets and their standard deviations, according to (3.13)–(3.15), are:

$$M(1) \cong 0.45 \text{ years, } M(2) \cong 0.37 \text{ years, } M(3) \cong 0.29 \text{ years,}$$

$$\sigma(1) \cong 0.27 \text{ years, } \sigma(2) \cong 0.20 \text{ years, } \sigma(3) \cong 0.15 \text{ years.}$$

Hence, from (3.17), the subsystem mean lifetimes in the states are:

$$\overline{M}(1) \cong 0.08 \text{ years, } \overline{M}(2) \cong 0.08 \text{ years, } \overline{M}(3) \cong 0.29 \text{ years.}$$

If a critical state is $r = 2$, then from (3.18) the subsystem risk function is given by

$$r(t) \cong 1 - \overline{R}'(t,2).$$

Hence, by (3.19), the moment when the subsystem risk exceeds the permitted level $\delta = 0.05$ is

$$\tau = r^{-1}(\delta) \cong 0.10 \text{ years.}$$

The system exact and approximate reliability functions are identical. Their behaviour and the system risk function are presented in Table 6.3 and Figure 6.6.

Table 6.3. The values of the multi-state reliability function and the risk function of the port bulk transportation system

t	$\overline{R}'(t,1)$	$\overline{R}'(t,2)$	$\overline{R}'(t,3)$	$r(t)$
0.00	1.00000	1.00000	1.00000	0.00000
0.10	0.96437	0.94542	0.91038	0.05458
0.20	0.86490	0.79891	0.68683	0.20109
0.30	0.72138	0.60339	0.42949	0.39661
0.40	0.55949	0.40727	0.22251	0.59273
0.50	0.40342	0.24558	0.09546	0.75442
0.60	0.27031	0.13223	0.03389	0.86777
0.70	0.16820	0.06351	0.00994	0.93649
0.80	0.09712	0.02719	0.00241	0.97281
0.90	0.05198	0.01036	0.00048	0.98964
1.00	0.02575	0.00351	0.00008	0.99649
1.10	0.01180	0.00105	0.00001	0.99895
1.20	0.00499	0.00028	0.00000	0.99972
1.30	0.00195	0.00007	0.00000	0.99993
1.40	0.00070	0.00001	0.00000	0.99999
1.50	0.00023	0.00000	0.00000	1.00000

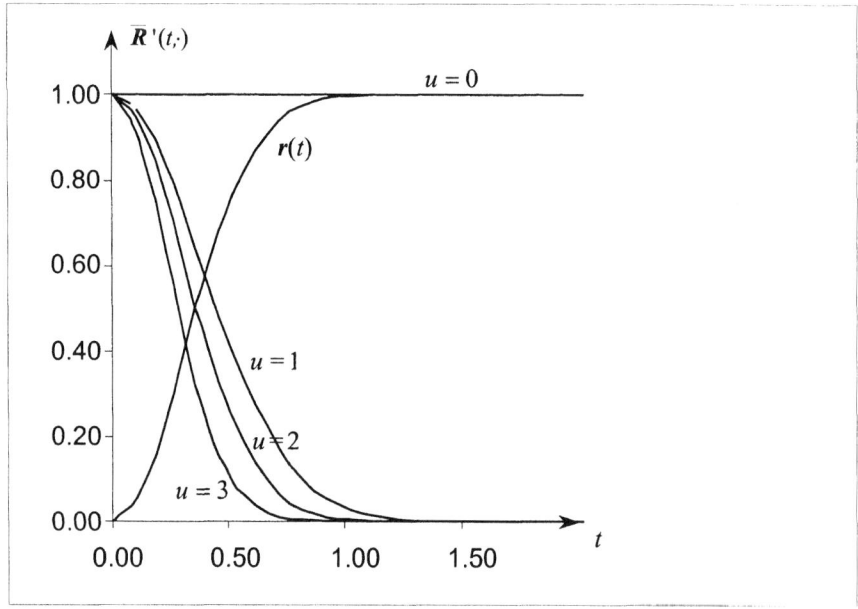

Fig. 6.6. Graphs of the multi-state reliability function and the risk function of the port bulk cargo transportation system

6.5. Reliability of a shipyard rope transportation system

Ship-rope elevators are used to dock and undock ships coming to shipyards for repairs. The elevator utilised in the Naval Shipyard in Gdynia, with the scheme presented in Figure 6.7, is composed of a steel platform-carriage placed in its syncline (hutch). The platform is moved vertically with 10 rope-hoisting winches fed by separate electric motors, each rope having a maximum load of 300 tonnes. During ship docking the platform, with the ship settled in special supporting carriages on the platform, is raised to the wharf level (upper position). During undocking, the operation is reversed. While the ship is moving into or out of the syncline and while stopped in the upper position the platform is held on hooks and the loads in the ropes are relieved. Since the platform-carriage and electric motors are highly reliable in comparison to the ropes, which work in extremely aggressive conditions (salt water, wind, pollution and so on), in our further analysis we will discuss the reliability of the rope system only.

The system under consideration is in order if all its ropes do not fail. Thus we may assume that it is a series system composed of 10 components. Each of the ropes is composed of 22 strands: 10 outer and 12 inner. The outer strands of ropes are composed of 26 steel wires. Their inner strands are composed of 19 steel wires and they form the rope's steel core, which is covered by a plastic layer. The cross-section of the rope is shown in Figure 6.8.

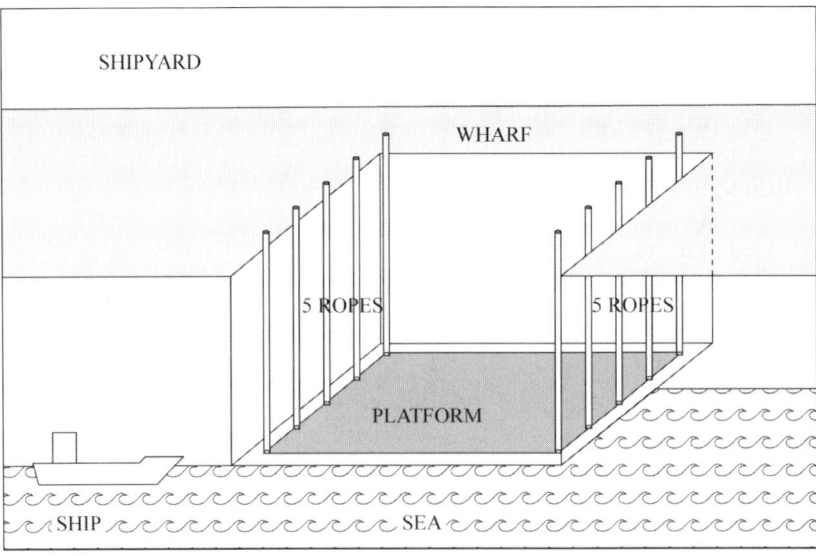

Fig. 6.7. The scheme of the ship-rope transportation system

Thus, considering the strands as basic components of the system, according to Definitions 3.14–3.16 we conclude that the rope elevator is a parallel-series system composed of $k_n = 10$ series-linked subsystems (ropes) with $l_n = 22$ parallel-linked components (strands).

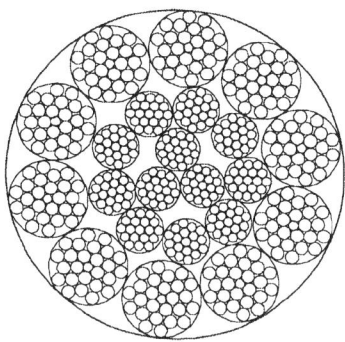

Fig. 6.8. The cross-section of the rope

According to safety standards and to the requirements for approximately comparable reliability (durability of outer and inner strands, after considering technical norms ([22], [102]) and expert opinion ([79]), the following reliability states of all strands have been distinguished:

state 3 – a strand is new, without any defects,

state 2 – the number of broken wires in the strand is greater than 0% and less than 25% of all its wires, or corrosion of wires is greater than 0% and less than 25%,

state 1 – the number of broken wires in the strand is greater than or equal to 25% and less than 50% of all its wires, or corrosion of wires is greater than or equal to 25% and less than 50%,

state 0 – otherwise (a strand is failed).

Considering these component reliability states, according to Definitions 3.14–3.16, we conclude that the rope system is a homogeneous regular four-state parallel-series system. Thus, from (3.30)–(3.31), its reliability function is defined by

$$\overline{\boldsymbol{R}}_{10,22}(t,\cdot) = [1, \overline{\boldsymbol{R}}_{10,22}(t,1), \overline{\boldsymbol{R}}_{10,22}(t,2), \overline{\boldsymbol{R}}_{10,22}(t,3)],$$

where

$$\overline{\boldsymbol{R}}_{10,22}(t,u) = [1 - [F(t,u)]^{22}]^{10}, \ t \in (-\infty,\infty), \ u = 1,2,3. \tag{6.7}$$

According to experts' opinions ([79]) the mean numbers of ship docking and undocking are equal to 80 per year. It means that the elevator is active 160 times per year. Moreover, the rope elevator system reliability depends strongly on the tonnage of the docking ships and therefore the following states of the elevator system loading have been distinguished:

state 4 – loading from 2250 to 1750 tonnes,

state 3 – loading from 1750 to 1250 tonnes,

state 2 – loading from 1250 to 750 tonnes,
state 1 – loading from 750 to 250 tonnes,
state 0 – without loading.

The loading states i, the frequencies n_i of loading states per year, the duration times t_i of the loading states and the total duration times T_i of the loading states per year of the rope elevator system are given in Table 6.4.

Table 6.4. The rope elevator loading states characteristics

i	n_i	$t_i(h)$	$T_i(h)$
4	20	8	160
3	80	6	480
2	40	3	120
1	20	2	40
0	-	-	7960
Total	160	-	8760

Using these data and the formula

$$p_i = \frac{T_i}{8760}, \, i = 0,1,2,3,4,$$

it is possible to evaluate the probabilities of the elevator loading states. They are as follows

$$p_0 = 0.9087, \, p_1 = 0.0046, \, p_2 = 0.0137, \, p_3 = 0.0548, \, p_4 = 0.0182. \quad (6.8)$$

Considering these particular elevator loading states we obtain the following formula for its reliability function

$$\overline{\boldsymbol{R}}'_{10,22}(t,\cdot) = [1, \overline{\boldsymbol{R}}'_{10,22}(t,1), \overline{\boldsymbol{R}}'_{10,22}(t,2), \overline{\boldsymbol{R}}'_{10,22}(t,3)], \quad (6.9)$$

where

$$\overline{\boldsymbol{R}}'_{10,22}(t,u) = p_0 \, \overline{\boldsymbol{R}}^{(0)}_{10,22}(t,u) + p_1 \, \overline{\boldsymbol{R}}^{(1)}_{10,22}(t,u) + p_2 \, \overline{\boldsymbol{R}}^{(2)}_{10,22}(t,u)$$

$$+ \, p_3 \, \overline{\boldsymbol{R}}^{(3)}_{10,22}(t,u) + p_4 \, \overline{\boldsymbol{R}}^{(4)}_{10,22}(t,u), \, t \in (-\infty,\infty), \, u = 1,2,3, \quad (6.10)$$

and according to (6.7)

$$\overline{\boldsymbol{R}}^{(i)}_{10,22}(t,u) = [1 - [F^{(i)}(t,u)]^{22}]^{10}, \, i = 0,1,2,3,4, \quad (6.11)$$

are its multi-state reliability functions in particular loading states, while

$$R^{(i)}(t,u) = 1 - F^{(i)}(t,u),\ t \in (-\infty,\infty),\ u = 1,2,3,\ i = 0,1,2,3,4,$$

are component (strand) multi-state reliability functions in particular elevator loading states.

According to rope reliability data given in their technical certificates ([22]) and experts' opinions ([79]) based on the nature of strand failures it has been assumed that the strands have multi-state Weibull reliability functions

$$R^{(i)}(t,u) = 1 \text{ for } t < 0,\ R^{(i)}(t,u) = \exp[-\beta_i(u)t^{\alpha_i(u)}] \text{ for } t \geq 0,$$
$$u = 1,2,3,\ i = 0,1,2,3,4,$$

with the following parameters

$$\alpha_i(u),\ \beta_i(u),\ i = 0,1,2,3,4,$$

in the particular elevator loading states:

$$\alpha_0(u) = 3,\quad \beta_0(u) = 0.003u,$$

$$\alpha_1(u) = 3,\quad \beta_1(u) = 0.006u,$$

$$\alpha_2(u) = 3,\quad \beta_2(u) = 0.008u,$$

$$\alpha_3(u) = 2,\quad \beta_3(u) = 0.090u,$$

$$\alpha_4(u) = 1.5,\quad \beta_4(u) = 0.250u,\ u = 1,2,3.$$

Hence, after considering (6.10), (6.11) and (6.8), the exact elevator reliability function is given by the formula (6.9) with

$$\overline{R}'_{10,22}(t,u) = 0.9087[1-[1-\exp[-0.003ut^3]]^{22}]^{10}$$

$$+ 0.0046[1-[1-\exp[-0.006ut^3]]^{22}]^{10}$$

$$+ 0.0137[1-[1-\exp[-0.008ut^3]]^{22}]^{10}$$

$$+ 0.0548[1-[1-\exp[-0.09ut^2]]^{22}]^{10}$$

$$+ 0.0182[1-[1-\exp[-0.25ut^{3/2}]]^{22}]^{10},\ u = 1,2,3. \tag{6.12}$$

Since the number of parallel subsystems in the system is $k_n = 10$ and the number of components in each subsystem is $l_n = 22$, then considering that

$$l_n = 22 >> \log k_n = \log 10 \cong 2.3$$

it seems reasonable to apply in finding the limit reliability function of the rope elevator system either Corollary 6.2 or Corollary 5.11. We will use both of them separately; first at different loading states for estimating the system reliability functions given by (6.11), and approximating the reliability function of the system by combining the results according to (6.12). Next, we will compare and comment on the achieved results.

Application of Corollary 6.2
According to (6.6), assuming normalising constants

$$b_n^{(i)}(u) = [\frac{1}{\beta_i(u)} \log \frac{l_n}{\log k_n}]^{1/\alpha_i(u)},$$

$$a_n^{(i)}(u) = (b_n^{(i)}(u))^{1-\alpha_i(u)} /(\alpha_i(u)\beta_i(u)\log k_n), \ u = 1,2,3, \ i = 0,1,2,3,4, \qquad (6.13)$$

we conclude that

$$\overline{\mathcal{R}}_3^{(i)}(t,\cdot) = [1, \overline{\mathcal{R}}_3^{(i)}(t,1), \overline{\mathcal{R}}_3^{(i)}(t,2), \overline{\mathcal{R}}_3^{(i)}(t,3)], \ t \in (-\infty,\infty),$$

where

$$\overline{\mathcal{R}}_3^{(i)}(t,u) = \exp[-\exp[t]] \text{ for } t \in (-\infty,\infty), \ u = 1.2,3, \ i = 0,1,2,3,4,$$

is the multi-state reliability function of the rope elevator system in the ith loading state. According to (6.13) and (3.49) for particular elevator loading states we have:

loading state 0

$$b_n^{(0)}(1) = 9.0950, \ a_n^{(0)}(1) = 0.5834,$$

$$b_n^{(0)}(2) = 7.2187, \ a_n^{(0)}(2) = 0.4630,$$

$$b_n^{(0)}(3) = 6.3061, \ a_n^{(0)}(3) = 0.4045,$$

$$\overline{R}_{10,22}^{(0)}(t,1) \cong \exp[-\exp[1.7141t - 15.5909]],$$

$$\overline{R}_{10,22}^{(0)}(t,2) \cong \exp[-\exp[2.1598t - 15.5909]],$$

$$\overline{\boldsymbol{R}}_{10,22}^{(0)}(t,3) \cong \exp[-\exp[2.4722t - 15.5909]],$$

loading state 1

$$b_n^{(1)}(1) = 7.2187, \ a_n^{(1)}(1) = 0.4630,$$

$$b_n^{(1)}(2) = 5.7295, \ a_n^{(1)}(2) = 0.3675,$$

$$b_n^{(1)}(3) = 5.0052, \ a_n^{(1)}(3) = 0.3210,$$

$$\overline{\boldsymbol{R}}_{10,22}^{(1)}(t,1) \cong \exp[-\exp[2.1598t - 15.5909]],$$

$$\overline{\boldsymbol{R}}_{10,22}^{(1)}(t,2) \cong \exp[-\exp[2.7211t - 15.5909]],$$

$$\overline{\boldsymbol{R}}_{10,22}^{(1)}(t,3) \cong \exp[-\exp[3.1153t - 15.5909]],$$

loading state 2

$$b_n^{(2)}(1) = 6.5587, \ a_n^{(2)}(1) = 0.4207,$$

$$b_n^{(2)}(2) = 5.2056, \ a_n^{(2)}(2) = 0.3339,$$

$$b_n^{(2)}(3) = 4.5475, \ a_n^{(2)}(3) = 0.2917,$$

$$\overline{\boldsymbol{R}}_{10,22}^{(2)}(t,1) \cong \exp[-\exp[2.3771t - 15.5909]],$$

$$\overline{\boldsymbol{R}}_{10,22}^{(2)}(t,2) \cong \exp[-\exp[2.9950t - 15.5909]],$$

$$\overline{\boldsymbol{R}}_{10,22}^{(2)}(t,3) \cong \exp[-\exp[3.4282t - 15.5909]],$$

loading state 3

$$b_n^{(3)}(1) = 5.0078, \ a_n^{(3)}(1) = 0.4818,$$

$$b_n^{(3)}(2) = 3.5410, \ a_n^{(3)}(2) = 0.3407,$$

$$b_n^{(3)}(3) = 2.8912, \ a_n^{(3)}(3) = 0.2782,$$

$$\overline{R}^{(3)}_{10,22}(t,1) \cong \exp[-\exp[2.0756t-10.3939]],$$

$$\overline{R}^{(3)}_{10,22}(t,2) \cong \exp[-\exp[2.9351t-10.3939]],$$

$$\overline{R}^{(3)}_{10,22}(t,3) \cong \exp[-\exp[3.5945t-10.3939]],$$

loading state 4

$$b_n^{(4)}(1)=4.3357, \ a_n^{(4)}(1)=0.5562,$$

$$b_n^{(4)}(2)=2.7313, \ a_n^{(4)}(2)=0.3504,$$

$$b_n^{(4)}(3)=2.0844, \ a_n^{(4)}(3)=0.2674,$$

$$\overline{R}^{(4)}_{10,22}(t,1) \cong \exp[-\exp[1.7979t-7.7954]],$$

$$\overline{R}^{(4)}_{10,22}(t,2) \cong \exp[-\exp[2.8539t-7.7954]],$$

$$\overline{R}^{(4)}_{10,22}(t,3) \cong \exp[-\exp[3.7397t-7.7954]].$$

Combining the achieved results of reliability approximation, according to (6.9) and (6.10), we get the approximate reliability function of the rope elevator system given by

$$\overline{R}'_{10,22}(t,\cdot) = [1, \overline{R}'_{10,22}(t,1), \overline{R}'_{10,22}(t,2), \overline{R}'_{10,22}(t,3)], \ t \in (-\infty,\infty), \qquad (6.14)$$

where

$$\overline{R}'_{10,22}(t,1) \cong 0.9087 \ \exp[-\exp[1.7141t-15.5909]]$$

$$+0.0046 \ \exp[-\exp[2.1598t-15.5909]]$$

$$+0.0137 \ \exp[-\exp[2.3771t-15.5909]]$$

$$+0.0548 \ \exp[-\exp[2.0756t-10.3939]]$$

$$+0.0182 \ \exp[-\exp[1.7979t-7.7954]],$$

$\overline{R}'_{10,22}(t,2) \cong 0.9087 \ \exp[-\exp[2.1598t-15.5909]]$

$+0.0046 \ \exp[-\exp[2.7211t-15.5909]]$

$+0.0137 \ \exp[-\exp[2.9950t-15.5909]]$

$+0.0548 \ \exp[-\exp[2.9351t-10.3939]]$

$+0.0182 \ \exp[-\exp[2.8539t-7.7954]]$,

$\overline{R}'_{10,22}(t,3) \cong 0.9087 \ \exp[-\exp[2.4722t-15.5909]]$

$+0.0046 \ \exp[-\exp[3.1153t-15.5909]]$

$+0.0137 \ \exp[-\exp[3.4282t-15.5909]]$

$+0.0548 \ \exp[-\exp[3.5945t-10.3939]]$

$+0.0182 \ \exp[-\exp[3.7397t-7.7954]].$

The approximate expected values of the system lifetimes $T(u)$ in the state subsets, according to (3.13), are ([73]):

$M(1) \cong 8.40$ years, $M(2) \cong 6.70$ years, $M(3) \cong 6.30$ years.

Hence, according to (3.17), the mean lifetimes of the system in the states are:

$\overline{M}(1) \cong 1.70$ years, $\overline{M}(2) \cong 0.40$ years, $\overline{M}(3) \cong 6.30$ years.

If a critical state is $r = 2$, then a system risk, by (3.18), is given by

$r(t) = 1 - \overline{R}'_{10,22}(t,2) \cong 1-0.9087 \ \exp[-\exp[2.1598t-15.5909]]$

$-0.0046 \ \exp[-\exp[2.7211t-15.5909]]$

$-0.0137 \ \exp[-\exp[2.9950t-15.5909]]$

$-0.0548 \ \exp[-\exp[2.9351t-10.3939]]$

$-0.0182 \ \exp[-\exp[2.8539t-7.7954]].$ \hfill (6.15)

The time when the risk exceeds a permitted level $\delta = 0.05$, according to (3.19), is

$\tau = r^{-1}(\delta) \cong 3.50$ years.

Application of Corollary 5.11
According to (5.75), letting

$$b_n^{(i)}(u) = [\frac{1}{\beta_i(u)} \log l_n]^{1/\alpha_i(u)},$$

$$a_n^{(i)}(u) = (b_n^{(i)}(u))^{1-\alpha_i(u)} / (\alpha_i(u)\beta_i(u)), u = 1,2,3, i = 0,1,2,3,4, \qquad (6.16)$$

we conclude that

$$\overline{\mathfrak{R}}_{10}^{(i)}(t,\cdot) = [1, \overline{\mathfrak{R}}_{10}^{(i)}(t,1), \overline{\mathfrak{R}}_{10}^{(i)}(t,2), \overline{\mathfrak{R}}_{10}^{(i)}(t,3)], t \in (-\infty,\infty),$$

where

$$\overline{\mathfrak{R}}_{10}^{(i)}(t,u) = [1 - \exp[-\exp[-t]]]^k \text{ for } t \in (-\infty,\infty), u = 1,2,3, i = 0,1,2,3,4,$$

is the multi-state limit reliability function of the rope elevator system in the *i*th loading state.
Hence, considering (6.16) and (3.4), for the particular states we have:

loading state 0

$$b_n^{(0)}(1) = 10.1002, \ a_n^{(0)}(1) = 1.0892,$$

$$b_n^{(0)}(2) = 8.0165, \ a_n^{(0)}(2) = 0.8645,$$

$$b_n^{(0)}(3) = 7.0031, \ a_n^{(0)}(3) = 0.7552,$$

$$\overline{R}_{10,22}^{(0)}(t,1) \cong [1 - \exp[-\exp[-0.9181t + 9.2731]]]^{10},$$

$$\overline{R}_{10,22}^{(0)}(t,2) \cong [1 - \exp[-\exp[-1.1568t + 9.2731]]]^{10},$$

$$\overline{R}_{10,22}^{(0)}(t,3) \cong [1 - \exp[-\exp[-1.3242t + 9.2731]]]^{10},$$

loading state 1

$$b_n^{(1)}(1) = 8.01650, \ a_n^{(1)}(1) = 0.8645,$$

$$b_n^{(1)}(2) = 6.3627, \ a_n^{(1)}(2) = 0.6861,$$

$$b_n^{(1)}(3) = 5.5583, \ a_n^{(1)}(3) = 0.5994,$$

$$\overline{R}_{10,22}^{(1)}(t,1) \cong [1 - \exp[-\exp[-1.1568t + 9.2731]]]^{10},$$

$$\overline{R}_{10,22}^{(1)}(t,2) \cong [1 - \exp[-\exp[-1.4574t + 9.2731]]]^{10},$$

$$\overline{R}_{10,22}^{(1)}(t,1) \cong [1 - \exp[-\exp[-1.6683t + 9.2731]]]^{10},$$

loading state 2

$$b_n^{(2)}(1) = 7.2835, \ a_n^{(2)}(1) = 0.7854,$$

$$b_n^{(2)}(2) = 5.7809, \ a_n^{(2)}(2) = 0.6234,$$

$$b_n^{(2)}(3) = 5.0501, \ a_n^{(2)}(3) = 0.5446,$$

$$\overline{R}_{10,22}^{(2)}(t,1) \cong [1 - \exp[-\exp[-1.2732t + 9.2731]]]^{10},$$

$$\overline{R}_{10,22}^{(1)}(t,1) \cong [1 - \exp[-\exp[-1.6041t + 9.2731]]]^{10},$$

$$\overline{R}_{10,22}^{(1)}(t,1) \cong [1 - \exp[-\exp[-1.8362t + 9.2731]]]^{10},$$

loading state 3

$$b_n^{(3)}(1) = 5.8605, \ a_n^{(3)}(1) = 0.9480,$$

$$b_n^{(3)}(2) = 4.1440, \ a_n^{(3)}(2) = 0.6703,$$

$$b_n^{(3)}(3) = 3.3835, \ a_n^{(3)}(3) = 0.5473,$$

$$\overline{R}_{10,22}^{(3)}(t,1) \cong [1 - \exp[-\exp[-1.0549t + 6.1821]]]^{10},$$

$$\overline{R}_{10,22}^{(3)}(t,2) \cong [1 - \exp[-\exp[-1.4918t + 6.1821]]]^{10},$$

$$\overline{R}_{10,22}^{(3)}(t,3) \cong [1 - \exp[-\exp[-1.8271t + 6.1821]]]^{10},$$

loading state 4

$$b_n^{(4)}(1) = 5.3470, \ a_n^{(4)}(1) = 1.1532,$$

$$b_n^{(4)}(2) = 3.3684, \ a_n^{(4)}(2) = 0.7265,$$

$$b_n^{(4)}(3) = 2.5706, \ a_n^{(4)}(3) = 0.5544,$$

$$\overline{R}_{10,22}^{(4)}(t,1) \cong [1 - \exp[-\exp[-0.8671t + 4.6366]]]^{10},$$

$$\overline{R}_{10,22}^{(4)}(t,2) \cong [1 - \exp[-\exp[-1.3765t + 4.6366]]]^{10},$$

$$\overline{R}_{10,22}^{(4)}(t,3) \cong [1 - \exp[-\exp[-1.8037t + 4.6366]]]^{10}.$$

Combining the achieved results of reliability approximation, from (6.9) and (6.10), we get the approximate reliability function of the rope elevator system given by

$$\overline{R}'_{10,22}(t,\cdot) = [1, \overline{R}'_{10,22}(t,1), \overline{R}'_{10,22}(t,2), \overline{R}'_{10,22}(t,3)], \tag{6.17}$$

where

$$\overline{R}'_{10,22}(t,1) \cong 0.9087 \ [1 - \exp[-\exp[-0.9181t + 9.2731]]]^{10}$$

$$+ 0.0046 \ [1 - \exp[-\exp[-1.1568t + 9.2731]]]^{10}$$

$$+ 0.0137 \ [1 - \exp[-\exp[-1.2732t + 9.2731]]]^{10}$$

$$+ 0.0548 \ [1 - \exp[-\exp[-1.0549t + 6.1821]]]^{10}$$

$$+ 0.0182 \ [1 - \exp[-\exp[-0.8671t + 4.6366]]]^{10},$$

$$\overline{R}'_{10,22}(t,2) \cong 0.9087 \ [1 - \exp[-\exp[-1.1568t + 9.2731]]]^{10}$$

$$+ 0.0046 \ [1 - \exp[-\exp[-1.4574t + 9.2731]]]^{10}$$

$$+ 0.0137 \ [1 - \exp[-\exp[-1.6041t + 9.2731]]]^{10}$$

$$+0.0548 \, [1-\exp[-\exp[-1.4918t+6.1821]]]^{10}$$

$$+0.0182 \, [1-\exp[-\exp[-1.3765t+4.6366]]]^{10} \, ,$$

$$\overline{R}'_{10,22} \, (t,3) \cong 0.9087 \, [1-\exp[-\exp[-1.3242t+9.2731]]]^{10}$$

$$+0.0046 \, [1-\exp[-\exp[-1.6683t+9.2731]]]^{10}$$

$$+0.0137 \, [1-\exp[-\exp[-1.8362t+9.2731]]]^{10}$$

$$+0.0548 \, [1-\exp[-\exp[-1.8271t+6.1821]]]^{10}$$

$$+0.0182 \, [1-\exp[-\exp[-1.8037t+4.6366]]]^{10} \, .$$

The approximate expected values of the system lifetimes $T(u)$ in the state subsets, according to (3.13), are ([73]):

$$M(1) \cong 8.70 \text{ years}, \; M(2) \cong 6.90 \text{ years}, \; M(3) \cong 6.40 \text{ years}.$$

Hence, according to (3.17), the mean lifetimes of the system in the states are:

$$\overline{M}(1) \cong 1.80 \text{ years}, \quad \overline{M}(2) \cong 0.50 \text{ years}, \quad \overline{M}(3) \cong 6.40 \text{ years}.$$

If a critical state is $r = 2$, then a system risk, according to (3.18), is given by

$$r(t) \cong 1 - \overline{R}'_{10,22} \, (t,2) \cong 1-0.9087 \, [1-\exp[-\exp[-1.1568t+9.2731]]]^{10}$$

$$-0.0046 \, [1-\exp[-\exp[-1.4574t+9.2731]]]^{10}$$

$$-0.0137 \, [1-\exp[-\exp[-1.6041t+9.2731]]]^{10}$$

$$-0.0548 \, [1-\exp[-\exp[-1.4918t+6.1821]]]^{10}$$

$$-0.0182 \, [1-\exp[-\exp[-1.3765t+4.6366]]]^{10} \, . \qquad (6.18)$$

The time when the risk exceeds a permitted level $\delta = 0.05$, according to (3.19), is

$$\tau = r^{-1}(\delta) \cong 3.60 \text{ years.}$$

Results comparison

The behaviour of the rope elevator's exact reliability function given by (6.12), and its approximate reliability functions given by (6.14) and (6.17) in the state subsets is shown in Tables 6.5–6.7 and Figures 6.9–6.11. The behaviour of the approximate risk functions determined by (6.15) and (6.18) is illustrated in Table 6.8 and Figure 6.12. Moreover in Tables 6.5–6.7 there are the differences between exact and approximate system reliability functions in the state subsets marked by $\varDelta_1(u)$ in the case where approximate formula (6.14) applies and by $\varDelta_2(u)$ where approximate formula (6.17) applies. From data presented in the tables and the graphs it can be noticed that the evaluation of the elevator system reliability given by (6.14) is pessimistic and that given by (6.17) is optimistic. Therefore, in fact, the elevator reliability characteristics are better than those calculated from (6.14) and worse than those calculated from (6.17). From these it is possible to find the following upper and lower bounds for the elevator mean sojourn times $T(u)$ in the state subsets ([73]):

$$8.40 < M(1) < 8.70 \text{ years}, \ 6.70 < M(2) < 6.90 \text{ years}, \ 6.30 < M(3) < 6.40 \text{ years},$$

and in particular states as well:

$$1.70 < \overline{M}(1) < 1.80 \text{ years}, \ 0.40 < \overline{M}(2) < 0.50 \text{ years}, \ 6.30 < \overline{M}(3) < 6.40 \text{ years}.$$

The moment when the risk exceeds a permitted level $\delta = 0.05$ falls in the interval

$$3.50 < \tau < 3.60 \text{ years}.$$

Table 6.5. The values of the rope elevator exact and approximate reliability functions in the state subset $u \geq 1$

t	$\overline{R}'_{10,22}(t,1)$				
	formula (6.12)	formula (6.14)	formula (6.17)	$\varDelta_1(1)$	$\varDelta_2(1)$
0.00	1.00000	0.99999	1.00000	+0.00001	0.00000
1.00	1.00000	0.99994	1.00000	+0.00006	0.00000
2.00	1.00000	0.99962	1.00000	+0.00038	0.00000
3.00	0.99983	0.99755	0.99991	+0.00228	–0.00008
4.00	0.99239	0.98579	0.99344	+0.00660	–0.00105
5.00	0.95025	0.94697	0.95073	+0.00328	–0.00048
6.00	0.92547	0.91902	0.92644	+0.00645	–0.00097
7.00	0.91191	0.88727	0.91268	+0.02464	–0.00077
8.00	0.86579	0.77990	0.89943	**+0.08589**	–0.03364
9.00	0.42655	0.38888	0.46813	+0.03767	**–0.04158**
10.00	0.01781	0.00817	0.01559	+0.00964	+0.00222
11.00	0.00002	0.00000	0.00003	+0.00002	–0.00001
12.00	0.00000	0.00000	0.00000	0.00000	0.00000

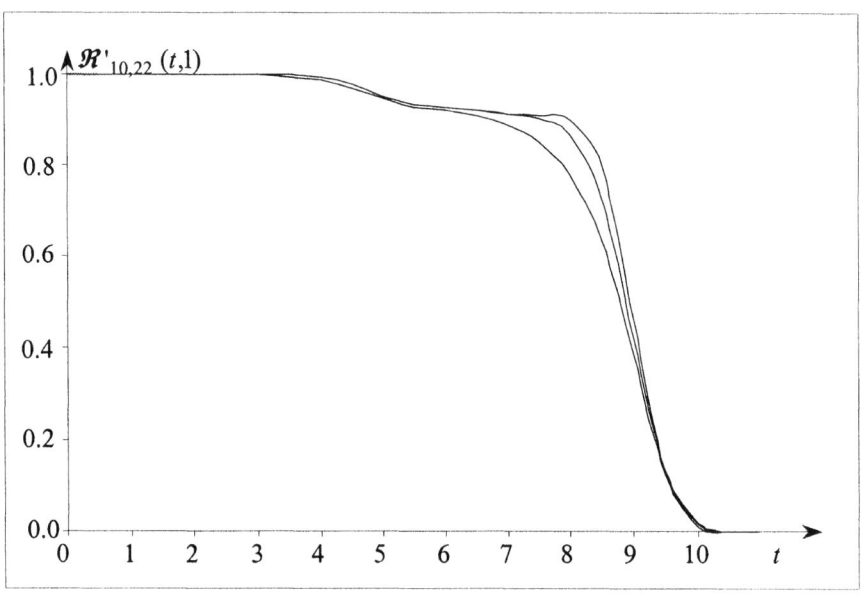

Fig. 6.9. Graphs of the rope elevator exact and approximate reliability functions in the state subset $u \geq 1$

Table 6.6. The values of the rope elevator exact and approximate reliability functions in the state subset $u \geq 2$

t	$\overline{R}'_{10,22}(t,2)$				
	formula (6.12)	formula (6.14)	formula (6.17)	$\Delta_1(2)$	$\Delta_2(2)$
0.00	1.00000	0.99999	1.00000	+0.00001	0.00000
1.00	1.00000	0.99984	1.00000	+0.00016	0.00000
2.00	0.99961	0.99727	0.99975	+0.00234	−0.00014
3.00	0.98009	0.97367	0.98184	+0.00642	−0.00175
4.00	0.92903	0.92690	0.92880	+0.00213	+0.00023
5.00	0.92215	0.91318	0.92335	+0.00897	−0.00120
6.00	0.90136	0.84621	0.90902	+0.05515	−0.00766
7.00	0.54698	0.48709	0.60936	**+0.05989**	**−0.06238**
8.00	0.01183	0.00408	0.01031	+0.00775	+0.00152
9.00	0.00000	0.00000	0.00000	0.00000	0.00000
10.00	0.00000	0.00000	0.00000	0.00000	0.00000
11.00	0.00000	0.00000	0.00000	0.00000	0.00000
12.00	0.00000	0.00000	0.00000	0.00000	0.00000

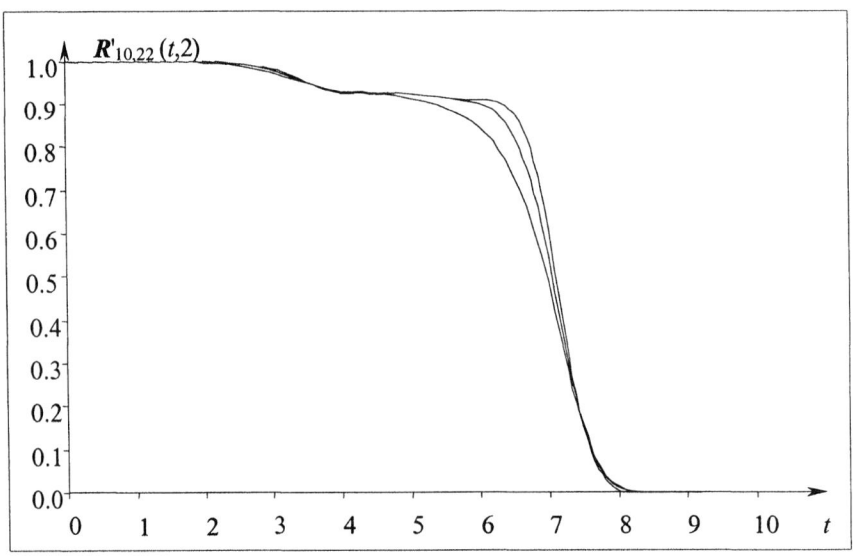

Fig. 6.10. Graphs of the rope elevator exact and approximate reliability functions in the state subset $u \geq 2$

Table 6.7. The values of the rope elevator exact and approximate reliability functions in the state $u = 3$

t	$\overline{R}'_{10,22}(t,3)$				
	formula (6.12)	formula (6.14)	formula (6.17)	$\Delta_1(3)$	$\Delta_2(3)$
0.00	1.00000	0.99999	1.00000	+0.00001	0.00000
1.00	1.00000	0.99962	1.00000	+0.00038	0.00000
2.00	0.99152	0.98838	0.99151	+0.00314	0.00001
3.00	0.94035	0.93919	0.94011	+0.00116	+0.00024
4.00	0.92634	0.92183	0.92686	+0.00451	−0.00052
5.00	0.90922	0.87529	0.91095	+0.03393	−0.00173
6.00	0.64678	0.56865	0.72020	**+0.07813**	**−0.07342**
7.00	0.01084	0.00352	0.00946	+0.00732	+0.00138
8.00	0.00000	0.00000	0.00000	0.00000	0.00000
9.00	0.00000	0.00000	0.00000	0.00000	0.00000
10.00	0.00000	0.00000	0.00000	0.00000	0.00000
11.00	0.00000	0.00000	0.00000	0.00000	0.00000
12.00	0.00000	0.00000	0.00000	0.00000	0.00000

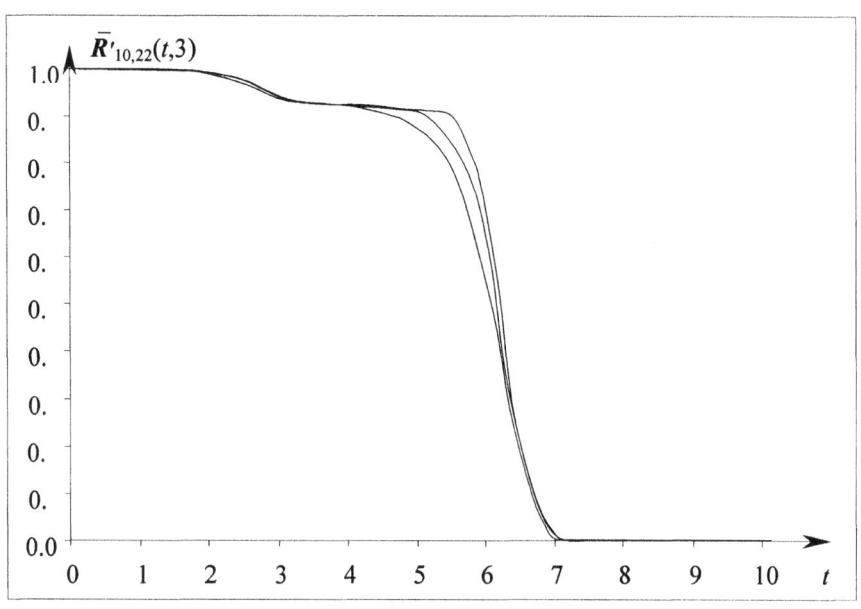

Fig. 6.11. Graphs of the rope elevator exact and approximate reliability functions in the state *u* = 3

Table 6.8. The approximate values of the rope elevator risk functions

t	$r(t)$	
	formula (6.15)	formula (6.18)
0.00	0.00001	0.00000
1.00	0.00016	0.00000
2.00	0.00273	0.00025
3.00	0.02633	0.01816
4.00	0.07310	0.07120
5.00	0.08682	0.07665
6.00	0.15379	0.09098
7.00	0.51291	0.39064
8.00	0.99592	0.98969
9.00	1.00000	1.00000
10.00	1.00000	1.00000

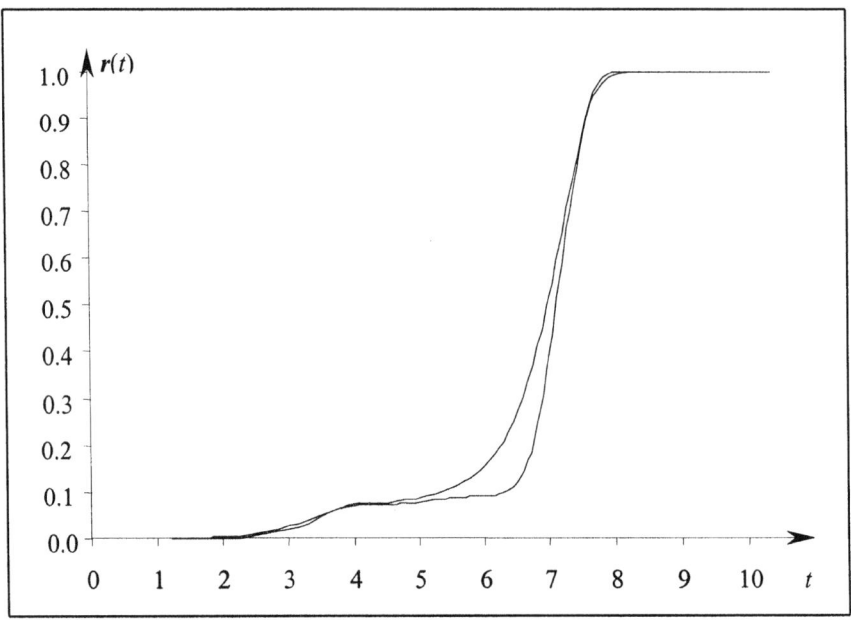

Fig. 6.12. Graphs of the approximate rope elevator risk functions

The values calculated from (6.12), (6.14) and (6.17) and the graphs illustrated in the figures allow us to state that the differences between exact and approximate reliability functions of the rope elevator are insignificant in practice. Thus replacing the exact reliability function by one of asymptotic form does not lead to large mistakes in the evaluation of system reliability characteristics. The accuracy of the rope elevator reliability evaluation testifies that the asymptotic approach to system reliability evaluation is sensible even in case when the numbers of the system's components are not large enough.

Limit reliability functions of the regular homogeneous parallel-series Weibull systems have been applied here to the reliability evaluation of the ship-rope elevator. The application of the asymptotic approach gives approximate formulae, which allow us to evaluate lower and upper limits for the system exact reliability function. The results of evaluation are also approximate because of the lack of exact reliability data for components. In cases when data are more exact it is possible to develop the methods to non-homogeneous systems assuming different reliability functions of the system components (strands). It is also possible to consider the system more particularly and try to approximate its reliability by analysing its basic components, i.e. particular wires of the strands. In this case it seems to be possible to construct an *"m* out of *n"*-series reliability model of the system.

CHAPTER 7

RELIABILITY OF LARGE MULTI-STATE EXPONENTIAL SYSTEMS

Limit reliability functions of multi-state series, parallel, "m out of n", series-parallel and parallel-series systems composed of components having exponential reliability functions are fixed. Next, the results are presented in the form of tables containing exact algorithms of the procedure while evaluating reliability characteristics of these systems' reliability in order to deliver the reliability practitioners a simple and convenient tool for everyday practice. The tables are composed of three parts, containing reliability data of the evaluated system, necessary calculations, and results of the system reliability evaluation. The way of using the algorithms is illustrated by several examples.

7.1. Auxiliary theorems

Results presented in this chapter are established and proved in [75].

Proposition 7.1
If components of the non-homogeneous multi-state series system have exponential reliability functions

$$R^{(i)}(t, \cdot) = [R^{(i)}(t,1), \ldots, R^{(i)}(t,z)], \, t \in (-\infty, \infty),$$

where

$$R^{(i)}(t,u) = 1 \text{ for } t < 0, \, R^{(i)}(t,u) = \exp[-\lambda_i(u)t] \text{ for } t \geq 0, \, i = 1,2,\ldots,a, \, u = 1,2,\ldots,z,$$

$$a_n(u) = \frac{1}{\lambda(n)n}, \, b_n(u) = 0, \, u = 1,2,\ldots,z,$$

where

$$\lambda(u) = \max_{1 \le i \le a}\{\lambda_i(u)\} \, , \, u = 1,2,...,z,$$

then

$$\overline{\mathcal{R}}'_2 (t,\cdot) = [1, \overline{\mathcal{R}}'_2 (t,1),..., \overline{\mathcal{R}}'_2 (t,z)], \, t \in (-\infty,\infty),$$

where

$$\overline{\mathcal{R}}'_2 (t,u) = 1 \text{ for } t < 0, \, \overline{\mathcal{R}}'_2 (t,u) = \exp[-\overline{d}(t,u) \, t] \text{ for } t \ge 0, \, u = 1,2,...,z,$$

and

$$\overline{d}(t,u) = \sum_{i=1}^{a} q_i \frac{\lambda_i(u)}{\lambda(u)} \text{ for } t \ge 0, \, u = 1,2,...,z,$$

is its limit reliability function.

Proposition 7.2
If components of the non-homogeneous multi-state parallel system have exponential reliability functions

$$R^{(i)}(t,\cdot) = [R^{(i)}(t,1),..., R^{(i)}(t,z)], \, t \in (-\infty,\infty),$$

where

$$R^{(i)}(t,u) = 1 \text{ for } t < 0, \, R^{(i)}(t,u) = \exp[-\lambda_i(u)t] \text{ for } t \ge 0, \, i = 1,2,...,a, \, u = 1,2,...,z,$$

$$a_n(u) = \frac{1}{\lambda(u)} \, , \, b_n(u) = \frac{1}{\lambda(u)} \log n \, , \, u = 1,2,...,z,$$

where

$$\lambda(u) = \min_{1 \le i \le a}\{\lambda_i(u)\} \, , \, u = 1,2,...,z,$$

then

$$\mathcal{R}'_3 (t,\cdot) = [1, \mathcal{R}'_3 (t,1),..., \mathcal{R}'_3 (t,z)], \, t \in (-\infty,\infty),$$

where

$$\mathcal{R}'_3 (t,u) = 1 - \exp[-d(t,u)\exp[-t]] \text{ for } t \in (-\infty,\infty), \, u = 1,2,...,z,$$

and

$$d(t,u) = \sum_{(i:\lambda_i(u)=\lambda(u))} q_i \quad \text{for } t \in (-\infty,\infty),\ u = 1,2,...,z,$$

is its limit reliability function.

Proposition 7.3
If components of the homogeneous multi-state "*m* out of *n*" system have exponential reliability functions

$$R(t,\cdot) = [R(t,1),..., R(t,z)],\ t \in (-\infty,\infty),$$

where

$$R(t,u) = 1 \text{ for } t < 0,\ R(t,u) = \exp[-\lambda(u)t] \text{ for } t \geq 0,\ u = 1,2,...,z,$$

and

Case 1. $m = $ constant ($m / n \to 0$ as $n \to \infty$),

$$a_n(u) = \frac{1}{\lambda(u)},\ b_n(u) = \frac{1}{\lambda(u)}\log n,\ u = 1,2,...,z,$$

then

$$\mathfrak{R}_3^{(0)}(t,\cdot) = [1, \mathfrak{R}_3^{(0)}(t,1),..., \mathfrak{R}_3^{(0)}(t,z)],\ t \in (-\infty,\infty),$$

where

$$\mathfrak{R}_3^{(0)}(t,u) = 1 - \sum_{i=0}^{m-1} \frac{[\exp[-it]}{i!}\exp[-\exp[-t]] \text{ for } t \in (-\infty,\infty),\ u = 1,2,..., z,$$

is its limit reliability function.

Case 2. $m/n \to \mu,\ 0 < \mu < 1$, as $n \to \infty$,

$$a_n(u) = \frac{1}{\lambda(u)}\sqrt{\frac{n-m+1}{(n+1)m}},\ b_n(u) = \frac{1}{\lambda(u)}\log\frac{n+1}{m},\ u = 1,2,...,z,$$

then

$$\mathfrak{R}_6^{(\mu)}(t,\cdot) = [1, \mathfrak{R}_6^{(\mu)}(t,1),..., \mathfrak{R}_6^{(\mu)}(t,z)],\ t \in (-\infty,\infty),$$

where

$$\mathcal{R}_6^{(\mu)}(t,u) = 1 - \frac{1}{\sqrt{2\pi}} \int_{-\infty}^{t} e^{-\frac{x^2}{2}} dx \text{ for } t \in (-\infty,\infty), u = 1,2,...,z,$$

is its limit reliability function.

Case 3. $n - m = \overline{m}$ = constant ($m/n \to 1$, as $n \to \infty$),

$$a_n(u) = \frac{1}{\lambda n}, b_n(u) = 0, \ u = 1,2,...,z,$$

then

$$\overline{\mathcal{R}}_9^{(1)}(t,\cdot) = [1, \overline{\mathcal{R}}_9^{(1)}(t,1),..., \overline{\mathcal{R}}_9^{(1)}(t,z)], t \in (-\infty,\infty),$$

where

$$\overline{\mathcal{R}}_9^{(1)}(t,u) = 1 \text{ for } t < 0, \ \overline{\mathcal{R}}_9^{(1)}(t,u) = \sum_{i=0}^{\overline{m}} \frac{t^i}{i!} \exp[-t] \text{ for } t \geq 0, \ u = 1,2,..., z,$$

is its limit reliability function.

Proposition 7.4
If components of the non-homogeneous regular multi-state series-parallel system have exponential reliability functions

$$R^{(i,j)}(t,\cdot) = [R^{(i,j)}(t,1),..., R^{(i,j)}(t,z)], t \in (-\infty,\infty),$$

where

$$R^{(i,j)}(t,u) = 1 \text{ for } t < 0, R^{(i,j)}(t,u) = \exp[-\lambda_{ij}(u)t] \text{ for } t \geq 0, i = 1,2,...,a, j = 1,2,...,e_i,$$
$$u = 1,2,...,z,$$

and

Case 1. $k_n = n, \ l_n > 0,$

$$a_n(u) = \frac{1}{\lambda(u)l_n}, b_n(u) = \frac{\log n}{\lambda(u)l_n}, \ u = 1,2,...,z,$$

where

$$\lambda_i(u) = \sum_{j=1}^{e_i} p_{ij} \lambda_{ij}(u) \, , \ \ \lambda(u) = \min_{1 \le i \le a}\{\lambda_i(u)\} \, , \ u = 1,2,...,z,$$

then

$$\mathcal{R'}_3(t,\cdot) = [1, \mathcal{R'}_3(t,1),..., \mathcal{R'}_3(t,z)\,], \ t \in (-\infty,\infty),$$

where

$$\mathcal{R'}_3(t,u) = 1 - \exp[-d(t,u)\exp[-t]] \ \text{for } t \in (-\infty,\infty), \ u = 1,2,...,z,$$

and

$$d(t,u) = \sum_{(i:\lambda_i(u)=\lambda(u))} q_i \ \ \text{for } t \in (-\infty,\infty), \ u = 1,2,...,z,$$

is its limit reliability function,

Case 2. $k_n \to k, \ l_n \to \infty,$

$$a_n(u) = \frac{1}{\lambda(u)l_n} \, , \ b_n(u) = 0, \ u = 1,2,...,z,$$

where

$$\lambda_i(u) = \sum_{j=1}^{e_i} p_{ij} \lambda_{ij}(u) \, , \ \ \lambda(u) = \min_{1 \le i \le a}\{\lambda_i(u)\} \, , \ u = 1,2,...,z,$$

then

$$\mathcal{R'}_9(t,\cdot) = [1, \mathcal{R'}_9(t,1),..., \mathcal{R'}_9(t,z)\,], \ t \in (-\infty,\infty),$$

where

$$\mathcal{R'}_9(t,u) = 1 \ \text{for } t < 0,$$

$$\mathcal{R'}_9(t,u) = 1 - \prod_{i=1}^{a}[1 - d_i(t,u)\exp[-t]]^{q_i k} \ \text{for } t \ge 0, \ u = 1,2,...,z,$$

and

$$d_i(t,u) = \exp[-(\frac{\lambda_i(u)}{\lambda(u)} - 1)t] \ \text{for } t \ge 0, \ i = 1,2,...,a, \ u = 1,2,...,z,$$

is its limit reliability function.

Proposition 7.5
If components of the non-homogeneous regular multi-state parallel-series system have exponential reliability function

$$R^{(i,j)}(t,\cdot) = [\, R^{(i,j)}(t,1)\,,...,\, R^{(i,j)}(t,z)\,],\ t \in (-\infty,\infty),$$

where

$$R^{(i,j)}(t,u) = 1 \text{ for } t < 0,\ R^{(i,j)}(t,u) = \exp[-\lambda_{ij}(u)t] \text{ for } t \geq 0,\ i = 1,2,...,a,\ j = 1,2,...,e_i,$$
$$u = 1,2,...,z,$$

and

Case 1. $k_n = n,\ l_n \to l,\ l \in (0,\infty),$

$$a_n(u) = \frac{1}{\lambda(u)n^{1/l_n}},\ b_n(u) = 0,\ u = 1,2,...,z,$$

where

$$\lambda_i(u) = \prod_{i=1}^{e_i} \lambda_{ij}^{p_{ij}}(u),\ \lambda(u) = \max_{1 \leq i \leq a}\{\lambda_i(u)\},\ u = 1,2,...,z,$$

then

$$\overline{\mathscr{R}}'_2(t,\cdot) = [1,\, \overline{\mathscr{R}}'_2(t,1)\,,...,\, \overline{\mathscr{R}}'_2(t,z)\,],\ t \in (-\infty,\infty),$$

where

$$\overline{\mathscr{R}}'_2(t,u) = 1 \text{ for } t < 0,\ \overline{\mathscr{R}}'_2(t,u) = \exp[-\overline{d}(t,u)\,t^l] \text{ for } t \geq 0,\ u = 1,2,...,z,$$

and

$$\overline{d}(t,u) = \sum_{i=1}^{a} q_i \left(\frac{\lambda_i(u)}{\lambda(u)}\right)^l \text{ for } t \geq 0,\ u = 1,2,...,z,$$

is its limit reliability function,

Case 2. $k_n = n,\ c \ll l_n,\ c\log n - l_n \gg s,\ c > 0,\ s > 0,$

$$a_n(u) = \frac{1}{(n^{1/l_n} - 1)\lambda(u)l_n} \; , \; \; b_n(u) = -\frac{1}{\lambda(u)} \log(1 - n^{-1/l_n}), u = 1,2,...,z,$$

where

$$\lambda_i(u) = \prod_{i=1}^{e_i} \lambda_{ij}^{p_{ij}}(u) \, , \; \lambda(u) = \max_{1 \le i \le a}\{\lambda_i(u)\} \, , \; u = 1,2,...,z,$$

then

$$\overline{\mathcal{R}}'_3(t,\cdot) = [1, \overline{\mathcal{R}}'_3(t,1),..., \overline{\mathcal{R}}'_3(t,z)], t \in (-\infty,\infty),$$

where

$$\overline{\mathcal{R}}'_3(t,u) = \exp[-\overline{d}(t,u)\exp[t]] \text{ for } t \in (-\infty,\infty), u = 1,2,...,z,$$

and

$$\overline{d}(t,u) = \sum_{(i:\lambda_i(u)=\lambda(u))} q_i \quad \text{for } t \in (-\infty,\infty), u = 1,2,...,z,$$

is its limit reliability function,

Case 3. $k_n = n, \; l_n - c \log n \sim s, c > 0, s \in (-\infty,\infty),$

$$a_n(u) = \frac{1}{\alpha(u)l_n} \, , \; n \, \beta(u)^{l_n} = 1, u = 1,2,...,z,$$

where

$$\beta_i(u) = \prod_{j=1}^{e_i}[1 - \exp[-\lambda_{ij}(u)b_n(u)]]^{p_{ij}} \, , \; \alpha_i(u) = \sum_{j=1}^{e_i}\frac{\exp[-\lambda_{ij}(u)b_n(u)]}{1 - \exp[-\lambda_{ij}(u)b_n(u)]}p_{ij}\lambda_{ij}(u) \, ,$$

$$\beta(u) = \max_{1 \le i \le a}\{\beta_i(u)\}, \; \alpha(u) = \max\{\alpha_i(u) : \beta_i(u) = \beta(u)\} \, , \; u = 1,2,...,z,$$

then

$$\overline{\mathcal{R}}'_3(t,\cdot) = [1, \overline{\mathcal{R}}'_3(t,1),..., \overline{\mathcal{R}}'_3(t,z)], t \in (-\infty,\infty),$$

where

$$\overline{\mathcal{R}}'_3(t,u) = \exp[-\overline{d}(t,u)\exp[t]] \text{ for } t \in (-\infty,\infty), u = 1,2,...,z,$$

and

$$\bar{d}(t,u) = \sum_{(i:\beta_i(u)=\beta(u))} q_i \exp[\frac{\alpha_i(u)-\alpha(u)}{\alpha(u)} t] \text{ for } t \in (-\infty,\infty), u = 1,2,...,z,$$

is its limit reliability function,

Case 3 . (for a homogeneous system) $k_n = n$, $l_n - c \log n \sim s$, $c > 0$, $s \in (-\infty,\infty)$,

$$a_n(u) = \frac{1}{\lambda(u)l_n(n^{1/l_n} - 1)}, \; b_n(u) = -\frac{1}{\lambda(u)} \log(1 - n^{-1/l_n}), u = 1,2,...,z,$$

then

$$\overline{\mathcal{R}}_3(t,\cdot) = [1, \overline{\mathcal{R}}_3(t,1),..., \overline{\mathcal{R}}_3(t,z)], t \in (-\infty,\infty),$$

where

$$\overline{\mathcal{R}}_3(t,u) = \exp[-\exp[t]] \text{ for } t \in (-\infty,\infty), u = 1,2,...,z,$$

is its limit reliability function,

Case 4. $k_n = n$, $l_n - \log n \gg s$, $c > 0$, $s > 0$,

$$a_n(u) = \frac{1}{\lambda(u)\log n}, \; b_n(u) = \frac{1}{\lambda(u)} \log(\frac{l_n}{\log n}), u = 1,2,...,z,$$

where

$$\lambda_i(u) = \max_{1 \le j \le e_i}\{\lambda_{ij}(u)\}, \; \lambda(u) = \max_{1 \le i \le a}\{\lambda_i(u)\}, u = 1,2,...,z,$$

then

$$\overline{\mathcal{R}}'_3(t,\cdot) = [1, \overline{\mathcal{R}}'_3(t,1),..., \overline{\mathcal{R}}'_3(t,z)], t \in (-\infty,\infty),$$

where

$$\overline{\mathcal{R}}'_3(t,u) = \exp[-\bar{d}(t,u)\exp[t]] \text{ for } t \in (-\infty,\infty), u = 1,2,...,z,$$

and

$$\overline{d}(t,u) = \sum_{(i:\lambda_i(u)=\lambda(u))} q_i \quad \text{for } t \in (-\infty,\infty),\ u = 1,2,...,z,$$

is its limit reliability function,

Case 5. $k_n \rightarrow k,\ l_n \rightarrow \infty,$

$$a_n(u) = \frac{1}{\lambda(u)}\ ,\ b_n(u) = \frac{1}{\lambda(u)}\log l_n\ ,\ u = 1,2,...,z,$$

where

$$\lambda_i(u) = \max_{1\le j\le e_i}\{\lambda_{ij}(u)\}\ ,\ \lambda(u) = \max_{1\le i\le a}\{\lambda_i(u)\}\ ,\ u = 1,2,...,z,$$

then

$$\overline{\mathscr{R}}'_{10}(t,\cdot) = [1, \overline{\mathscr{R}}'_{10}(t,1),..., \overline{\mathscr{R}}'_{10}(t,z)],\ t \in (-\infty,\infty),$$

where

$$\overline{\mathscr{R}}'_{10}(t,u) = \prod_{(i:\lambda_i(u)=\lambda(u))}[1-\exp[-\exp[-t]]]^{q_ik}\quad \text{for } t \in (-\infty,\infty),\ u = 1,2,...,z,$$

is its limit reliability function.

7.2. Algorithms for reliability evaluation of multi-state exponential systems

The results of section 7.1 are presented here in the form of tables containing exact algorithms of the procedure, while evaluating reliability characteristics of these systems' reliability in order to deliver reliability practitioners a simple and convenient tool for everyday practice. The tables are composed of three parts containing reliability data of the evaluated system, necessary calculations and results of the system reliability evaluation. The way of using of the algorithms is illustrated by several examples. These algorithms are also the basis of the computer program ([71]), allowing us to accelerate the system reliability evaluation.

Table 7.1. Algorithm of reliability evaluation of a series system

System type	A series system
Data	number of components n number of component kinds a fractions of different component kinds $q_1,...,q_a$ number of component and system states z transition rates between components' state subsets $\lambda_1(u),...,\lambda_a(u), u = 1,2,...,z$ system critical reliability state r admissible system risk level δ

Calculations	$A(u) = n\sum\limits_{i=1}^{a} q_i \lambda_i(u), u = 1,2,...,z$

Results	system reliability function $\overline{\boldsymbol{R}}'_n(t,\cdot) = \overline{\boldsymbol{\mathcal{R}}}'_2((t - b_n(u))/a_n(u),\cdot)$ $\quad\quad = [1,\exp[-A(1)t],...,\exp[-A(z)t]]$ system mean lifetimes in state subsets $M(u) = 1/A(u), u = 1,2,...,z$ standard deviations of system lifetimes in state subsets $\sigma(u) = 1/A(u), u = 1,2,...,z$ system mean lifetimes in particular states $\overline{M}(u) = M(u) - M(u + 1), u = 1,2,...,z - 1, \overline{M}(z) = M(z)$ system risk function $r(t) = 1 - \exp[-A(r)t]$ exceeding moment of admissible system risk level δ $\tau = -(\log(1 - \delta))/A(r)$

Table 7.2. Algorithm of reliability evaluation of a parallel system

System type	A parallel system
Data	number of components n number of component kinds a fractions of different component kinds $q_1,...,q_a$ number of component and system states z transition rates between components' state subsets $\lambda_1(u),...,\lambda_a(u), u = 1,2,...,z$ system critical reliability state r admissible system risk level δ

| Calculations | $\lambda(u) = \min_{1\leq i\leq a}\{\lambda_i(u)\}$, $u = 1,2,...,z$

 $d(t,u) = \sum_{(i:\lambda_i(u)=\lambda(u))} q_i, u = 1,2,..., z$
 $A(u) = \lambda(u)$, $B(u) = \log(nd(t,u))$, $u = 1,2,...,z$ |

| Results | system reliability function
$\boldsymbol{R'}_n(t,\cdot) \cong \boldsymbol{\mathscr{R}'}_3((t - b_n(u))/a_n(u),\cdot)$
$\quad = [1,1 - \exp[-\exp[-A(1)t + B(1)]],...,$
$\qquad\qquad 1 - \exp[-\exp[-A(z)t + B(z)]]]$
system mean lifetimes in state subsets
$M(u) = [0.5772 + B(u)]/A(u), u = 1,2,...,z$
standard deviations of system lifetimes in state subsets
$\sigma(u) = \pi /(A(u)\sqrt{6})$, $u = 1,2,...,z$
system mean lifetimes in particular states
$\overline{M}(u) = M(u) - M(u + 1), u = 1,2,...,z - 1, \overline{M}(z) = M(z)$
system risk function
$r(t) = \exp[-\exp[-A(r)t + B(r)]]$
exceeding moment of admissible system risk level δ
$\tau = [B(r) - \log(-\log \delta)]/ A(r)$ |

Table 7.3. Algorithm of reliability evaluation of an "*m* out of *n*" system

System type	An "*m* out of *n*" system
Data	number of components *n* threshold number *m* number of component and system states *z* transition rates between components' state subsets $\lambda(u), u = 1,2,..., z$ system critical reliability state *r* admissible system risk level δ

| Calculations | **Case 1.** $m = $ constant $(m / n \rightarrow 0 \text{ as } n \rightarrow \infty)$
$A(u) = \lambda(u)$, $B(u) = \log n$, $u = 1,2,...,z$
Case 2. $(m / n \rightarrow \mu,\ 0 < \mu < 1,\ \text{as } n \rightarrow \infty)$
$$A(u) = \lambda(u)\sqrt{\frac{(n+1)m}{(n-m+1)}}\ ,\ B(u) = \sqrt{\frac{(n+1)m}{(n-m+1)}}\ \log\frac{n+1}{m}$$
$u = 1,2,...,z$
Case 3. $n - m = \overline{m} = $ constant $(m / n \rightarrow 1 \text{ as } n \rightarrow \infty)$
$A(u) = n\lambda(u), u = 1,2,..., z$ |

Results	**Case 1.** m = constant $(m/n \to 0$ *as* $n \to \infty)$
	system reliability function

$$R_n^{(m)}(t,\cdot) \cong \mathscr{R}_3^{(0)}((t-b_n(u))/a_n(u),\cdot)$$

$$= [1,\ 1 - \sum_{i=0}^{m-1} \frac{\exp[-iA(1)t+iB(1)]}{i!} \exp[\exp[-A(1)t+B(1)]],\dots,$$

$$1 - \sum_{i=0}^{m-1} \frac{\exp[-iA(z)t+iB(z)]}{i!} \exp[\exp[-A(1)t+B(1)]]]$$

system mean lifetimes in state subsets
$M(u) =$ (a geometric integration), $u = 1,2,\dots,z$
standard deviations of system lifetimes in state subsets
$\sigma(u) =$ (a geometric integration), $u = 1,2,\dots,z$
system mean lifetimes in particular states
$\overline{M}(u) = M(u) - M(u+1),\ u = 1,2,\dots,z-1,\ \overline{M}(z) = M(z)$
system risk function

$$r(t) = \sum_{i=0}^{m-1} \frac{\exp[-iA(r)t+iB(r)]}{i!} \exp[\exp[-A(r)t+B(r)]]$$

exceeding moment of admissible system risk level δ
$\tau = r^{-1}(t)$

Results	**Case 2.** $m/n \to \mu,\ (0 < \mu < 1$ as $n \to \infty)$
	system reliability function

$$R_n^{(m)}(t,\cdot) \cong \mathscr{R}_6^{(\mu)}((t-b_n(u))/a_n(u),\cdot)$$

$$= [1,\ 1 - \frac{1}{\sqrt{2\pi}} \int_{-\infty}^{A(1)t-B(1)} e^{-\frac{x^2}{2}} dx,\dots,$$

$$1 - \frac{1}{\sqrt{2\pi}} \int_{-\infty}^{A(z)t-B(z)} e^{-\frac{x^2}{2}} dx]$$

system mean lifetimes in state subsets
$M(u) = B(u)/A(u),\ u = 1,2,\dots,z,$
system mean lifetimes in particular states
$\overline{M}(u) = M(u) - M(u+1),\ u = 1,2,\dots,z-1,\ \overline{M}(z) = M(z)$
system risk function

$$r(t) = \frac{1}{\sqrt{2\pi}} \int_{-\infty}^{A(r)t-B(r)} e^{-\frac{x^2}{2}} dx$$

exceeding moment of admissible system risk level δ
$\tau = r^{-1}(t)$

Results	**Case 3.** $n - m = \overline{m}$ = constant $(m/n \to 1$ *as* $n \to \infty)$
	system reliability function

$$\overline{R}_n^{(\overline{m})}(t,\cdot) \cong \overline{\mathcal{R}}_9^{(1)}((t - b_n(u))/a_n(u),\cdot)$$

$$= [1, \sum_{i=0}^{\overline{m}} \frac{[A(1)]^i t^i}{i!} \exp[-A(1)t],$$

$$\sum_{i=0}^{\overline{m}} \frac{[A(2)]^i t^i}{i!} \exp[-A(2)t], \dots,$$

$$\sum_{i=0}^{\overline{m}} \frac{[A(z)]^i t^i}{i!} \exp[-A(z)t] \text{ for } t \geq 0$$

system mean lifetimes in state subsets

$M(u) = $ (a geometric integration), $u = 1,2,\dots,z$

standard deviations of system lifetimes in state subsets

$\sigma(u) = $ (a geometric integration), $u = 1,2,\dots,z$

system mean lifetimes in particular states

$\overline{M}(u) = M(u) - M(u + 1), u = 1,2,\dots,z - 1, \overline{M}(z) = M(z)$

system risk function

$$r(t) = 1 - \sum_{i=0}^{\overline{m}} \frac{[A(r)]^i t^i}{i!} \exp[-A(r)t]$$

exceeding moment of admissible system risk level δ

$\tau = r^{-1}(t)$

Table 7.4. Algorithm of reliability evaluation of a series-parallel system

System type	A series-parallel system
Data	number of series subsystems linked in parallel k_n number of components in series subsystems l_n number of types of series subsystems a fractions of series subsystems of particular types $q_1,...,q_a$ numbers of types of components in series subsystems $e_1,...,e_a$ fractions of components of particular types in series subsystems $p_{11},..., p_{1e_1}$ $...$ $p_{a1},..., p_{ae_a}$ number of component and system states z component transition rates between state subsets $\lambda_{11}(u),..., \lambda_{1e_1}(u),$ $...$ $\lambda_{a1}(u),..., \lambda_{ae_a}(u), u = 1,2,..., z$ system critical state r system risk admissible level δ

Calculations	**Case 1.** system shape: $k_n \to \infty$, $l_n > 0$ $\lambda_i(u) = \sum\limits_{j=1}^{e_i} p_{ij}\lambda_{ij}(u), i = 1,2,...,a$, $\lambda(u) = \min\limits_{1\le i\le a}\{\lambda_i(u)\}$, $u = 1,2,...,z$ $d(t,u) = \sum\limits_{(i:\lambda_i(u)=\lambda(u))} q_i$, $u = 1,2,..., z$ $A(u) = \lambda(u)l_n$, $B(u) = \log(nd(t,u))$, $u = 1,2,...,z$

Calculations	**Case 2.** system shape: $k_n \to k$, $l_n \to \infty$ $\lambda_i(u) = \sum\limits_{j=1}^{e_i} p_{ij}\lambda_{ij}(u), i = 1,2,...,a$ $A_i(u) = \lambda_i(u)l_n$, $i = 1,2,...a, u = 1,2,...,z$

Results	**Case 1.** system shape: $k_n \to \infty$, $l_n > 0$

system reliability function

$R'_{k_n l_n}(t,\cdot) \cong \mathcal{R}'_3(t - b_n(u))/a_n(u),\cdot)$

$$= [1, 1 - \exp[-\exp[-A(1)t + B(1)]], ...,$$
$$1 - \exp[-\exp[-A(z)t + B(z)]]]$$

system mean lifetimes in state subsets

$M(u) = [0.5772 + B(u)]/A(u)$, $u = 1,2,...,z$

standard deviations of system lifetimes in state subsystems

$\sigma(u) = \pi/(A(u)\sqrt{6})$, $u = 1,2,...,z$

system mean lifetimes in particular states

$\overline{M}(u) = M(u) - M(u+1)$, $u = 1,2,...,z-1$, $\overline{M}(z) = M(z)$

system risk function

$r(t) = \exp[-\exp[-A(r)t + B(r)]]$

exceeding moment of admissible system risk level δ

$\tau = [B(r) - \log(-\log \delta)]/A(r)$

Results	**Case 2.** system shape: $k_n \to k$, $l_n \to \infty$

system reliability function

$R'_{k_n l_n}(t,\cdot) = \mathcal{R}'_9((t - b_n(u))/a_n(u),\cdot)$

$$= [1, 1 - \prod_{i=1}^{a}[1 - \exp[-A_i(1)t]]^{q_i k}, ...,$$

$$1 - \prod_{i=1}^{a}[1 - \exp[-A_i(z)t]]^{q_i k}]$$

system mean lifetimes in state subsystems

$$M(u) = \int_0^\infty [1 - \prod_{i=1}^{a}[1 - \exp[-A_i(u)t]]^{q_i k}] dt, u = 1,2,..., z$$

standard deviations of system lifetimes in state subsystems

$\sigma(u) = \sqrt{N(u) - M^2(u)}$,

$$N(u) = 2\int_0^\infty t[1 - \prod_{i=1}^{a}[1 - \exp[-A_i(u)t]]^{q_i k}] dt, u = 1,2,...,z$$

system mean lifetimes in particular states

$\overline{M}(u) = M(u) - M(u+1)$, $u = 1,2,...,z-1$, $\overline{M}(z) = M(z)$

system risk function

$$r(t) = \prod_{i=1}^{a}[1 - \exp[-A_i(r)t]]^{q_i k}$$

exceeding moment of admissible system risk level δ

$\tau = r^{-1}(\delta)$

Table 7.5. Algorithm of reliability evaluation of a parallel-series system

System type	A parallel-series system
Data	number of parallel subsystems linked in series k_n number of components in parallel subsystems l_n number of types of parallel subsystems a fractions of parallel subsystems of particular types $q_1,...,q_a$ numbers of types of components in parallel subsystems $e_1,...,e_a$ fractions of components of particular types in parallel subsystems $p_{11},..., p_{1e_1}$ $...$ $p_{a1},..., p_{ae_a}$ number of component and system states z component transition rates between state subsets $\lambda_{11}(u),..., \lambda_{1e_1}(u),$ $...$ $\lambda_{a1}(u),..., \lambda_{ae_a}(u), u = 1,2,..., z$ system critical state r system risk admissible level δ

| Calculations | **Case 1.** system shape: $k_n \to \infty$, $l_n \to l$

$\lambda_i(u) = \prod\limits_{j=1}^{e_i} \lambda_{ij}^{p_{ij}}(u), i = 1,2,..., a$, $u = 1,2,...,z$

$A(u) = n \sum\limits_{i=1}^{a} q_i [\lambda_i(u)]^l, u = 1,2,..., z$ |

Calculations	**Case 2.** system shape: $k_n \to \infty$, $c << l_n$, $c \log n - l_n >> s$, $c > 0$, $s > 0$
	$\lambda_i(u) = \sum_{j=1}^{e_i} \lambda_{ij}^{p_{ij}}(u)$, $i = 1, 2, ..., a$, $\lambda(u) = \max_{1 \le i \le a}\{\lambda_i(u)\}$
	$\bar{d}(t, u) = \sum_{(i : \lambda_i(u) = \lambda(u))} q_i$, $u = 1, 2, ..., z$
	$A(u) = (n^{1/l_n} - 1)\lambda(u)l_n$
	$B(u) = (n^{1/l_n} - 1)l_n \log(1 - n^{-1/l_n}) + \log \bar{d}(t, u)$

Calculations	**Case 3.** system shape: $k_n = n$, $l_n - c \log n \sim s$, $c > 0$, $s \in (-\infty, \infty)$
	$\beta_i(u) = \prod_{j=1}^{e_i}[1 - \exp[-\lambda_{ij}(u)b_n(u)]]^{p_{ij}}$
	$\alpha_i(u) = \sum_{j=1}^{e_i} \frac{\exp[-\lambda_{ij}(u)b_n(u)]}{1 - \exp[-\lambda_{ij}(u)b_n(u)]} p_{ij}\lambda_{ij}(u)$
	$\beta(u) = \max_{1 \le i \le a}\{\beta_i(u)\}$,
	$n\,\beta(u)^{l_n} = 1$, $\alpha(u) = \max\{\alpha_i(u) : \beta_i(u) = \beta(u)\}$
	$A_i(u) = \alpha_i(u)l_n$, $B_i(u) = -b_n(u)\alpha_i(u)l_n$, $i = 1, 2, ..., a$

Calculations	**Case 3'.** (a homogeneous system) system shape: $k_n = n$, $l_n - c \log n \sim s$, $c > 0$, $s \in (-\infty, \infty)$
	$A(u) = (n^{1/l_n} - 1)\lambda(u)l_n$, $B(u) = (n^{1/l} - 1)l_n \log(1 - n^{-1/l_n})$

Calculations	**Case 4.** system shape: $k_n \to \infty$, $l_n - c \log n >> s$, $c > 0$, $s > 0$
	$\lambda_i(u) = \max_{1 \le j \le e_i}\{\lambda_{ij}(u)\}$, $\lambda(u) = \max_{1 \le i \le a}\{\lambda_i(u)\}$
	$\bar{d}(t, u) = \sum_{(i : \lambda_i(u) = \lambda(u))} q_i$
	$A(u) = \lambda(u) \cdot \log n$, $B(u) = -\log n \cdot \log(l_n / \log n) + \log \bar{d}(t, u)$

Calculations	**Case 5.** system shape: $k_n \to k$, $l_n \to \infty$
	$\lambda_i(u) = \max_{1 \le j \le e_i}\{\lambda_{ij}(u)\}$, $\lambda(u) = \max_{1 \le i \le a}\{\lambda_i(u)\}$
	$A(u) = \lambda(u)$, $B(u) = \log l_n$

Results	**Case 1.** system shape: $k_n \to \infty$, $l_n \to l$
	system reliability function
	$\bar{R}'_{k_n l_n}(t, \cdot) \cong \bar{\mathcal{R}}'_2((t - b_n(u))/a_n(u), \cdot)$
	$= [1, \exp[-A(1)t^l], \dots, \exp[-A(z)t^l]$
	mean system sojourn times in state subsets
	$M(u) = \Gamma((l+1)/l)[A(u)]^{-1/l}$, $u = 1, 2, \dots, z$
	standard deviations of system sojourn times in state subsets
	$\sigma(u) = \sqrt{\Gamma((l+2)/l) - \Gamma^2((l+1)/l)}[A(u)]^{-1/l}$, $u = 1, 2, \dots, z$
	mean system sojourn times in particular states
	$\bar{M}(u) = M(u) - M(u+1)$, $u = 1, 2, \dots, z-1$, $\bar{M}(z) = M(z)$
	system risk function
	$r(t) \cong 1 - \exp[-A(r)t^l]$
	exceeding moment of admissible system risk level δ
	$\tau \cong [-(\log(1-\delta))/A(r)]^{1/l}$

Results	**Case 2.** system shape: $k_n \to \infty$, $c \ll l_n$, $c \log n - l_n \gg s$,
	$c > 0$, $s > 0$
	system reliability function
	$\bar{R}'_{k_n l_n}(t, \cdot) \cong \bar{\mathcal{R}}'_3((t - b_n(u))/a_n(u), \cdot)$
	$= [1, \exp[-\exp[A(1)t + B(1)]], \dots,$
	$\exp[-\exp[A(z)t + B(z)]]]$
	mean system sojourn times in state subsets
	$M(u) = [-0.5772 - B(u)]/A(u)$, $u = 1, 2, \dots, z$
	standard deviations of system sojourn times in state subsets
	$\sigma(u) = \pi/(A(u)\sqrt{6})$, $u = 1, 2, \dots, z$
	mean system sojourn times in particular states
	$\bar{M}(u) = M(u) - M(u+1)$, $u = 1, 2, \dots, z-1$, $\bar{M}(z) = M(z)$
	system risk function
	$r(t) \cong 1 - \exp[-\exp[A(r)t + B(r)]$
	exceeding moment of admissible system risk level δ
	$\tau \cong [\log(-\log(1-\delta)) - B(r)]/A(r)$

Results	**Case 3.** system shape: $k_n = n$, $l_n - c \log n \sim s$, $c > 0$, $s \in (-\infty,\infty)$ system reliability function $\overline{R}'_{k_n l_n}(t,\cdot) = \overline{\mathcal{R}}'_3((t - b_n(u))/a_n(u),\cdot)$ $= [1, \exp[-\sum_{(i:\beta_i(1)=\beta(1))} q_i \exp[A_i(1)t + B_i(1)]], \ldots,$ $\exp[-\sum_{(i:\beta_i(z1)=\beta(z))} q_i \exp[A_i(z)t + B_i(z)]]]$ mean system sojourn times in state subsets $M(u) = \int\limits_0^\infty \exp[-\sum_{(i:\beta_i(u)=\beta(u))} q_i \exp[A_i(u)t + B_i(u)]]dt$, $u = 1,2,\ldots,z$ standard deviations of system sojourn times in state subsets $\sigma(u) = \sqrt{N(u) - M^2(u)}$ $N(u) = 2\int\limits_0^\infty t \exp[-\sum_{(i:\beta_i(u)=\beta(u))} q_i \exp[A_i(u)t + B_i(u)]]dt$, $u = 1,2,\ldots,z$ mean system sojourn times in particular states $\overline{M}(u) = M(u) - M(u+1)$, $u = 1,2,\ldots,z-1$, $\overline{M}(z) = M(z)$ a system risk function $r(t) = 1 - \exp[-\sum_{(i:\beta_i(r)=\beta(r))} \exp[A_i(r)t + B_i(r)]]$, $u = 1,2,\ldots,z$ exceeding moment of admissible system risk level δ $\tau = r^{-1}(\delta)$

Results	**Case 3'.** (a homogeneous system) system shape: $k_n = n$, $l_n - c \log n \sim s$, $c > 0$, $s \in (-\infty,\infty)$ system reliability function $\overline{R}_{k_n l_n}(t,\cdot) \cong \overline{\mathcal{R}}_3((t - b_n(u))/a_n(u),\cdot)$ $= [1, \exp[-\exp[A(1)t + B(1)]], \ldots,$ $\exp[-\exp[A(z)t + B(z)]]]$ mean system sojourn times in state subsets $M(u) = [-0.5772 - B(u)]/A(u)$, $u = 1,2,\ldots,z$ standard deviations of system sojourn times in state subsets $\sigma(u) = \pi /(A(u)\sqrt{6})$, $u = 1,2,\ldots,z$, mean system sojourn times in particular states $\overline{M}(u) = M(u) - M(u+1)$, $u = 1,2,\ldots,z-1$, $\overline{M}(z) = M(z)$ system risk function $r(t) \cong 1 - \exp[-\exp[A(r)t + B(r)]]$ exceeding moment of admissible system risk level δ $\tau \cong [\log(-\log(1-\delta)) - B(r)] / A(r)$

Results	**Case 4.** system shape: $k_n \to \infty$, $l_n - c \log n >> s, c > 0, s > 0$
	system reliability function
	$\overline{R}'_{k_n l_n}(t,\cdot) \cong \overline{\mathscr{R}}'_3((t - b_n(u))/a_n(u),\cdot)$
	$\qquad = [1, \exp[-\exp[A(1)t + B(1)]],...,$
	$\qquad\qquad \exp[-\exp[A(z)t + B(z)]]]$
	mean system sojourn times in state subsets
	$M(u) = [-0.5772 - B(u)]/A(u), u = 1,2,...,z$
	standard deviations of system sojourn times in state subsets
	$\sigma(u) = \pi/(A(u)\sqrt{6}), u = 1,2,...,z$
	mean system sojourn times in particular states
	$\overline{M}(u) = M(u) - M(u + 1), u = 1,2,...,z - 1, \overline{M}(z) = M(z)$
	system risk function
	$r(t) \cong 1 - \exp[-\exp[A(r)t + B(r)]$
	exceeding moment of admissible system risk level δ
	$\tau \cong [\log(-\log(1 - \delta)) - B(r)]/A(r)$

Results	**Case 5.** system shape: $k_n \to k$, $l_n \to \infty$
	system reliability function
	$\overline{R}'_{k_n l_n}(t,\cdot) = \overline{\mathscr{R}}'_{10}((t - b_n(u))/a_n(u),\cdot)$
	$\qquad = [1, \prod_{(i:\lambda_i(1)=\lambda(1))}[1 - \exp[-\exp[-A(1)t + B(1)]]]^{q_ik},...,$
	$\qquad\qquad \prod_{(i:\lambda_i(z)=\lambda(z))}[1 - \exp[-\exp[-A(z)t + B(z)]]]^{q_ik}]$
	system mean sojourn times in state subsets
	$M(u) = \int_0^\infty [\prod_{(i:\lambda_i(1)=\lambda(1))}[1 - \exp[-\exp[-A(u)t + B(u)]]]^{q_ik}]dt, u = 1,2,...,z$
	standard deviations of system sojourn times in state subsets
	$\sigma(u) = \sqrt{N(u) - M^2(u)}$
	$N(u) = 2\int_0^\infty t[\prod_{(i:\lambda_i(u)=\lambda(u))}[1 - \exp[-\exp[A(u)t + B(u)]]]^{q_ik}]dt, u = 1,2,...,z$
	mean system sojourn times in particular states
	$\overline{M}(u) = M(u) - M(u + 1), u = 1,2,...,z - 1, \overline{M}(z) = M(z)$
	system risk function
	$r(t) = 1 - \prod_{(i:\lambda_i(r)=\lambda(r))}[1 - \exp[-\exp[-A(r)t + B(r)]]]^{q_ik}$
	exceeding moment of admissible system risk level δ
	$\tau = r^{-1}(\delta)$

7.3. Algorithms application to reliability evaluation of exponential systems

The examples of the system reliability evaluation presented here are an illustration of the usage of the algorithms, addressed directly to reliability practitioners.

Example 7.1. (*a piping system*)
Let us consider a piping system composed of $n = 80$ four-state pipe segments of four types such that 20 pipe segments have exponential reliability functions with transition rates between state subsets

$$\lambda_1(u) = 4^{u-4}, u = 1,2,3,$$

20 pipe segments have exponential reliability functions with transition rates between state subsets

$$\lambda_2(u) = (4.5)^{u-4}, u = 1,2,3,$$

10 pipe segments have exponential reliability functions with transition rates between state subsets

$$\lambda_3(u) = (6.5)^{u-4}, u = 1,2,3,$$

and 30 pipe segments have exponential reliability functions with transition rates between state subsets

$$\lambda_4(u) = 8^{u-4}, u = 1,2,3.$$

According to Definition 3.17, the considered piping is a non-homogeneous multi-state series system with parameters

$$n = 80, a = 4, q_1 = 2/8, q_2 = 2/8, q_3 = 1/8, q_4 = 3/8.$$

Its reliability evaluation may be performed using the algorithm presented in Table 7.1. Sequential steps of the procedure are given in Table 7.6.

Table 7.6. Reliability evaluation of the piping system

System type	A non-homogeneous series system
Data	number of system components $n = 80$

	number of component types
	$a = 4$
	fractions of different component types
	$q_1 = 2/8$, $q_2 = 2/8$, $q_3 = 1/8$, $q_4 = 3/8$
	number of component and system states
	$z = 3$
	component transition rates between state subsets
	$\lambda_1(1) = 4^{-3}$, $\lambda_2(1) = (4.5)^{-3}$, $\lambda_3(1) = (6.5)^{-3}$, $\lambda_4(1) = 8^{-3}$
	$\lambda_1(2) = 4^{-2}$, $\lambda_2(2) = (4.5)^{-2}$, $\lambda_3(2) = (6.5)^{-2}$, $\lambda_4(2) = 8^{-2}$
	$\lambda_1(3) = 4^{-1}$, $\lambda_2(3) = (4.5)^{-1}$, $\lambda_3(3) = (6.5)^{-1}$, $\lambda_4(3) = 8^{-1}$
	system reliability critical state
	$r = 2$
	admissible level of system risk
	$\delta = 0.05$

Calculations	$A(1) = 80[(2/8)4^{-3} + (2/8)(4.5)^{-3} + (1/8)(6.5)^{-3} + (3/8)8^{-3}]$
	$= 0.627$
	$A(2) = 80[(2/8)4^{-2} + (2/8)(4.5)^{-2} + (1/8)(6.5)^{-2} + (3/8)8^{-2}]$
	$= 2.943$
	$A(3) = 80[(2/8)4^{-1} + (2/8)(4.5)^{-1} + (1/8)(6.5)^{-1} + (3/8)8^{-1}]$
	$= 14.733$

Results	system approximate reliability function
	$\boldsymbol{R}'_{80}(t,\cdot) = \boldsymbol{\mathcal{R}}'_2((t - b_n(u))/a_n(u),\cdot)$
	$= [1,\exp[-0.627t],\exp[-2.943t],\exp[-14.733t]]$ for $t \geq 0$
	system mean lifetimes in state subsets
	$M(1) = 1/0.627 \cong 1.59$, $M(2) = 1/2.943 \cong 0.34$,
	$M(3) = 1/14.733 \cong 0.07$
	standard deviations of system lifetimes in state subsets
	$\sigma(1) \cong 1.59$, $\sigma(2) \cong 0.34$, $\sigma(3) \cong 0.07$
	system mean lifetimes in particular states
	$\overline{M}(1)) \cong 1.59 - 0.34 \cong 1.25$, $\overline{M}(2) \cong 0.34 - 0.07 \cong 0.27$,
	$\overline{M}(3) \cong 0.07$
	system risk function
	$r(t) = 1 - \exp[-A(2)t] = 1 - \exp[-2.943t]$
	exceeding moment of admissible system risk level δ
	$\tau = -(\log(1 - 0.05))/2.94 \cong 0.017$

The behaviour of the multi-state reliability function and risk function of the piping system is illustrated in Table 7.7 and Figure 7.1.

Table 7.7. The values of the multi-state reliability function and the risk function of the piping system

t	$\bar{\mathcal{R}}'_2(\dfrac{t-b_n(1)}{a_n(1)},1)$	$\bar{\mathcal{R}}'_2(\dfrac{t-b_n(2)}{a_n(2)},2)$	$\bar{\mathcal{R}}'_2(\dfrac{t-b_n(3)}{a_n(3)},3)$	$r(t)$
0.0	1.00000	1.00000	1.00000	0.00000
0.2	0.88214	0.55510	0.05252	0.44456
0.4	0.77818	0.30814	0.00276	0.69149
0.6	0.68647	0.17105	0.00014	0.82864
0.8	0.60556	0.09495	0.00001	0.90482
1.0	0.53419	0.05271	0.00000	0.94713
1.2	0.47123	0.02926	0.00000	0.97064
1.4	0.41570	0.01624	0.00000	0.98369
1.6	0.36670	0.00902	0.00000	0.99094
1.8	0.32349	0.00500	0.00000	0.99497
2.0	0.28536	0.00278	0.00000	0.99721
2.2	0.25173	0.00154	0.00000	0.99845
2.4	0.22206	0.00086	0.00000	0.99914
2.6	0.19589	0.00048	0.00000	0.99952
2.8	0.17280	0.00026	0.00000	0.99974

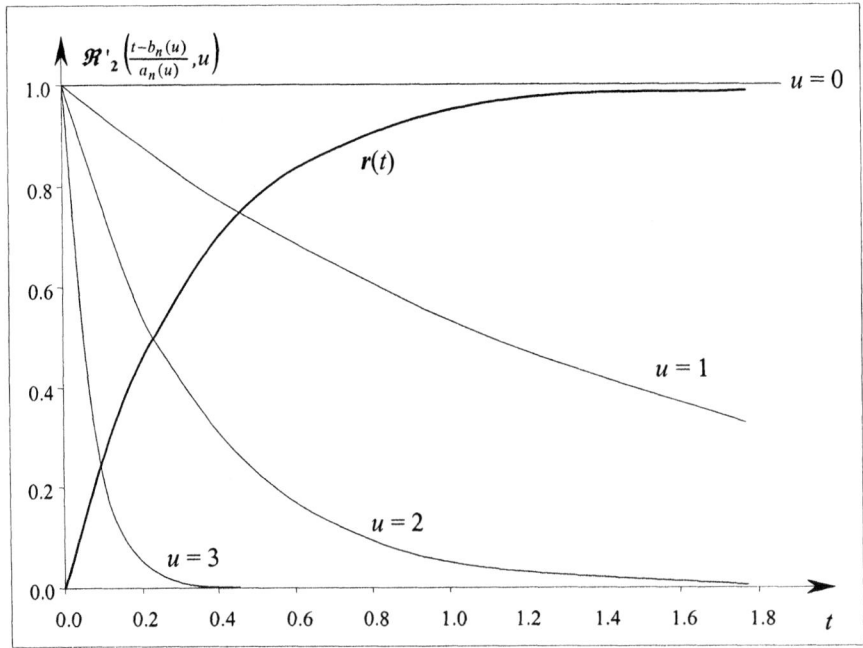

Fig. 7.1. Graphs of the multi-state reliability function and the risk function of the piping system

Example 7.2. (*a model parallel system*)

Let us consider a homogeneous six-state parallel system such that

$$n = 30, \ \lambda(u) = 10^{u-6}h^{-1}, \ u = 1,2,3,4,5.$$

We will perform its reliability evaluation based on Table 7.2. The procedure is given in Table 7.8.

Table 7.8. Reliability evaluation of the homogeneous parallel system

System type	A homogeneous parallel system
Data	number of system components $n = 30$ number of different component types $a = 1$ fractions of different component types $q_1 = 1$ number of component and system states $z = 5$ component transition rates between state subsets $\lambda_1(1) = 10^{-5}, \ \lambda_1(2) = 10^{-4}, \ \lambda_1(3) = 10^{-3}, \ \lambda_1(4) = 10^{-2},$ $\lambda_1(5) = 10^{-1}$ system reliability critical state $r = 2$ admissible critical level of system risk $\delta = 0.05$

| Calculations | $\lambda(1) = 10^{-5}, \ \lambda(2) = 10^{-4}, \ \lambda(3) = 10^{-3}, \ \lambda(4) = 10^{-2},$
$\lambda(5) = 10^{-1}$
$d(u) = 1, \ u = 1,2,3,4,5$
$A(1) = 10^{-5}, \ A(2) = 10^{-4}, \ A(3) = 10^{-3}, \ A(4) = 10^{-2}, \ A(5) = 10^{-1}$
$B(u) = \log 80 \cong 3.4012, \ u = 1,2,...,z$ |

Results	system reliability function
	$\boldsymbol{R}_{30}(t,\cdot) \cong \boldsymbol{\mathfrak{R}}_3((t-b_n(u))/a_n(u),\cdot)$
	$\quad = [1, 1 - \exp[-\exp[-0.00001t + 3.4012]],$
	$\qquad 1 - \exp[-\exp[-0.0001t + 3.4012]],$
	$\qquad 1 - \exp[-\exp[-0.001t + 3.4012]],$
	$\qquad 1 - \exp[-\exp[-0.01t + 3.4012]],$
	$\qquad 1 - \exp[-\exp[-0.1t + 3.4012]]]$ for $t \in (-\infty, \infty)$
	system mean lifetimes in state subsets
	$M(1) \cong 397840,\ M(2) \cong 39784,\ M(3) \cong 3978.4,\ M(4) \cong 397.84,$
	$M(5) \cong 39.784,$
	standard deviations of system lifetimes in state subsets
	$\sigma(1) \cong 128190,\ \sigma(2) \cong 12819,\ \sigma(3) \cong 1281.9,\ \sigma(4) \cong 128.19,$
	$\sigma(5) \cong 12.819,$
	system mean lifetimes in particular states
	$\overline{M}(1) \cong 358056,\ \overline{M}(2) \cong 35806,\ \overline{M}(3) \cong 3581,\ \overline{M}(4) \cong 358,$
	$\overline{M}(5) \cong 40$
	system risk function
	$r(t) \cong \exp[-\exp[-0.0001t + 3.4012]],$
	exceeding moment of admissible system risk level δ
	$\tau = [3.4012 - \log(-\log 0.05)]/0.0001 \cong 23040$

Example 7.3. (*a piping system*)
Let us consider a piping system composed of $k_n = 3$ pipelines linked in parallel, each of them composed of $l_n = 100$ six-state pipe segments of two types linked in series. Two of the pipelines consist of 40 pipe segments that have exponential reliability functions with transition rates between state subsets

$$\lambda_{11}(u) = (2.5)^{u-6},\ u = 1,2,3,4,5,$$

and 60 pipe segments that have exponential reliability functions with transition rates between state subsets

$$\lambda_{12}(u) = 2^{u-6},\ u = 1,2,3,4,5,$$

The third pipeline consists of 50 pipe segments that have exponential reliability functions with transition rates between state subsets

$$\lambda_{21}(u) = (1.9)^{u-6},\ u = 1,2,3,4,5,$$

and 50 pipe segments that have exponential reliability functions with transition rates between state subsets

$$\lambda_{22}(u) = (2.1)^{u-6},\ u = 1,2,3,4,5.$$

Thus the considered piping system is a non-homogeneous regular six-state series-parallel system with parameters

$k_n = k = 3$, $l_n = 100$, $a = 2$, $q_1 = 2/3$, $q_2 = 1/3$.

$q_1 = 2$, $p_{11} = 0.4$, $p_{12} = 0.6$,

$e_2 = 2$, $p_{21} = 0.5$, $p_{22} = 0.5$.

We will perform its reliability evaluation according to the procedure given in Table 7.4 (Case 2). This procedure is presented in Table 7.9.

Table 7.9. Reliability evaluation of the piping system

System type	A non-homogeneous regular series-parallel system
Data	system shape $k_n \rightarrow k, l_n \rightarrow \infty$ number of series subsystems linked in parallel $k_n = 3$ number of components in series subsystems $l_n = 100$ number of subsystem types $a = 2$ fractions of different subsystem types $q_1 = 2/3$, $q_2 = 1/3$ numbers of component types in series subsystems $e_1 = 2$, $e_2 = 2$ fractions of different component types in series subsystems $p_{11} = 0.4$, $p_{12} = 0.6$ $p_{21} = 0.5$, $p_{22} = 0.5$ number of component and system states $z = 5$ component transition rates between state subsets $\lambda_{11}(u) = (2.5)^{u-6}$, $\lambda_{12}(u) = 2^{u-6}$ $\lambda_{21}(u) = (1.9)^{u-6}$, $\lambda_{22}(u) = (2.1)^{u-6}$, $u = 1,2,3,4,5$ system reliability critical state $r = 2$ admissible level of system risk $\delta = 0.05$

Calculations	**Case 2.** $k_n \rightarrow k$, $l_n \rightarrow \infty$
	$\lambda_1(u) = 0.4\cdot(2.5)^{u-6} + 0.6\cdot 2^{u-6}$
	$\lambda_2(u) = 0.5\cdot(1.9)^{u-6} + 0.5\cdot(2.1)^{u-6}$
	$A_1(u) = 40\cdot(2.5)^{u-6} + 60\cdot 2^{u-6}$
	$A_2(u) = 50\cdot(1.9)^{u-6} + 50\cdot(2.1)^{u-6}$, $u = 1,2,3,4,5$

Results	**Case 2.** $k_n \rightarrow k$, $l_n \rightarrow \infty$
	system reliability function
	$\mathbf{R'}_{3,100}(t,\cdot) = \mathfrak{R'}_9((t - b_n(u))/a_n(u),\cdot)$
	$\quad = [1,1 - [1 - \exp[-2.285t]]^2[1 - \exp[-3.244t]],$
	$\qquad 1 - [1 - \exp[-4.774t]]^2[1 - \exp[-6.408t]],$
	$\qquad 1 - [1 - \exp[-10.060t]]^2[1 - \exp[-19.096t]],$
	$\qquad 1 - [1 - \exp[-21.400t]]^2[1 - \exp[-25.188t]],$
	$\qquad 1 - [1 - \exp[-46.000t]]^2[1 - \exp[-50.125t]]]$
	for $t \geq 0$
	system mean lifetimes in state subsets
	$M(u) = \int_0^\infty [1 - [1 - \exp[-A_1(u)t]]^2[1 - \exp[-A_2(u)]]]dt$
	$\quad = 3/[2A_1(u)] + 1/A_2(u) + 1/[2A_1(u) + A_2(u)] - 2/[A_1(u) + A_2(u)],$
	i.e.
	$M(1) = 0.74$, $M(2) = 0.35$, $M(3) = 0.16$, $M(4) = 0.05$, $M(5) = 0.04$
	standard deviations of system lifetimes in state subsets
	$\sigma(u) = \sqrt{N(u) - M^2(u)}$ (*)
	$N(u) = 2\int_0^\infty t[1 - [1 - \exp[-A_1(u)t]]^2[1 - \exp[-A_2(u)t]]]dt,$
	i.e.
	$\sigma(1) = 0.43$, $\sigma(2) = 0.20$, $\sigma(3) = 0.10$, $\sigma(4) = 0.05$, $\sigma(5) = 0.02$,
	system mean lifetimes in particular states
	$\overline{M}(1) = 0.39$, $\overline{M}(2) = 0.19$, $\overline{M}(3) = 0.11$, $\overline{M}(4) = 0.01$,
	$\overline{M}(5) = 0.04$
	system risk function
	$r(t) = [1 - \exp[-4.774t]]^2[1 - \exp[-6.408t]]$
	exceeding moment of admissible system risk level δ
	$\tau = r^{-1}(\delta) = 0.09$ years

(*)

$$\sigma = \sqrt{5/[4A_1]^2 + 1/[A_2]^2 + 1/[2A_1 + A_2]^2 + 4/[(A_1 + A_2)(2A_1 + A_2)] - 8/[A_1 + A_2]^2}$$

The behaviour of the system multi-state reliability function and risk function is illustrated in Table 7.10 and Figure 7.2.

Table 7.10. The values of the piping system multi-state reliability function and its risk function

t	$\mathscr{R}'_9\,((t-b_n(u))/a_n(u),u)$					$r(t)$
	$u=1$	$u=2$	$u=3$	$u=4$	$u=5$	
0.00	1.0000	1.0000	1.0000	1.0000	1.0000	0.0000
0.05	0.9983	0.9876	0.9039	0.6909	0.2565	0.0124
0.10	0.9884	0.9318	0.6572	0.2842	0.0265	0.0682
0.20	0.9358	0.7267	0.2660	0.0338	0.0002	0.2733
0.25	0.8948	0.6123	0.1623	0.0113	0.0000	0.3877
0.30	0.8468	0.5053	0.0983	0.0038	0.0000	0.4947
0.35	0.7943	0.4108	0.0594	0.0013	0.0000	0.5892
0.40	0.7391	0.3303	0.0359	0.0004	0.0000	0.6697
0.45	0.6832	0.2634	0.0217	0.0001	0.0000	0.7366
0.50	0.6279	0.2088	0.0131	0.0000	0.0000	0.7912
0.55	0.5741	0.1649	0.0079	0.0000	0.0000	0.8351
0.60	0.5228	0.1298	0.0048	0.0000	0.0000	0.8702
0.65	0.4743	0.1020	0.0029	0.0000	0.0000	0.8980
0.70	0.4289	0.0800	0.0017	0.0000	0.0000	0.9200

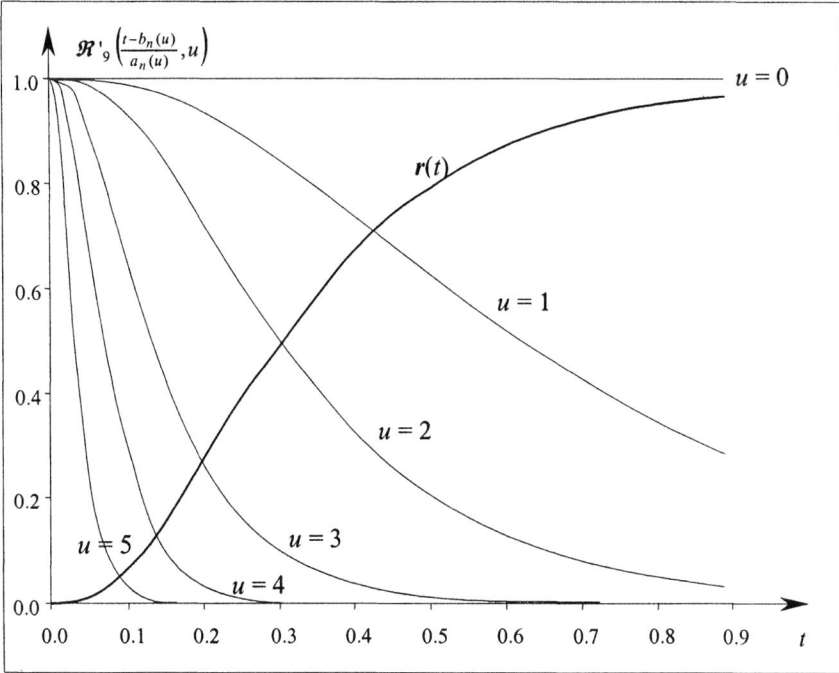

Fig. 7.2. The graphs of the multi-state reliability function of the piping system and its risk function

Example 7.4. (*a bus transportation system*)
A bus transportation company has $k_n = 100$ transportation lines. In each of the lines passengers have at their disposal $l_n = 3$ buses on which they may travel. Buses have six-state reliability functions, i.e. $z = 5$. Forty of the lines in the considered system have two buses that have exponential reliability functions with transition rates between state subsets

$$\lambda_{11}(u) = (2.5)^{(2u - 12)/3}, \, u = 1,2,3,4,5,$$

and one bus that has exponential reliability functions with transition rates between state subsets

$$\lambda_{12}(u) = 2^{(2u - 12)/3}, \, u = 1,2,3,4,5.$$

The remaining 60 lines have two buses that have exponential reliability functions with transition rates between state subsets

$$\lambda_{21}(u) = (1.9)^{(2u - 12)/3}, \, u = 1,2,3,4,5,$$

and one bus that has exponential reliability functions with transition rates between state subsets

$$\lambda_{22}(u) = (2.1)^{(2u - 12)/3}, \, u = 1,2,3,4,5.$$

The bus transportation system is able to perform its transportation tasks if at least one bus on each of the lines is not failed. Thus, according to Definition 3.21, it is a multi-state non-homogeneous regular parallel-series system with parameters

$$k_n = n = 100, \, l_n = l = 3, \, a = 2, \, q_1 = 0.4, \, q_2 = 0.6,$$

$$e_1 = 2, \, p_{11} = 2/3, \, p_{12} = 1/3, \, e_2 = 2, \, p_{21} = 2/3, \, p_{22} = 1/3.$$

Its reliability evaluation is performed in Table 7.11, according to the algorithm given in Table 7.5 (Case 1).

Table 7.11. Reliability evaluation of the bus transportation system

System type	A non-homogeneous regular parallel-series system
Data	system shape: number of parallel subsystems linked in series $k_n = 100$ number of components in parallel subsystems $l_n = 3$ number of types of parallel subsystems $a = 2$

	fractions of parallel subsystems of particular types
	$q_1 = 0.4$, $q_2 = 0.6$
	numbers of component types in parallel subsystems
	$e_1 = 2$, $e_2 = 2$
	fractions of components of particular types in parallel subsystems
	$p_{11} = 2/3$, $p_{12} = 1/3$
	$p_{21} = 2/3$, $p_{22} = 1/3$
	number of component and system states
	$z = 5$
	component transitions rates between state subsets
	$\lambda_{11}(u) = (2.5)^{(2u-12)/3}$, $\lambda_{12}(u) = 2^{(2u-12)/3}$
	$\lambda_{21}(u) = (1.9)^{(2u-12)/3}$, $\lambda_{22}(u) = (2.1)^{(2u-12)/3}$, $u = 1,2,3,4,5$
	a system critical state
	$r = 2$
	system risk admissible level
	$\delta = 0.05$

Calculations	***Case 1.*** system shape: $k_n \to \infty$, $l_n \to l$
	$\lambda_1(u) = (2.5)^{(2u-12)2/9} \cdot 2^{(2u-12)1/9}$
	$\lambda_2(u) = (1.9)^{(2u-12)2/9} \cdot (2.1)^{(2u-12)1/9}$
	$A(u) = 40(2.5)^{(2u-12)2/3} \cdot 2^{(2u-12)/3} + 60(1.9)^{(2u-12)2/3} \cdot (2.1)^{(2u-12)/3}$

Results	***Case 1.*** system shape: $k_n \to \infty$, $l_n \to l$
	system reliability function
	$\overline{R}'_{100,3}(t,\cdot) \cong \overline{\mathcal{R}}'_2((t - b_n(u))/a_n(u),\cdot)$
	$\quad = [1,\exp[-0.079t^3],\exp[-0.318t^3],\exp[-1.300t^3],$
	$\quad\quad \exp[-5.408t^3],\exp[-22.974t^3]]$ for $t \geq 0$
	mean system sojourn times in state subsets
	$M(u) = \Gamma(4/3)[A(u)]^{-1/3}$, $u = 1,2,...,5$, i.e.
	$M(1) = 2.08$, $M(2) = 1.31$, $M(3) = 0.82$, $M(4) = 0.51$, $M(5) = 0.31$
	standard deviations of system sojourn times in state subsets
	$\sigma(u) = [\Gamma(5/3) - \Gamma^2(4/3)]^{1/2}[A(u)]^{-1/3}$, $u = 1,2,...,5$, i.e.
	$\sigma(1) = 0.57$, $\sigma(2) = 0.23$, $\sigma(3) = 0.09$, $\sigma(4) = 0.03$, $\sigma(5) = 0.01$,
	mean system sojourn times in particular states
	$\overline{M}(1) = 0.77$, $\overline{M}(2) = 0.49$, $\overline{M}(3) = 0.31$, $\overline{M}(4) = 0.20$,
	$\overline{M}(5) = 0.31$
	system risk function
	$r(t) = 1 - \exp[-0.318t^3]$
	exceeding moment of admissible system risk level δ
	$\tau = r^{-1}(\delta) = [(-\log(1-\delta))/0.318]^{1/3} = 0.54$ years

The behaviour of the approximate multi-state reliability function components and the risk function of the bus transportation system is presented in Table 7.12 and Figure 7.3.

Table 7.12. Values of the multi-state reliability function and the risk function of the bus transportation system

	$\overline{\mathcal{R}}'_2\,((t-b_n(u))/a_n(u),\cdot)$					
t	$u=1$	$u=2$	$u=3$	$u=4$	$u=5$	$r(t)$
0.0	1.0000	1.0000	1.0000	1.0000	1.0000	0.0000
0.4	0.9950	0.9799	0.9222	0.7092	0.8317	0.0191
0.6	0.9811	0.9336	0.7609	0.3136	0.2289	0.0629
0.8	0.9604	0.8497	0.5233	0.0640	0.0069	0.1428
1.0	0.9240	0.7276	0.2822	0.0047	0.0000	0.2599
1.2	0.8723	0.5772	0.1124	0.0001	0.0000	0.4056
1.4	0.8051	0.4179	0.0311	0.0000	0.0000	0.5622
1.6	0.7236	0.2718	0.0056	0.0000	0.0000	0.7086
1.8	0.6308	0.1565	0.0006	0.0000	0.0000	0.8272
2.0	0.5315	0.0786	0.0000	0.0000	0.0000	0.9100
2.2	0.4312	0.0338	0.0000	0.0000	0.0000	0.9594
2.4	0.3355	0.0123	0.0000	0.0000	0.0000	0.9844
2.6	0.2494	0.0037	0.0000	0.0000	0.0000	0.9950

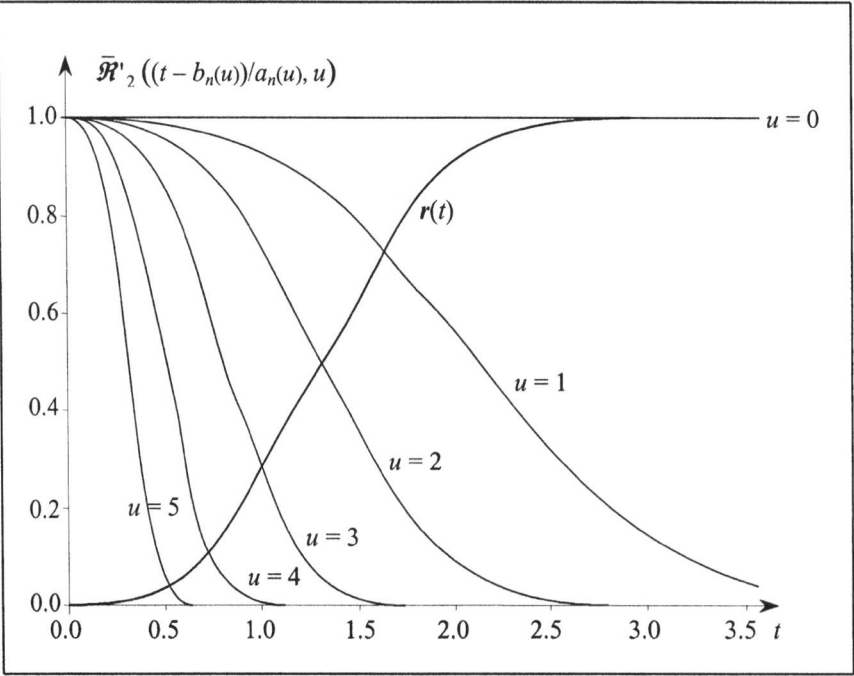

Fig. 7.3. Graphs of the multi-state reliability function and the risk function of the bus transportation system

CHAPTER 8

RELATED AND OPEN PROBLEMS

Domains of attraction for limit reliability functions of two-state systems are introduced. They are understood as the conditions, that the reliability functions of the particular components of the system have to satisfy in order that the system limit reliability function is one of the limit reliability functions from the previously fixed class for this system. Exemplary theorems concerned with domains of attraction for limit reliability functions of homogeneous series systems are presented and the application of one of them is illustrated. A practically important problem of accuracy of the asymptotic approach to large systems reliability evaluation concerned with the speed of convergence of system reliability sequence is discussed. This problem is illustrated by analysing the speed of convergence of the homogeneous series-parallel system reliability sequences to its limit reliability function. Series-"m out of n" systems and "m out of n"-series systems are defined and exemplary theorems on their limit reliability functions are presented and applied to the reliability evaluation of an illumination system and a rope elevator. Hierarchical series-parallel and parallel-series systems of any order are defined, their reliability functions are determined and limit theorems on their reliability functions are applied to reliability evaluation of exemplary hierarchical systems of order two. Applications of the asymptotic approach in large series systems reliability improvement are also presented. The chapter is completed by showing the possibility of applying the asymptotic approach to the reliability analysis of large systems placed in their operation processes. In this scope, the asymptotic approach to reliability evaluation for a large port grain transportation system related to its operation process is performed.

8.1. Domains of attraction for system limit reliability functions

The problem of domains of attraction for the limit reliability functions of two-state systems considered in this book is solved completely in [23], [71]–[72] and [80]–[81]. We will illustrate this problem partly for two-state series homogeneous systems only.

From Theorem 4.1 given in Chapter 4 it follows that the class of limit reliability functions for a homogeneous series system is composed of three functions, $\overline{\mathfrak{R}}_i(t)$, $i = 1,2,3$, defined by (4.3)–(4.5). Now we will determine domains of attraction $D_{\overline{\mathfrak{R}}_i}$ for these fixed functions, i.e. we will determine the conditions which the reliability functions $R(t)$ of the particular components of the homogeneous series system have to satisfy in order that the system limit reliability function is one of the reliability functions $\overline{\mathfrak{R}}_i(t)$, $i = 1,2,3$.

Proposition 8.1
If $R(t)$ is a reliability function of the homogeneous series system components, then

$$R(t) \in D_{\overline{\mathfrak{R}}_1}$$

if and only if

$$\lim_{r \to -\infty} \frac{1 - R(r)}{1 - R(rt)} = t^{\,\alpha} \text{ for } t > 0.$$

Proposition 8.2
If $R(t)$ is a reliability function of the homogeneous series system components, then

$$R(t) \in D_{\overline{\mathfrak{R}}_2}$$

if and only if

(i) $\exists \, y \in (-\infty, \infty) \ R(y) = 1$ and $R(y + \varepsilon) < 1$ for $\varepsilon > 0$,

(ii) $\lim_{r \to 0^+} \frac{1 - R(rt + y)}{1 - R(r + y)} = t^{\,\alpha} \text{ for } t > 0.$

Proposition 8.3
If $R(t)$ is a reliability function of the homogeneous series system components, then

$$R(t) \in D_{\overline{\mathfrak{R}}_3}$$

if and only if

$$\lim_{n \to \infty} n[1 - R(a_n t + b_n)] = e^{\,t} \text{ for } t \in (-\infty, \infty)$$

with

$$b_n = \inf\{t : R(t+0) \leq 1 - \frac{1}{n} \leq R(t-0)\},$$

$$a_n = \inf\{t : R(t(1+0)+b_n) \leq 1 - \frac{e}{n} \leq R(t(1-0)+b_n)\}.$$

Example 8.1
If components of the homogeneous series system have reliability functions

$$R(t) = \begin{cases} 1, & t < 0 \\ 1-t, & 0 \leq t < 1 \\ 0, & t \geq 1, \end{cases}$$

then

$$R(t) \in D_{\overline{\mathfrak{R}}_2}.$$

Motivation: Since

(i) $R(0) = 1$ and $R(\varepsilon) < 1$ for each $\varepsilon > 0$,

(ii) $\displaystyle \lim_{r \to 0^+} \frac{1 - R(rt + y)}{1 - R(r + y)} = \lim_{r \to 0^+} \frac{1 - (1 - rt)}{1 - (1 - r)} = t$ for $t > 0$,

then by Proposition 8.2, $R(t) \in D_{\overline{\mathfrak{R}}_2}$, where

$$\overline{\mathfrak{R}}_2(t) = \begin{cases} 1, & t < 0 \\ \exp[-t], & t \geq 0. \end{cases}$$

The results of the analysis on domains of attraction for limit reliability functions of two-state systems may automatically be transmitted to multi-state systems. To do this, it is sufficient to apply theorems about two-state systems such as the ones presented here to each vector co-ordinate of the multi-state reliability functions ([71]).

8.2. Speed of convergence of system reliability function sequences

A practically important problem of accuracy of the asymptotic approach to large systems reliability evaluation is concerned with the speed of convergence of the system reliability sequence to its limit reliability function. This problem is progressively solved for two-state systems in [29], [108]–[110] and [71], and we will illustrate it by analysing the speed of convergence of the homogeneous series-parallel system reliability sequence to its limit reliability function ([71]).

Proposition 8.4
If

$$\lim_{n\to\infty} k_n = \infty, \quad \lim_{n\to\infty} l_n = l,$$

and a component reliability function $R(t)$ of the regular homogeneous two-state series-parallel system satisfies the inequality

$$0 < [R(a_n t + b_n)]^l < \frac{1}{2},$$

then for the limit reliability function $\mathfrak{R}_3(t)$ defined by (4.54) and for the system exact reliability function $R_{k_n,l_n}(t)$ given by (2.5), we have

$$\left| \mathfrak{R}_3(t) - R_{k_n,l_n}(a_n t + b_n) + \frac{1}{2}[k_n R^l(a_n t + b_n)]^2 \exp[-k_n R^l(a_n t + b_n)] \right|$$

$$\leq \left| R^{l_n}(a_n t + b_n) - R^l(a_n t + b_n) \right| \cdot k_n + \frac{1}{k_n^2} C(k_n, R^l(a_n t + b_n)) + \left| \int_{k_n R^l(a_n t + b_n)}^{V_{3}(t)} e^{-x} dx \right|,$$

where

$$C(k_n, R^l(a_n t + b_n)) = \frac{1}{2} \pi \exp[2k_n R^l(a_n t + b_n)] \cdot [k_n R^l(a_n t + b_n)]^3$$

$$\cdot \left[\frac{4}{3[1 - 2R^l(a_n t + b_n)]} + [\frac{16}{9} \frac{n[R^l(a_n t + b_n)]^3}{[1 - 2R^l(a t + b_n)]^2} + \frac{8}{3} \frac{n[R^l(a_n t + b_n)]^2}{[1 - 2R^l(a t + b_n)]} \right.$$

$$\left. + k_n R^l(a_n t + b_n)] \cdot \exp[2k_n [R^l(a_n t + b_n)]^2 + \frac{8}{3} \frac{n[R^l(a_n t + b_n)]^3}{[1 - 2R^l(a t + b_n)]}]].$$

Example 8.2
A homogeneous series-parallel system is composed of $k_n = 100$ series subsystems linked in parallel. Each series subsystem is composed of $l_n = 4$ components with Rayleigh reliability function

$$R(t) = \begin{cases} 1, & t \leq 0 \\ \exp[-t^2], & t > 0. \end{cases} \qquad (8.1)$$

Assuming

$$a_n = \frac{1}{2\sqrt{l_n \log k_n}}, \quad b_n = \sqrt{\frac{1}{l_n} \log k_n},$$

and considering the relationships

$$\lim_{n \to \infty} k_n = \infty, \quad \lim_{n \to \infty} l_n = l,$$

for $t \in (-\infty, \infty)$ and sufficiently large n, we have

$$a_n t + b_n = \frac{t}{2\sqrt{l_n \log k_n}} + \sqrt{\frac{1}{l_n} \log k_n} > 0.$$

Hence and from (8.1) we have

$$V(t) = \lim_{n \to \infty} k_n [R(a_n t + b_n)]^{l_n}$$

$$= \lim_{n \to \infty} k_n \exp[-l_n \left(\frac{t}{2\sqrt{l_n \log k_n}} + \sqrt{\frac{1}{l_n} \log k_n} \right)^2]$$

$$= \lim_{n \to \infty} k_n \exp[-l_n \left(\frac{1}{4 l_n \log k_n} t^2 + \frac{1}{l_n} t + \frac{1}{l_n} \log k_n \right)]$$

$$= \lim_{n \to \infty} k_n \exp[-\frac{1}{4 \log k_n} t^2 - t - \log k_n]$$

$$= \lim_{n \to \infty} \exp[-\frac{1}{4 \log k_n} t^2 - t] = \exp[-t] \text{ for } t \in (-\infty, \infty)$$

Thus, considering Lemma 4.12, the system limit reliability function is given by

$$\mathfrak{R}_3(t) = 1 - \exp[-\exp[-t]] \text{ for } t \in (-\infty, \infty).$$

Moreover, from Proposition 8.4, we have

$$\left| \mathfrak{R}_3(t) - \mathbf{R}_{100,4}(a_{100}t + b_{100}) + \frac{1}{2}[100R^4(a_{100}t + b_{100})]^2 \exp[-100R^4(a_{100}t + b_{100})] \right|$$

$$\leq \frac{1}{10000} C(100, R^4(a_{100}t + b_{100})) + \left| \int_{100R^4(a_{100}t+b_{100})}^{V_3(t)} e^{-x} \, dx \right|,$$

where for $t > 0$, we have

$$C(100, R^4(a_{100}t + b_{100}))$$

$$= \frac{1}{2}\pi \exp[2\exp[-\frac{1}{4\log 100}t^2 - t]] \, [\exp[-\frac{1}{4\log 100}t^2 - t]]^3$$

$$\cdot [\frac{200}{3[50 - \exp[-\frac{1}{4\log 100}t^2 - t]]} + [\frac{16}{9}\frac{50[\exp[-\frac{1}{4\log 100}t^2 - t]]^3}{[50 - \exp[-\frac{1}{4\log 100}t^2 - t]]^2}$$

$$+ \frac{8}{3}\frac{50[\exp[-\frac{1}{4\log 100}t^2 - t]]^2}{50 - \exp[-\frac{1}{4\log 100}t^2 - t]} + \exp[-\frac{1}{4\log 100}t^2 - t]]$$

$$\cdot \exp[2[\exp[-\frac{1}{4\log 100}t^2 - t]]^2 + \frac{8}{3}\frac{50[\exp[-\frac{1}{4\log 100}t^2 - t]]^3}{50 - \exp[-\frac{1}{4\log 100}t^2 - t]}]].$$

These evaluations, i.e. the system limit reliability function and its lower and upper bounds for $k_n = 100$ and $l_n = 4$, are presented in Table 8.3 and Figure 8.3. Additionally, to illustrate the speed of convergence the evaluations for $k_n = 10$ and $l_n = 4$ and for $k_n = 50$ and $l_n = 4$ are presented in Tables 8.1–8.2 and in Figures 8.1–8.2.

The results of the analysis of the speed of reliability sequences' convergence to limit reliability functions for two-state systems may automatically be transferred to multi-state systems. To do this, it is sufficient to apply theorems on two-state systems like the one presented here to each vector co-ordinate of the multi-state reliability functions ([71]).

Table 8.1. The evaluation of the speed of convergence of reliability function sequences for a homogeneous series-parallel system ($k_n = 10$, $l_n = 4$)

t	$\mathfrak{R}_3(t)$	Δ_1	Δ_2	$\mathfrak{R}_3(t) + \Delta_1$	$\mathfrak{R}_3(t) + \Delta_2$	$R_{10}(a_{10}t + b_{10})$
0.00	0.6321	−0.3911	0.3948	0.2410	1.0269	0.6513
0.25	0.5410	−0.0930	0.0957	0.4481	0.6368	0.5530
0.50	0.4548	−0.0318	0.0337	0.4229	0.4885	0.4558
0.75	0.3765	−0.0237	0.0250	0.3527	0.4015	0.3653
1.00	0.3078	−0.0283	0.0291	0.2795	0.3369	0.2851
1.25	0.2491	−0.0347	0.0352	0.2144	0.2843	0.2171
1.50	0.2000	−0.0397	0.0400	0.1603	0.2399	0.1616
1.75	0.1595	−0.0423	0.0425	0.1172	0.2020	0.1179
2.00	0.1266	−0.0426	0.0427	0.0839	0.1693	0.0843
2.25	0.1000	−0.0410	0.0410	0.0590	0.1411	0.0592
2.50	0.0788	−0.0380	0.0380	0.0408	0.1168	0.0409
2.75	0.0619	−0.0342	0.0342	0.0277	0.0961	0.0278
3.00	0.0486	−0.0300	0.0300	0.0186	0.0786	0.0186
3.25	0.0380	−0.0258	0.0258	0.0122	0.0638	0.0122
3.50	0.0297	−0.0218	0.0218	0.0080	0.0515	0.0080
3.75	0.0232	−0.0181	0.0181	0.0051	0.0414	0.0051

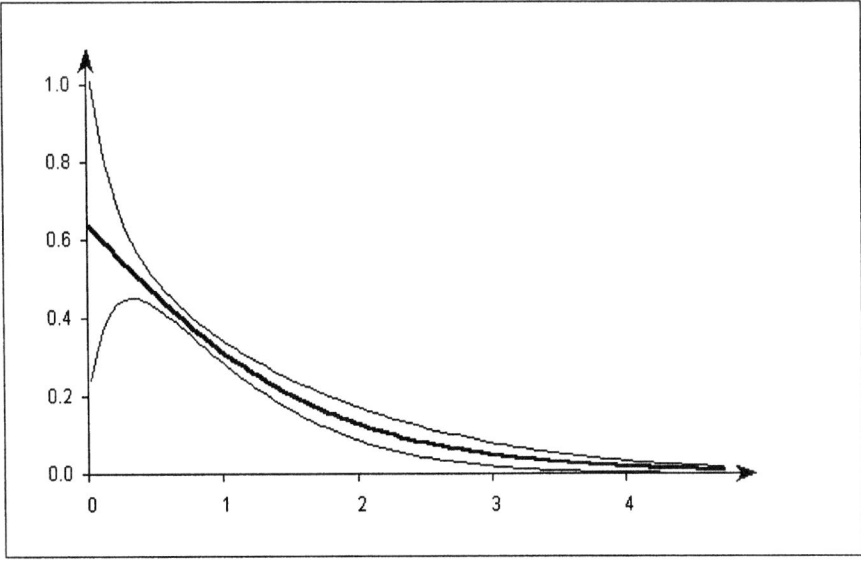

Fig. 8.1. The graphs of the limit reliability function and their lower and upper evaluations for a homogeneous series-parallel system ($k_n = 10$, $l_n = 4$)

Table 8.2. The evaluation of the speed of convergence of reliability function sequences for a homogeneous series-parallel system ($k_n = 50$, $l_n = 4$)

t	$\mathcal{R}_3(t)$	Δ_1	Δ_2	$\mathcal{R}_3(t) + \Delta_1$	$\mathcal{R}_3(t) + \Delta_2$	$R_{50}(a_{50}t + b_{50})$
0.00	0.6321	−0.0115	0.0116	0.6206	0.6438	0.6358
0.25	0.5410	−0.0044	0.0045	0.5366	0.5456	0.5424
0.50	0.4548	−0.0061	0.0062	0.4487	0.4609	0.4515
0.75	0.3765	−0.0107	0.0108	0.3657	0.3873	0.3673
1.00	0.3078	−0.0160	0.0160	0.2918	0.3238	0.2927
1.25	0.2491	−0.0207	0.0208	0.2284	0.2699	0.2289
1.50	0.2000	−0.0243	0.0243	0.1757	0.2243	0.1760
1.75	0.1595	−0.0264	0.0264	0.1331	0.1859	0.1333
2.00	0.1266	−0.0271	0.0271	0.0995	0.1537	0.0996
2.25	0.1000	−0.0266	0.0266	0.0734	0.1266	0.0735
2.50	0.0788	−0.0252	0.0252	0.0536	0.1040	0.0536
2.75	0.0619	−0.0233	0.0233	0.0387	0.0852	0.0387
3.00	0.0486	−0.0209	0.0209	0.0276	0.0695	0.0276
3.25	0.0380	−0.0185	0.0185	0.0195	0.0565	0.0196
3.50	0.0297	−0.0160	0.0160	0.0137	0.0458	0.0137
3.75	0.0232	−0.0137	0.0137	0.0095	0.0370	0.0095

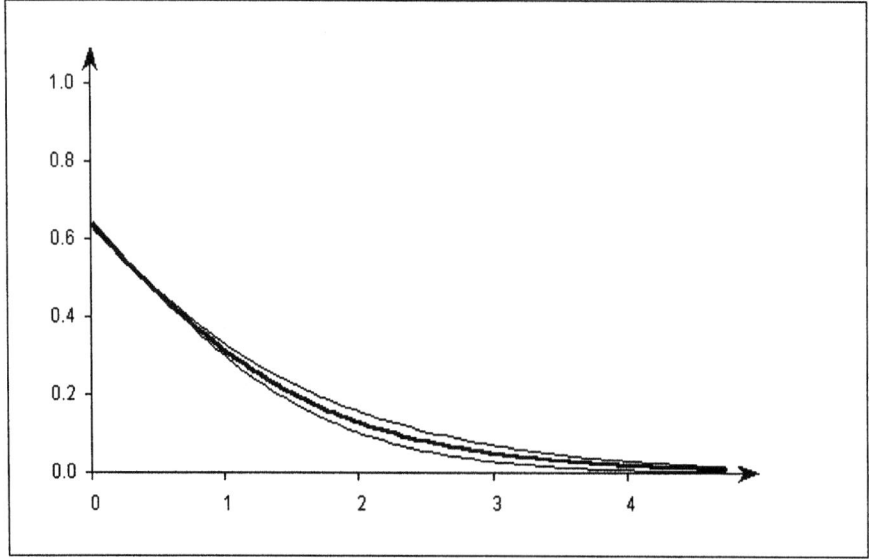

Fig. 8.2. The graphs of the limit reliability function and their lower and upper evaluations for a homogeneous series-parallel system ($k_n = 50$, $l_n = 4$)

Table 8.3. The evaluation of the speed of convergence of reliability function sequences for a homogeneous series-parallel system ($k_n = 100$, $l_n = 4$)

t	$\mathfrak{R}_3(t)$	Δ_1	Δ_2	$\mathfrak{R}_3(t) + \Delta_1$	$\mathfrak{R}_3(t) + \Delta_2$	$R_{100}(a_{100}t + b_{100})$
0.00	0.6321	−0.0028	0.0028	0.6293	0.6349	0.6340
0.25	0.5410	−0.0019	0.0020	0.5391	0.5430	0.5412
0.50	0.4548	−0.0047	0.0047	0.4501	0.4595	0.4513
0.75	0.3765	−0.0090	0.0090	0.3675	0.3855	0.3682
1.00	0.3078	−0.0136	0.0136	0.2942	0.3214	0.2946
1.25	0.2491	−0.0177	0.0177	0.2314	0.2668	0.2317
1.50	0.2000	−0.0208	0.0208	0.1792	0.2208	0.1794
1.75	0.1595	−0.0227	0.0227	0.1368	0.1822	0.1369
2.00	0.1266	−0.0234	0.0234	0.1032	0.1500	0.1033
2.25	0.1000	−0.0231	0.0231	0.0770	0.1231	0.0770
2.50	0.0788	−0.0220	0.0220	0.0568	0.1008	0.0568
2.75	0.0619	−0.0204	0.0204	0.0415	0.0823	0.0415
3.00	0.0486	−0.0185	0.0185	0.0301	0.0671	0.0301
3.25	0.0380	−0.0164	0.0164	0.0216	0.0544	0.0216
3.50	0.0297	−0.0143	0.0143	0.0154	0.0441	0.0154
3.75	0.0232	−0.0123	0.0123	0.0109	0.0356	0.0109

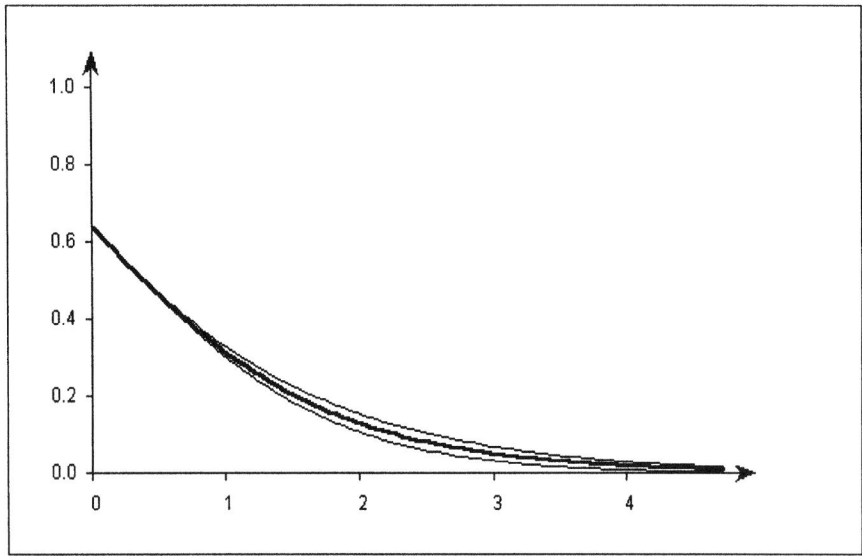

Fig. 8.3. The graphs of the limit reliability function and their lower and upper evaluations for a homogeneous series-parallel system ($k_n = 100$, $l_n = 4$)

8.3. Reliability of large series-"*m* out of *n*" systems

Definition 8.1
A two-state system is called a series-"*m* out of k_n" system if its lifetime *T* is given by

$$T = T_{(k_n-m+1)}, \quad m = 1,2,...,k_n,$$

where $T_{(k_n-m+1)}$ is the *m*th maximal order statistic in the set of random variables

$$T_i = \min_{1 \le j \le l_i} \{T_{ij}\}, \quad i = 1,2,...,k_n.$$

The above definition means that the series-"*m* out of k_n" system is composed of k_n series subsystems and it is not failed if and only if at least *m* out of its k_n series subsystems are not failed.
The series-"*m* out of k_n" system is a series-parallel for *m* = 1 and it becomes a series system for *m* = k_n.
The diagram of a series-"*m* out of k_n" system is given in Figure 8.4, where i_1, i_2, ..., $i_{k_n} \in \{1,2,...,k_n\}$ and $i_j \ne i_k$ for $j \ne k$.

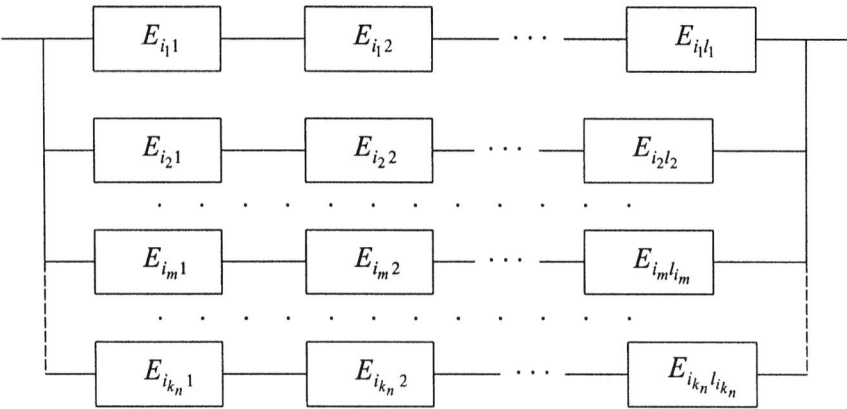

Fig. 8.4. The scheme of a series-"*m* out of k_n" system

The reliability function of the two-state series-"*m* out of k_n" system is given either by

$$R^{(m)}_{k_n,l_1,l_2,...,l_{k_n}}(t) = 1 - \sum_{\substack{r_1,r_2,...,r_{k_n}=0 \\ r_1+r_2+...+r_{k_n} \le m-1}}^{1} \prod_{i=1}^{k_n} [\prod_{j=1}^{l_i} R_{ij}(t)]^{r_i} [1 - \prod_{j=1}^{l_i} R_{ij}(t)]^{1-r_i}, \quad t \in (-\infty,\infty),$$

or by

$$\overline{R}^{(\overline{m})}_{k_n,l_1,l_2,\dots,l_{k_n}}(t) = \sum_{\substack{r_1,r_2,\dots,r_{k_n}=0 \\ r_1+r_2+\dots+r_{k_n}\le m}}^{1} \prod_{i=1}^{k_n}[1 - \prod_{j=1}^{l_i} R_{ij}(t)]^{r_i}[\prod_{j=1}^{l_i} R_{ij}(t)]^{1-r_i}, \quad t \in (-\infty,\infty),$$

where $\overline{m} = k_n - m$.

Definition 8.2
The series-"*m* out of k_n" system is called regular if

$$l_1 = l_2 = \dots = l_{k_n} = l_n, \quad l_n \in N.$$

The diagram of a regular series-"*m* out of k_n" system is given in Figure 8.5.

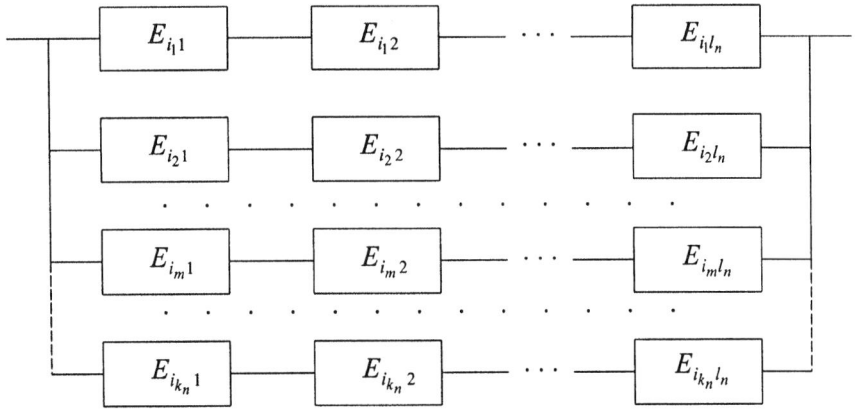

Fig. 8.5. The scheme of a regular series-"*m* out of k_n" system

Definition 8.3
The series-"*m* out of k_n" system is called homogeneous if its component lifetimes T_{ij} have an identical distribution function

$$F(t) = P(T_{ij} \le t), t \in (-\infty,\infty), i = 1,2,\dots,k_n, j = 1,2,\dots l_i,$$

i.e. if its components E_{ij} have the same reliability function

$$R(t) = 1 - F(t), t \in (-\infty,\infty).$$

From the above definitions it follows that the reliability function of the homogeneous and regular series-"*m* out of k_n" system is given either by

$$R_{k_n,l_n}^{(m)}(t) = 1 - \sum_{i=0}^{m-1} \binom{k_n}{i} [R^{l_n}(t)]^i [1 - R^{l_n}(t)]^{k_n-i}, \quad t \in (-\infty,\infty),$$

or by

$$\overline{R}_{k_n,l_n}^{(\overline{m})}(t) = \sum_{i=0}^{\overline{m}} \binom{k_n}{i} [1 - R^{l_n}(t)]^i [R^{l_n}(t)]^{k_n-i}, \quad t \in (-\infty,\infty), \ \overline{m} = k_n - m.$$

where k_n is the number of series subsystems in the "m out of k_n" system and l_n is the number of components of the series subsystems.

Corollary 8.1
If components of the homogeneous and regular two-state series-"m out of k_n" system have Weibull reliability function

$$R(t) = 1 \text{ for } t < 0, \quad R(t) = \exp[-\beta t^\alpha] \text{ for } t \geq 0, \ \alpha > 0, \ \beta > 0,$$

then its reliability function is given either by

$$R_{k_n,l_n}^{(m)}(t) = 1 \text{ for } t < 0,$$

$$R_{k_n,l_n}^{(m)}(t) = 1 - \sum_{i=0}^{m-1} \binom{k_n}{i} [\exp[-il_n\beta t^\alpha]][1 - \exp[-l_n\beta t^\alpha]]^{k_n-i} \text{ for } t \geq 0$$

or by

$$\overline{R}_{k_n,l_n}^{(\overline{m})}(t) = 1 \text{ for } t < 0,$$

$$\overline{R}_{k_n,l_n}^{(\overline{m})}(t) = \sum_{i=0}^{\overline{m}} \binom{k_n}{i} [1 - \exp[-l_n\beta t^\alpha]]^i [\exp[-(k_n-i)l_n\beta t^\alpha]], \ t \geq 0, \ \overline{m} = k_n - m. \quad (8.2)$$

Proposition 8.5
If components of the two-state homogeneous and regular series-"m out of k_n" system have Weibull reliability function

$$R(t) = 1 \text{ for } t < 0, \quad R(t) = \exp[-\beta t^\alpha] \text{ for } t \geq 0, \ \alpha > 0, \ \beta > 0,$$

and

$$k_n = n, \quad n - m = \overline{m} = constant \ (m/n \to 1 \text{ as } n \to \infty),$$

$$a_n = (\beta l_n n)^{-\frac{1}{\alpha}}, \quad b_n = 0,$$

then

$$
\overline{\mathscr{R}}_9^{(1)}(t) = \begin{cases} 1 & \text{for } t < 0 \\ \displaystyle\sum_{i=0}^{\overline{m}} \frac{t^{i\alpha}}{i!} \exp[-t^\alpha] & \text{for } t \ge 0, \alpha > 0, \end{cases}
$$

(8.3)

is its limit reliability function.

Motivation: We will use Lemma 4.11 with condition (4.34) modified into the form

$$
\overline{V}(t) = \lim_{n\to\infty} k_n [1 - R^{l_n}(a_n t + b_n)] \text{ for } t \in C_{\overline{V}}.
$$

Since

$$
a_n t + b_n = (\beta l_n n)^{-\frac{1}{\alpha}} t < 0 \text{ for } t < 0
$$

and

$$
a_n t + b_n = (\beta l_n n)^{-\frac{1}{\alpha}} t \ge 0 \text{ for } t \ge 0,
$$

then

$$
R(a_n t + b_n) = 1 \text{ for } t < 0 \quad \text{and} \quad R(a_n t + b_n) = \exp[-\beta[(\beta l_n n)^{-1/\alpha} t]^\alpha] \text{ for } t \ge 0.
$$

Hence

$$
\overline{V}(t) = \lim_{n\to\infty} k_n [1 - R^{l_n}(a_n t + b_n)] = 0 \text{ for } t < 0
$$

and

$$
\overline{V}(t) = \lim_{n\to\infty} k_n [1 - R^{l_n}(a_n t + b_n)]
$$

$$
= \lim_{n\to\infty} n[1 - \exp[-l_n \beta (a_n t + b_n)^\alpha]]
$$

$$
= \lim_{n\to\infty} n[1 - \exp[-\frac{1}{n} t^\alpha]]
$$

$$
= \lim_{n\to\infty} n[\frac{1}{n} t^\alpha + o(\frac{1}{n})] = t^\alpha \text{ for } t \ge 0.
$$

Thus, from Lemma 4.11, $\overline{\mathfrak{R}}_9^{(1)}(t)$ given by (8.3) is the system limit reliability function.

□

Example 8.3
An illumination of a sports hall is constructed of $k_n = 30$ rows of glow-lamps. Each row is composed of $l_n = 10$ series-connected glow-lamps. The lighting is good enough if at least 25 of the rows work. This lighting may be considered as a series-"25 out of 30" system. If the reliability functions of system components are Weibull with parameters $\alpha = 2$, $\beta = 0.0001$, i.e. if they are given by the formula

$$R(t) = \begin{cases} 1, & t < 0 \\ \exp[-0.0001t^2], & t \geq 0, \end{cases}$$

then according to (8.2), the reliability function of the system is given by

$$\overline{R}_{30,10}^{(5)}(t) = \sum_{i=0}^{5} \binom{30}{i} [1 - R^{10}(t)]^i [R^{10}(t)]^{30-i} .$$

$$= \sum_{i=0}^{5} \binom{30}{i} (1 - \exp[-10^{-3}t^2])^i \exp[-10^{-3}t^2(30-i)] .$$

Assuming

$$a_n = \frac{1}{(\beta l_n k_n)^{1/\alpha}} = 10\sqrt{3}^{-1} , \quad b_n = 0 ,$$

from Proposition 8.5 and using (1.1), we get

$$\overline{R}_{30,10}^{(5)}(t) \cong \overline{\mathfrak{R}}_9^{(1)}(10^{-1}\sqrt{3}t) = \begin{cases} 1 & \text{for } t < 0 \\ \sum_{i=0}^{5} \dfrac{3^i t^{2i}}{10^{2i} i!} \exp[-10^{-2}3t^2] & \text{for } t \geq 0, \end{cases}$$

The graphs of the system exact reliability function $\overline{R}_{30,10}^{(5)}(t)$ and its approximate reliability function $\overline{\mathfrak{R}}_9^{(1)}(10^{-1}\sqrt{3}t)$ are given in Figure 8.6 ([95]).

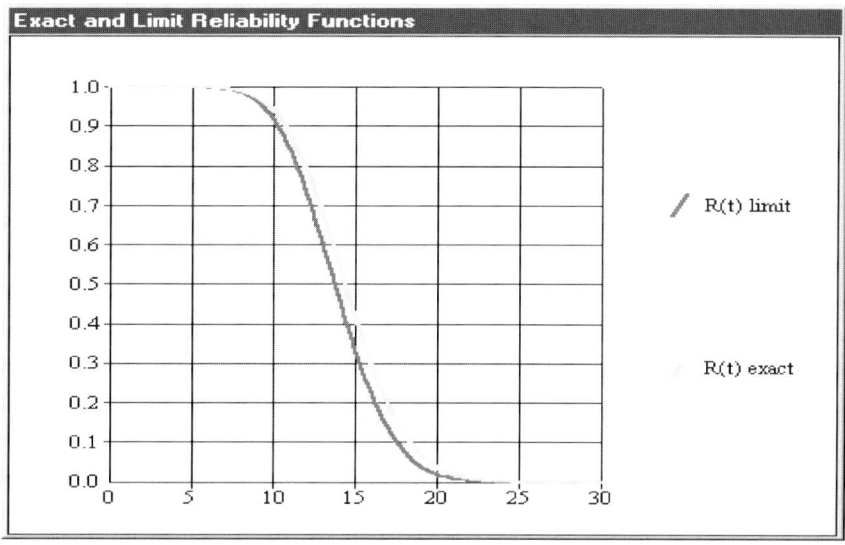

Fig. 8.6. The graphs of the reliability functions $\overline{R}_{30,10}^{(5)}(t)$ **and** $\overline{\mathcal{R}}_9^{(1)}(10^{-1}\sqrt{3}t)$

8.4. Reliability of large "*m* out of *n*"-series systems

Definition 8.4

A two-state system is called an "m_i out of l_i"-series system if its lifetime T is given by

$$T = \min_{1 \le i \le k_n} T_{(l_i - m_i + 1)} , \quad m_i = 1,2,...,l_i ,$$

where $T_{(l_i - m_i + 1)}$ is the m_ith maximal order statistic in the set of random variables

$$T_{i1}, T_{i2}, ..., T_{il_i}, \quad i = 1,2,...,k_n .$$

The above definition means that the "m_i out of l_i"-series system is composed of k_n subsystems that are "m_i out of l_i" systems and it is not failed if all its "m_i out of l_i" subsystems are not failed.

The "m_i out of l_i"-series system is a parallel-series system if $m_1 = m_2 = \ldots = m_{k_n} = 1$ and it becomes a series system if $m_i = l_i$ for all $i = 1,2, ..., k_n$.

The diagram of an "m_i out of l_i"-series system is given in Figure 8.7, where j_1, $j_2, \cdots, j_{l_i} \in \{1,2,...,l_i\}$, for $i = 1,2,...,k_n$ and $i_j \neq i_k$ for $j \neq k$.

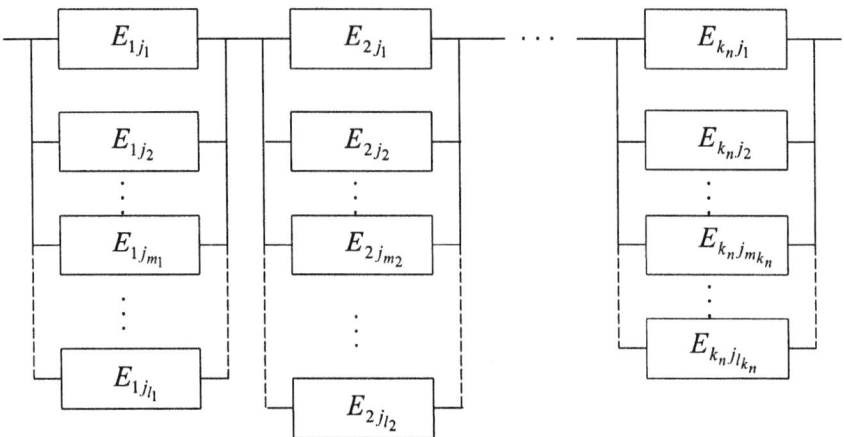

Fig. 8.7. The scheme of an "m_i out of l_i"-series system

The reliability function of the two-state "m_i out of l_i"-series system is given either by

$$\overline{R}_{k_n,l_1,l_2,...,l_{k_n}}^{(m_1,m_2,...,m_{k_n})}(t) = \prod_{i=1}^{k_n}[1 - \sum_{\substack{r_1,r_2,...,r_{l_i}=0 \\ r_1+r_2+...+r_{l_i} \le m_i-1}}^{1} [\prod_{j=1}^{l_i} R_{ij}(t)]^{r_i}[1 - \prod_{j=1}^{l_i} R_{ij}(t)]^{1-r_i}], \; t \in (-\infty,\infty),$$

or by

$$\overline{R}_{k_n,l_1,l_2,...,l_{k_n}}^{(\overline{m}_1,\overline{m}_2,...,\overline{m}_{k_n})}(t) = \prod_{i=1}^{k_n}[\sum_{\substack{r_1,r_2,...,r_{l_i}=0 \\ r_1+r_2+...+r_{l_i} \le \overline{m}_i}}^{1} [1 - \prod_{j=1}^{l_i} R_{ij}(t)]^{r_i}[\prod_{j=1}^{l_i} R_{ij}(t)]^{1-r_i}], \; t \in (-\infty,\infty),$$

where $\overline{m}_i = l_i - m_i, \; i = 1,2,...,k_n$.

Definition 8.5
The two-state "m_i out of l_i"-series system is called homogeneous if its component lifetimes T_{ij} have an identical distribution function

$$F(t) = P(T_{ij} \le t), \; t \in (-\infty,\infty), \; i = 1,2,...,k_n, \; j=1,2,...l_i,$$

i.e. if its components E_{ij} have the same reliability function

$$R(t) = 1 - F(t), \; t \in (-\infty,\infty).$$

Definition 8.6
The " m_i out of l_i "-series system is called regular if

$$l_1 = l_2 = \ldots = l_{k_n} = l_n \text{ and } m_1 = m_2 = \ldots = m_{k_n} = m, \text{ where } l_n, m \in N, \quad m \leq l_n.$$

The diagram of a regular " m out of l_n "-series system is given in Figure 8.8.

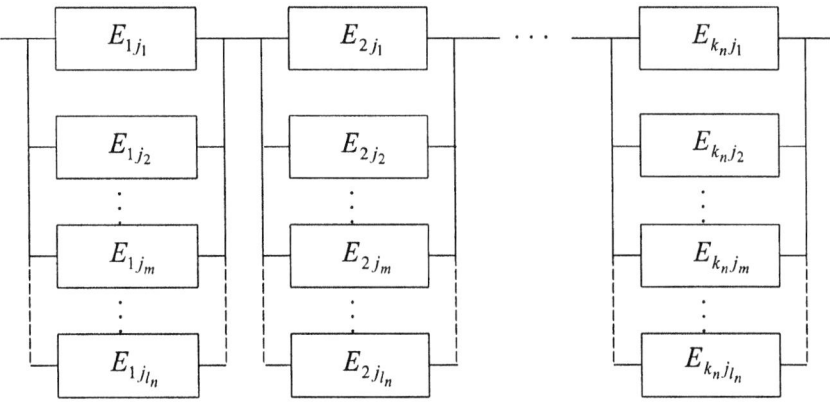

Fig. 8.8. The scheme of a regular "m out of l_n"-series system

The reliability function of the two-state homogeneous and regular „ m out of l_n "-series system is given either by

$$\overline{R}^{(m)}_{k_n,l_n}(t) = [1 - \sum_{i=0}^{m-1} \binom{l_n}{i} [R(t)]^i [1 - R(t)]^{l_n - i}]^{k_n} , \quad t \in (-\infty, \infty),$$

or by

$$\overline{R}^{(\overline{m})}_{k_n,l_n}(t) = [\sum_{i=0}^{\overline{m}} \binom{l_n}{i} [1 - R(t)]^i [R(t)]^{l_n - i}]^{k_n} , \quad t \in (-\infty, \infty), \quad \overline{m} = l_n - m.$$

where k_n is the number of "m out of l_n" subsystems linked in series and l_n is the number of components in the "m out of l_n" subsystems.

Corollary 8.2
If the components of the two-state homogeneous and regular "m out of l_n"-series system have Weibull reliability function

$$R(t) = 1 \text{ for } t < 0, \quad R(t) = \exp[-\beta t^\alpha] \text{ for } t \geq 0, \ \alpha > 0, \ \beta > 0,$$

then its reliability function is given either by

$$\overline{R}_{k_n,l_n}^{(m)}(t) = 1 \ \text{ for } \ t < 0,$$

$$\overline{R}_{k_n,l_n}^{(m)}(t) = [1 - \sum_{i=0}^{m-1} \binom{l_n}{i} \exp[-i\beta t^\alpha][1 - \exp[-\beta t^\alpha]]^{l_n - i}]^{k_n} \ \text{ for } \ t \geq 0, \tag{8.4}$$

or by

$$\overline{R}_{k_n,l_n}^{(\overline{m})}(t) = 1 \ \text{ for } \ t < 0,$$

$$\overline{R}_{k_n,l_n}^{(\overline{m})}(t) = [\sum_{i=0}^{\overline{m}} \binom{l_n}{i} [1 - \exp[-\beta t^\alpha]]^i \exp[-(l_n - i)\beta t^\alpha]]^{k_n} \ \text{ for } \ t \geq 0, \ \overline{m} = l_n - m.$$

Proposition 8.6

If components of the two-state homogeneous and regular "*m* out of l_n"-series system have Weibull reliability function

$$R(t) = 1 \ \text{ for } \ t < 0, \ \ R(t) = \exp[-\beta t^\alpha] \ \text{ for } \ t \geq 0, \ \alpha > 0, \ \beta > 0,$$

and

$$k_n \to k, \ \ l_n = n, \ \ m = \text{constant} \ (m/n \to 0 \text{ as } n \to \infty),$$

$$a_n = \frac{b_n}{\alpha \log n}, \ \ b_n = [\frac{\log n}{\beta}]^{\frac{1}{\alpha}},$$

then

$$[\overline{\Re}_3^{(0)}(t)]^k = [1 - \exp[-\exp[-t]] \sum_{i=0}^{m-1} \frac{\exp[-it]}{i!}]^k, \ \ t \in (-\infty, \infty), \tag{8.5}$$

is its limit reliability function.

Motivation: We will use Lemma 4.9 with unmodified condition (4.30) in the form

$$V(t) = \lim_{n \to \infty} n[R(a_n t + b_n)] \ \text{ for } \ t \in C_V.$$

Since for sufficiently large n, we have

$$a_n t + b_n = [\frac{\log n}{\beta}]^{\frac{1}{\alpha}} [1 + \frac{t}{\alpha \log n}] > 0 \ \text{ for } \ t \in (-\infty, \infty),$$

then

$$R(a_n t + b_n) = \exp[-\beta(a_n t + b_n)^\alpha]$$

$$= \exp[-\log n[1 + \frac{t}{\alpha \log n}]^\alpha]$$

$$= \exp[-\log n - t - o(1)] \text{ for } t \in (-\infty, \infty).$$

Hence

$$V(t) = \lim_{n \to \infty} n[R(a_n t + b_n)]$$

$$= \lim_{n \to \infty} n \exp[-\log n - t - o(1)]$$

$$= \lim_{n \to \infty} \exp[-t - o(1)] = \exp[-t] \text{ for } t \in (-\infty, \infty).$$

Thus, from Lemma 4.9, $[\overline{\mathfrak{R}}_3^{(0)}(t)]^k$ given by (8.5) is the system limit reliability function. □

Example 8.4

Let us consider the ship-rope elevator used to dock and undock ships coming in to shipyards for repairs. The elevator is composed of a steel platform-carriage placed in its syncline (hutch). The platform is moved vertically with 10 rope hoisting winches fed by separate electric motors. During ship docking the platform, with the ship settled in special supporting carriages on the platform, is raised to the wharf level (upper position). During undocking, the operation is reversed. While the ship is moving into or out of the syncline and while stopped in the upper position the platform is held on hooks and the loads in the ropes are relieved. In our further analysis we will discuss the reliability of the rope system only. The system under consideration is in order if all its ropes do not fail. Thus we may assume that it is a series system composed of 10 components (ropes). Each of the ropes is composed of 22 strands. Thus, considering the strands as basic components of the system and assuming that each of the ropes is not failed if at least $m = 5$ of its strands are not failed, according to Definitions 8.4–8.5, we conclude that the rope elevator is the two-state homogeneous and regular „5 out of 22"-series system. It is composed of $k_n = 10$ series-linked "5 out of 22" subsystems (ropes) with $l_n = 22$ components (strands). Assuming additionally that strands have Weibull reliability functions with parameters $\alpha = 2$, $\beta = 0.05$, i.e.

$$R(t) = \exp[-0.05t^2] \text{ for } t \geq 0,$$

from (8.4), we conclude that the elevator reliability function is given by

$$\overline{R^{(5)}_{10,22}}(t) = [1 - \sum_{i=0}^{4}(^{22}_{i})\exp[-i0.05t^2][1 - \exp[-0.05t^2]]^{22-i}]^{10} \text{ for } t \geq 0.$$

Next, applying Proposition 8.6 with

$$a_n = \frac{7.8626}{2\log 22} \cong 1.2718, \quad b_n = [\frac{\log 22}{0.05}]^{\frac{1}{2}} \cong 7.8626,$$

and (1.1) we get the following approximate formula for the elevator reliability function

$$\overline{R^{(5)}_{10,22}}(t) \cong [\overline{\mathfrak{R}^{(0)}_3}(0.7863t - 6.1821)]^{10}$$

$$= [1 - \exp[-\exp[-0.7863t + 6.1821]]\sum_{i=0}^{4}\frac{\exp[-0.7863it + 6.1821i]}{i!}]^{10}$$

for $t \in (-\infty, \infty)$,

8.5. Reliability of large hierarchical systems

Prior to defining the hierarchical systems of any order we once again consider a series-parallel system presented in Figure 8.9. This system here is called a series-parallel system of order 1.

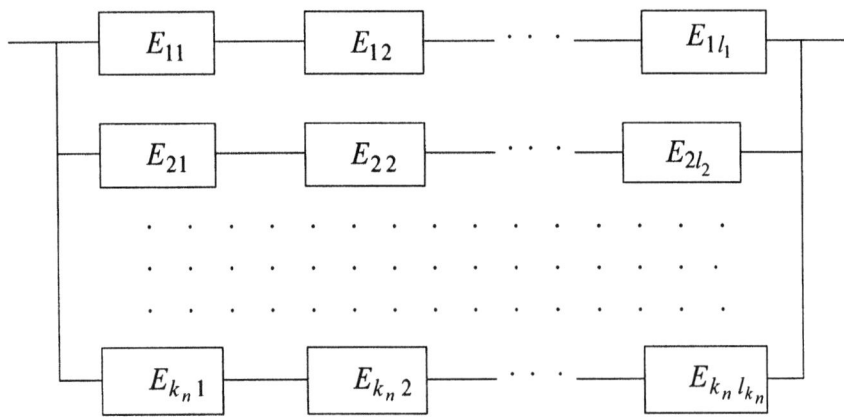

Fig. 8.9. The scheme of a series-parallel system of order 1

It is made up of components

$$E_{i_1 j_1}, \quad i_1 = 1,2,...,k_n, \quad j_1 = 1,2,...,l_{i_1},$$

with the lifetimes respectively

$$T_{i_1 j_1}, \quad i_1 = 1,2,...,k_n, \quad j_1 = 1,2,...,l_{i_1}.$$

Its lifetime is given by

$$T = \max_{1 \le i_1 \le k_n} \{ \min_{1 \le j_1 \le l_{i_1}} \{T_{i_1 j_1}\} \}. \tag{8.6}$$

Now we assume that each component

$$E_{i_1 j_1}, \quad i_1 = 1,2,...,k_n, \quad j_1 = 1,2,...,l_{i_1},$$

of the series-parallel system of order 1 is a subsystem composed of components (see Figure 8.10)

$$E_{i_1 j_1 i_2 j_2}, \quad i_2 = 1,2,...,k_n^{(i_1 j_1)}, \quad j_2 = 1,2,...,l_{i_2}^{(i_1 j_1)},$$

and has a series-parallel structure.
This means that each subsystem lifetime $T_{i_1 j_1}$ is given by

$$T_{i_1 j_1} = \max_{1 \le i_2 \le k_n^{(i_1 j_1)}} \{ \min_{1 \le j_2 \le l_{i_2}^{(i_1 j_1)}} \{T_{i_1 j_1 i_2 j_2}\} \}, \quad i_1 = 1,2,...,k_n, \quad j_1 = 1,2,...,l_{i_1}, \tag{8.7}$$

where

$$T_{i_1 j_1 i_2 j_2}, \quad i_2 = 1,2,...,k_n^{(i_1 j_1)}, \quad j_2 = 1,2,...,l_{i_2}^{(i_1 j_1)},$$

are the lifetimes of the subsystem components $E_{i_1 j_1 i_2 j_2}$.
The system defined this way is called a hierarchical series-parallel system of order 2. Its lifetime, from (8.6) and (8.7), is given by the formula

$$T = \max_{1 \le i_1 \le k_n} \{ \min_{1 \le j_1 \le l_{i_1}} [\max_{1 \le i_2 \le k_n^{(i_1 j_1)}} (\min_{1 \le j_2 \le l_{i_2}^{(i_1 j_1)}} T_{i_1 j_1 i_2 j_2})] \},$$

where k_n is the number of series systems linked in parallel and composed of series-parallel subsystems $E_{i_1 j_1}$, l_{i_1} are the numbers of series-parallel subsystems $E_{i_1 j_1}$ in these series systems, $k_n^{(i_1 j_1)}$ are the numbers of series systems in the series-parallel subsystems $E_{i_1 j_1}$ linked in parallel, and $l_{i_2}^{(i_1 j_1)}$ are the numbers of components in these series systems of the series-parallel subsystems $E_{i_1 j_1}$.

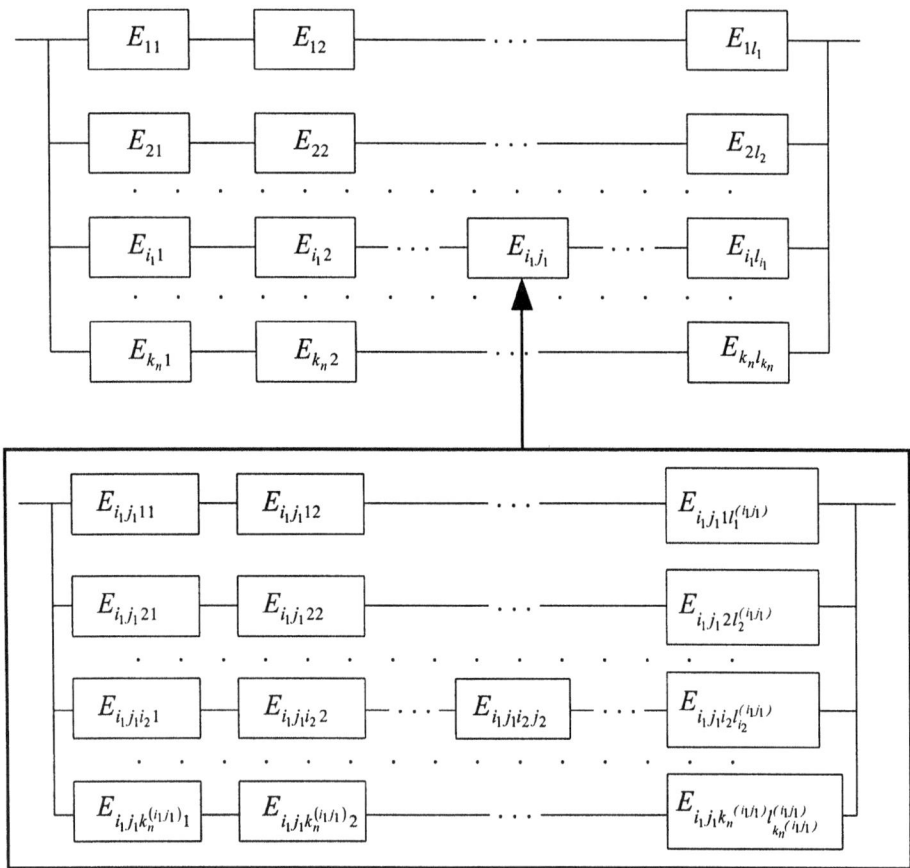

Fig. 8.10. The scheme of a series-parallel system of order 2

In an analogous way it is possible to define two-state parallel-series systems of order 2. Generally, in order to define hierarchical series-parallel and parallel-series systems of any order r, $r \geq 1$, we assume that

$$E_{i_1 j_1 \ldots i_r j_r} ,$$

where

$$i_1 = 1,2,\ldots,k_n , \; j_1 = 1,2,\ldots,l_{i_1} , \; i_2 = 1,2,\ldots,k_n^{(i_1 j_1)} , \; j_2 = 1,2,\ldots,l_{i_2}^{(i_1 j_1)} , \; \ldots,$$

$$i_r = 1,2,...,k_n^{(i_1j_1...i_{r-1}j_{r-1})}, \quad j_r = 1,2,...,l_{i_r}^{(i_1j_1...i_{r-1}j_{r-1})}$$

and

$$k_n, \; l_{i_1}, \; k_n^{(i_1j_1)}, \; l_{i_2}^{(i_1j_1)}, \; ..., \; k_n^{(i_1j_1...i_{r-1}j_{r-1})}, \; l_{i_r}^{(i_1j_1...i_{r-1}j_{r-1})} \in N,$$

are two-state components having reliability functions

$$R_{i_1j_1...i_rj_r}(t) = P(T_{i_1j_1...i_rj_r} > t), \; t \in (-\infty,\infty),$$

and random variables

$$T_{i_1j_1...i_rj_r},$$

where

$$i_1 = 1,2,...,k_n, \; j_1 = 1,2,...,l_{i_1}, \; i_2 = 1,2,...,k_n^{(i_1j_1)}, \; j_2 = 1,2,...,l_{i_2}^{(i_1j_1)}, \; ...,$$

$$i_r = 1,2,...,k_n^{(i_1j_1...i_{r-1}j_{r-1})}, \; j_r = 1,2,...,l_{i_r}^{(i_1j_1...i_{r-1}j_{r-1})},$$

are independent random variables with distribution functions

$$F_{i_1j_1...i_rj_r}(t) = P(T_{i_1j_1...i_rj_r} \leq t), \; t \in (-\infty,\infty),$$

representing the lifetimes of the components $E_{i_1j_1...i_rj_r}$.

Definition 8.7
A two-state system is called a series-parallel system of order r if its lifetime T is given by

$$T = \max_{1 \leq i_1 \leq k_n} \{ \min_{1 \leq j_1 \leq l_{i_1}} \{ \max_{1 \leq i_2 \leq k_n^{(i_1j_1)}} \{ \min_{1 \leq j_2 \leq l_{i_2}^{(i_1j_1)}} ... \max_{1 \leq i_r \leq k_n^{(i_1j_1...i_{r-1}j_{r-1})}} (\min_{1 \leq j_r \leq l_{i_r}^{(i_1j_1...i_{r-1}j_{r-1})}} T_{i_1j_1...i_rj_r})]...\}\}\},$$

where $k_n, \; k_n^{(i_1j_1)}, \; ..., \; k_n^{(i_1j_1i_2j_2...i_{r-1}j_{r-1})}$ are the numbers of suitable series systems of the system composed of series-parallel subsystems and linked in parallel, $l_{i_1}, \; l_{i_2}^{(i_1j_1)}, \; ...,$ $l_{i_{r-1}}^{(i_1j_1i_2j_2...i_{r-2}j_{r-2})}$ are the numbers of suitable series-parallel subsystems in these series systems, and $l_{i_r}^{(i_1j_1i_2j_2...i_{r-1}j_{r-1})}$ are the numbers of components in the series systems of the series-parallel subsystems.

Definition 8.8

A two-state series-parallel system of order r is called homogeneous if its component lifetimes $T_{i_1 j_1 \ldots i_r j_r}$ have an identical distribution function

$$F(t) = P(T_{i_1 j_1 \ldots i_r j_r}^{i} \le t), \; t \in (-\infty, \infty),$$

where

$$i_1 = 1, 2, \ldots, k_n, \; j_1 = 1, 2, \ldots, l_{i_1}, \; i_2 = 1, 2, \ldots, k_n^{(i_1 j_1)}, \; j_2 = 1, 2, \ldots, l_{i_2}^{(i_1 j_1)}, \ldots,$$

$$i_r = 1, 2, \ldots, k_n^{(i_1 j_1 \ldots i_{r-1} j_{r-1})}, \; j_r = 1, 2, \ldots, l_{i_r}^{(i_1 j_1 \ldots i_{r-1} j_{r-1})},$$

i.e. if its components $E_{i_1 j_1 \ldots i_r j_r}$ have the same reliability function

$$R(t) = 1 - F(t), \; t \in (-\infty, \infty).$$

Definition 8.9

A two-state series-parallel system of order r is called regular if

$$l_{i_1} = l_{i_2}^{(i_1 j_1)} = \ldots = l_{i_r}^{(i_1 j_1 \ldots i_{r-1} j_{r-1})} = l_n \text{ and } k_n^{(i_1 j_1)} = \ldots = k_n^{(i_1 j_1 \ldots i_{r-1} j_{r-1})} = k_n$$

where k_n is the number of series systems in the series-parallel subsystems and l_n are the numbers of series-parallel subsystems or respectively the numbers of components in these series systems.

Using mathematical induction it is possible to prove that the reliability function of the homogeneous and regular two-state hierarchical series-parallel system of order r is given by

$$\boldsymbol{R}_{k,k_n,l_n}(t) = 1 - [1 - [\boldsymbol{R}_{k-1,k_n,l_n}(t)]^{l_n}]^{k_n} \text{ for } k = 2, 3, \ldots, r$$

and

$$R_{1,k_n,l_n}(t) = 1 - [1 - [R(t)]^{l_n}]^{k_n}, \; t \in (-\infty, \infty),$$

where k_n and l_n are the numbers defined in Definition 8.9.

Corollary 8.3

If components of the homogeneous and regular two-state hierarchical series-parallel system of order r have an exponential reliability function

$R(t) = 1$ for $t < 0$, $R(t) = \exp[-\lambda t]$ for $t \geq 0$, $\lambda > 0$,

then its reliability function is given by

$$\boldsymbol{R}_{k,k_n,l_n}(t) = 1 \text{ for } t < 0, \quad \boldsymbol{R}_{k,k_n,l_n}(t) = 1 - [1 - [\boldsymbol{R}_{k-1,k_n,l_n}(t)]^{l_n}]^{k_n} \text{ for } t \geq 0$$

for $k = 2,3,...,r$ and

$$\boldsymbol{R}_{1,k_n,l_n}(t) = 1 - [1 - \exp[-\lambda l_n t]]^{k_n}, \quad t \in (-\infty,\infty).$$

Theorem 8.1 ([24])
If

(i) $\mathscr{R}(t) = 1 - \exp[-V(t)]$, $t \in (-\infty,\infty)$, is a non-degenerate reliability function,

(ii) $\lim\limits_{n\to\infty} l_n^{r-1} k_n^{-\frac{1}{l_n}} = 0$ for $r \geq 1$,

(iii) $\lim\limits_{n\to\infty} k_n^{l_n^{r-1}+...+1} [R(a_n t + b_n)]^{l_n^r} = V(t)$ for $t \in C_V$, $r \geq 1$, $t \in (-\infty,\infty)$,

then

$$\lim\limits_{n\to\infty} \boldsymbol{R}_{r,k_n,l_n}(a_n t + b_n) = \mathscr{R}(t) \text{ for } t \in C_{\mathscr{R}}, r \geq 1, t \in (-\infty,\infty).$$

Proposition 8.7
If components of the homogeneous and regular two-state hierarchical series-parallel system of order r have an exponential reliability function

$R(t) = 1$ for $t < 0$, $R(t) = \exp[-\lambda t]$ for $t \geq 0$, $\lambda > 0$,

$$\lim\limits_{n\to\infty} l_n^{r-1} k_n^{-\frac{1}{l_n}} = 0 \text{ for } r \geq 1,$$

and

$$a_n = \frac{1}{\lambda l_n^r}, \quad b_n = \frac{1}{\lambda}(\frac{1}{l_n} + \frac{1}{l_n^2} + ... + \frac{1}{l_n^r})\log k_n,$$

then

$$\mathcal{R}_3(t) = 1 - \exp[-\exp[-t]] \text{ for } t \in (-\infty, \infty),\tag{8.8}$$

is its limit reliability function.

Motivation: Since for sufficiently large n, we have

$$a_n t + b_n = \frac{t + (l_n^{r-1} + \dots + 1)\log k_n}{\lambda l_n^r} > 0 \text{ for } t \in (-\infty, \infty),$$

then

$$R(a_n t + b_n) = \exp[-\frac{t + (l_n^{r-1} + \dots + 1)\log k_n}{l_n^r}] \text{ for } t \in (-\infty, \infty)$$

and further

$$V(t) = \lim_{n \to \infty} k_n^{l_n^{r-1} + \dots + 1} R^{l_n^r}(a_n t + b_n)$$

$$= \lim_{n \to \infty} k_n^{l_n^{r-1} + \dots + 1} \exp[-l_n^r \frac{t + (l_n^{r-1} + \dots + 1)\log k_n}{l_n^r}] = \exp[-t] \text{ for } t \in (-\infty, \infty).$$

Hence from Theorem 8.1, $\mathcal{R}_3(t)$ given by (8.8) is the system limit reliability function.
□

Example 8.5

A hierarchical regular series-parallel homogeneous system of order $r = 2$ is such that $k_n = 200$, $l_n = 3$. The system components have identical exponential reliability functions with the failure rate $\lambda = 0.01$.

Under these assumptions its exact reliability function, according to Corollary 8.3, is given by

$$R_{2,200,3}(t) = 1, \ t < 0, \ R_{2,200,3}(t) = 1 - [1 - [1 - [1 - \exp[-0.01 \cdot 3t]]^{200}]^3]^{200}, \ t \geq 0.$$

Next applying Proposition 8.7 with normalising constants

$$a_n = \frac{1}{0.01 \cdot 9} = 11.1, \ b_n = \frac{1}{0.01}(\frac{1}{3} + \frac{1}{9})\log 200 = 235.5,$$

we conclude that the system limit reliability function is given by

$$\mathcal{R}_3(t) = 1 - \exp[-\exp[-t]] \text{ for } t \in (-\infty, \infty),$$

and from (1.1), the following approximate formula is valid

$$\boldsymbol{R}_{2,200,3}(t) \cong \mathcal{R}_3(0.09t - 21.2) = 1 - \exp[-\exp[-0.09t + 21.2]] \text{ for } t \in (-\infty, \infty).$$

The accuracy of this approximation is illustrated in Table 8.4 and Figure 8.11.

Table 8.4. Values of exact and approximate reliability functions of a hierarchical regular series-parallel homogeneous system of order 2

t	$\boldsymbol{R}_{2,200,3}(t)$	$\mathcal{R}_3(0.09t - 21.2)$	$\Delta = \boldsymbol{R}_{2,200,3}(t) - \mathcal{R}_3(0.09t - 21.2)$
200	0.999996	1.000000	−0.000004
210	0.997281	0.999953	−0.002672
220	0.934547	0.982668	−0.048121
230	0.705338	0.807704	−0.102366
240	0.414313	0.488455	**−0.074142**
250	0.205450	0.238551	−0.033101
260	0.092891	0.104885	−0.011994
270	0.040069	0.044050	−0.003981
280	0.016873	0.018149	−0.001276
290	0.007014	0.007419	−0.000405
300	0.002895	0.003023	−0.000128
310	0.001189	0.001230	−0.000041
320	0.000487	0.000500	−0.000013
330	0.000199	0.000203	−0.000116
340	0.000081	0.000083	−0.000002

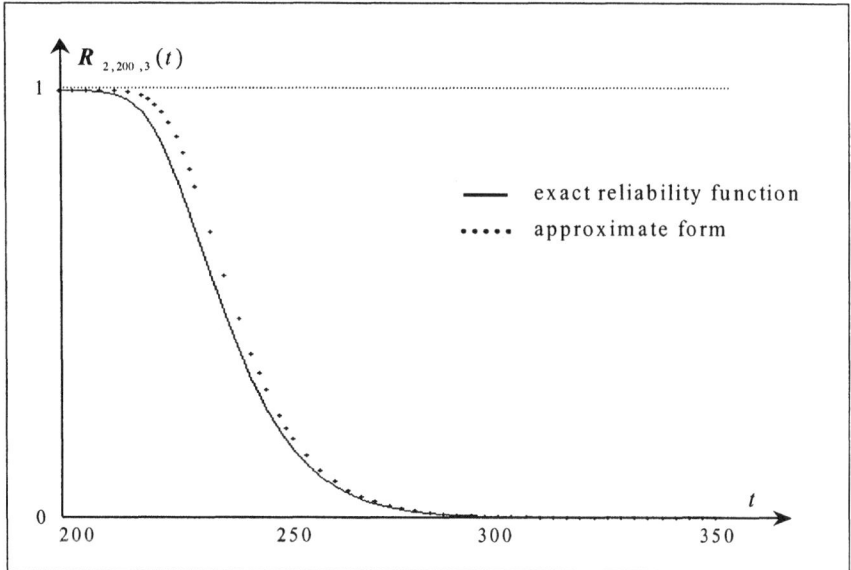

Fig. 8.11. Graphs of exact and approximate reliability functions of a hierarchical regular series-parallel homogeneous system of order 2

Definition 8.10
A two-state system is called a parallel-series system of order r if its lifetime T is given by

$$T = \min_{1 \le i_1 \le k_n} \{ \max_{1 \le j_1 \le l_{i_1}} \{ \min_{1 \le i_2 \le k_n^{(i_1 j_1)}} \{ \max_{1 \le j_2 \le l_{i_2}^{(i_1 j_1)}} \dots [\min_{1 \le i_r \le k_n^{(i_1 j_1 \dots i_{r-1} j_{r-1})}} (\max_{1 \le j_r \le l_{i_r}^{(i_1 j_1 \dots i_{r-1} j_{r-1})}} T_{i_1 j_1 \dots i_r j_r})] \dots \} \} \},$$

where k_n, $k_n^{(i_1 j_1)}$, ..., $k_n^{(i_1 j_1 i_2 j_2 \dots i_{r-1} j_{r-1})}$ are the numbers of suitable parallel systems of the system composed of parallel-series subsystems and linked in series, l_{i_1}, $l_{i_2}^{(i_1 j_1)}$, ..., $l_{i_{r-1}}^{(i_1 j_1 i_2 j_2 \dots i_{r-2} j_{r-2})}$ are the numbers of suitable parallel-series subsystems in these parallel systems, and $l_{i_r}^{(i_1 j_1 i_2 j_2 \dots i_{r-1} j_{r-1})}$ are the numbers of components in the parallel systems of the parallel-series subsystems.

Definition 8.11
A two-state parallel-series system of order r is called homogeneous if its component lifetimes $T_{i_1 j_1 \dots i_r j_r}$ have an identical distribution function

$$F(t) = P(T_{i_1 j_1 \dots i_r j_r} \le t),$$

where

$$i_1 = 1,2,\dots, k_n, \quad j_1 = 1,2,\dots, l_{i_1}, \quad i_2 = 1,2,\dots, k_n^{(i_1 j_1)}, \quad j_2 = 1,2,\dots, l_{i_2}^{(i_1 j_1)}, \dots,$$

$$i_r = 1,2,\dots, k_n^{(i_1 j_1 \dots i_{r-1} j_{r-1})}, \quad j_r = 1,2,\dots, l_{i_r}^{(i_1 j_1 \dots i_{r-1} j_{r-1})},$$

i.e. if its components $E_{i_1 j_1 \dots i_r j_r}$ have the same reliability function

$$R(t) = 1 - F(t),\ t \in (-\infty,\infty).$$

Definition 8.12
A two-state parallel-series system of order r is called regular if

$$l_{i_1} = l_{i_2}^{(i_1 j_1)} = \dots = l_{i_r}^{(i_1 j_1 \dots i_{r-1} j_{r-1})} = l_n \text{ and } k_n^{(i_1 j_1)} = \dots = k_n^{(i_1 j_1 \dots i_{r-1} j_{r-1})} = k_n$$

where k_n is the number of parallel systems in the parallel-series subsystems and l_n are the numbers of parallel-series subsystems or, respectively, the numbers of components in these parallel systems.

Using mathematical induction it is possible to prove that the reliability function of the homogeneous and regular two-state hierarchical parallel-series system of order r is given by

$$\overline{R}_{k,k_n,l_n}(t) = [1-[1-\overline{R}_{k-1,k_n,l_n}(t)]^{l_n}]^{k_n} \quad \text{for } k = 2,3,...,r$$

and

$$\overline{R}_{1,k_n,l_n}(t) = [1-[F(t)]^{l_n}]^{k_n}, \ t \in (-\infty,\infty),$$

where k_n and l_n are the numbers defined in Definition 8.12.

Corollary 8.4

If components of the homogeneous and regular two-state hierarchical parallel-series system of order r have an exponential reliability function

$$R(t) = 1 \text{ for } t < 0, \ R(t) = \exp[-\lambda t] \text{ for } t \ge 0, \ \lambda > 0,$$

then its reliability function is given by

$$\overline{R}_{k,k_n,l_n}(t) = [1-[1-\overline{R}_{k-1,k_n,l_n}(t)]^{l_n}]^{k_n} \quad \text{for } k = 2,3,...,r$$

and

$$\overline{R}_{1,k_n,l_n}(t) = [1-[1-\exp[-\lambda t]]^{l_n}]^{k_n}, \ t \in (-\infty,\infty),$$

Theorem 8.2 ([24])

If

(i) $\overline{\mathfrak{R}}(t) = \exp[-\overline{V}(t)], \ t \in (-\infty,\infty)$, is a non-degenerate reliability function,

(ii) $\lim_{n \to \infty} l_n^{r-1} k_n^{-\frac{1}{l_n}} = 0 \text{ for } r \ge 1,$

(iii) $\lim_{n \to \infty} k_n^{l_n^{r-1}+...+1} [F(a_n t + b_n)]^{l_n^r} = \overline{V}(t) \text{ for } t \in C_V, \ r \ge 1, \ t \in (-\infty,\infty),$

then

$$\lim_{n \to \infty} \overline{R}_{r,k_n,l_n}(a_n t + b_n) = \overline{\mathfrak{R}}(t) \text{ for } t \in C_{\overline{\mathfrak{R}}}, \ r \ge 1, \ t \in (-\infty,\infty). \tag{8.9}$$

Proposition 8.8
If components of the homogeneous and regular two-state hierarchical parallel-series
system of order r have an exponential reliability function

$$R(t) = 1 \text{ for } t < 0, \quad R(t) = \exp[-\lambda t] \text{ for } t \geq 0, \quad \lambda > 0,$$

$$\lim_{n \to \infty} l_n^{r-1} k_n^{-\frac{1}{l_n}} = 0 \text{ for } r \geq 1, \quad \lim_{n \to \infty} l_n = l, \quad l \in N,$$

and

$$a_n = \frac{1}{\lambda} \frac{1}{\frac{1}{k_n^{l_n}} + \dots + \frac{1}{l_n^r}}, \quad b_n = 0,$$

then

$$\overline{\mathfrak{R}}_2(t) = 1 \text{ for } t < 0 \text{ and } \overline{\mathfrak{R}}_2(t) = \exp[-t^{l^r}] \text{ for } t \geq 0 \qquad (8.10)$$

is its limit reliability function.
Motivation: Since for sufficiently large n, we have

$$a_n t + b_n = \frac{1}{\lambda} \frac{t}{\frac{1}{k_n^{l_n}} + \dots + \frac{1}{l_n^r}} < 0 \text{ for } t < 0$$

and

$$a_n t + b_n = \frac{1}{\lambda} \frac{t}{\frac{1}{k_n^{l_n}} + \dots + \frac{1}{l_n^r}} \geq 0 \text{ for } t \geq 0,$$

then

$$F(a_n t + b_n) = 0 \text{ for } t < 0$$

and

$$F(a_n t + b_n) = 1 - \exp[-\frac{t}{\frac{1}{k_n^{l_n}} + \dots + \frac{1}{l_n^r}}] \text{ for } t \geq 0,$$

and further

$$\overline{V}(t) = \lim_{n\to\infty} k_n^{l_n^{r-1}+\dots+1} F^{l_n^r}(a_n t + b_n) = 0 \text{ for } t < 0$$

and

$$\overline{V}(t) = \lim_{n\to\infty} k_n^{l_n^{r-1}+\dots+1} F^{l_n^r}(a_n t + b_n)$$

$$= \lim_{n\to\infty} k_n^{l_n^{r-1}+\dots+1} [1 - \exp[-\frac{t}{k_n^{\frac{1}{l_n}+\dots+\frac{1}{l_n^r}}}]]^{l_n^r}$$

$$= \lim_{n\to\infty} k_n^{l_n^{r-1}+\dots+1} [\frac{t^{l_n^r}}{k_n^{l_n^{r-1}+\dots+1}} + o(\frac{1}{k_n^{l_n^{r-1}+\dots+1}})] = t^{l^r} \text{ for } t \ge 0.$$

Hence from Theorem 8.2, $\overline{\mathscr{R}}_2(t)$ given by (8.10) is the system limit reliability function.
□

Example 8.6
We consider a hierarchical regular parallel-series homogeneous system of order $r = 2$ such that $k_n = 200$, $l_n = 3$, whose components have identical exponential reliability functions with the failure rate $\lambda = 0.01$.
Its exact reliability function, according to Corollary 8.4, is given by

$$\overline{R}_{2,200,3}(t) = 1 \text{ for } t < 0$$

and

$$\overline{R}_{2,200,3}(t) = [1 - [1 - [\ 1 - [1 - \exp[-0.01t]]^3\ ]^{200}\ ]^3]^{200} \text{ for } t \ge 0.$$

Next applying Proposition 8.8 with normalising constants

$$a_n = \frac{1}{0.01} \cdot \frac{1}{200^{1/3+1/9}} = 9.4912, \ b_n = 0,$$

we conclude that

$$\overline{\mathscr{R}}_2(t) = 1 \text{ for } t < 0 \text{ and } \overline{\mathscr{R}}_2(t) = \exp[-t^9] \text{ for } t \ge 0$$

is the system limit reliability function, and from (1.1), the following approximate formula is valid

$$\overline{R}_{2,200,3}(t) \cong \overline{\mathfrak{R}}_2(0.1054t) = \exp[-(0.1054t)^9] \text{ for } t \geq 0.$$

The accuracy of this approximation is illustrated in Table 8.5 and Figure 8.12.

Table 8.5. Values of exact and approximate reliability functions of a hierarchical regular parallel-series homogeneous system of order 2

t	$\overline{R}_{2,200,3}(t)$	$\overline{\mathfrak{R}}_2(0.1054t)$	$\Delta = \overline{R}_{2,200,3}(t) - \overline{\mathfrak{R}}_2(0.1054t)$
0	1.000000	1.000000	0.000000
2	0.999999	0.999999	0.000000
4	0.999656	0.999579	0.000077
6	0.988445	0.983952	0.004493
8	0.876948	0.806167	0.070781
10	0.451036	0.200822	**0.250214**
12	0.040806	0.000253	0.040553
14	0.000070	0.000000	0.000070
16	0.000000	0.000000	0.000000

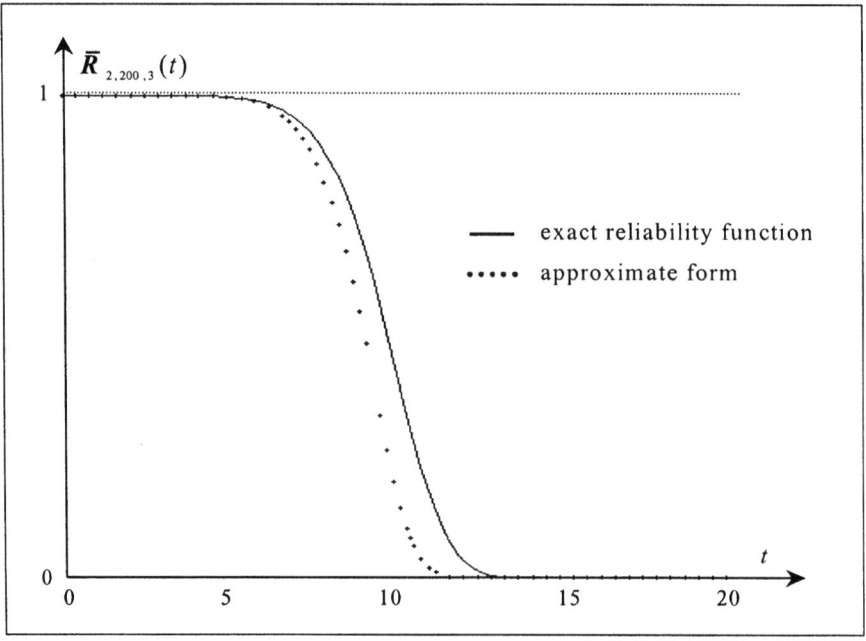

Fig. 8.12. Graphs of exact and approximate reliability functions of a hierarchical regular parallel-series homogeneous system of order 2

8.6. Asymptotic approach to systems reliability improvement

We first consider the homogeneous series system illustrated in Figure 8.13.

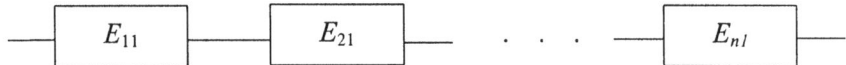

Fig. 8.13. The scheme of a series system

It is composed of n components E_{i1}, $i = 1,2,...,n$, having lifetimes T_{i1}, $i = 1,2,...,n$, and exponential reliability functions

$$R(t) = 1 \text{ for } t < 0, \quad R(t) = \exp[-\lambda t] \text{ for } t \geq 0, \quad \lambda > 0.$$

Its lifetime and its reliability function respectively are given by

$$T = \min_{1 \leq i \leq n}\{T_{i1}\},$$

$$R_n(t) = [R(t)]^n = \exp[-\lambda n t], \quad t \geq 0.$$

In order to improve of the reliability of this series system the following exemplary methods can be used:
– improving the reliability of its components by reducing their failure rates by a factor $\rho, 0 < \rho < 1$,
– a warm duplication (single reservation) of system components,
– a cold duplication of system components,
– a mixed duplication of system components,
– a hot system duplication,
– a cold system duplication.
It is supposed here that the reserve components are identical to the basic ones.
The results of these methods of system reliability improvement are briefly presented below, giving the system schemes, lifetimes and reliability functions.

Case 1. Improving the reliability of the system components by reducing their failure rates by a factor $\rho, 0 < \rho <1$,

$$T = \min_{1 \leq i \leq n}\{T_{i1}\},$$

$$R_n^{(1)}(t) = [R(\rho t)]^n = \exp[-\rho \lambda n t], \quad t \geq 0.$$

Case 2. A hot reservation of the system components

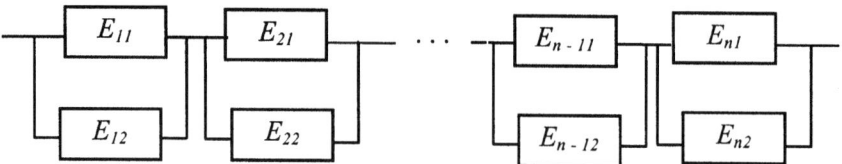

Fig. 8.14. The scheme of a series system with components having hot reservation

$$T = \min_{1 \le i \le n} \{ \max_{1 \le j \le 2} \{ T_{ij} \} \},$$

$$R_n^{(2)}(t) = [1 - [F(t)]^2]^n = [1 - [1 - \exp[-\lambda t]]^2]^n, \quad t \ge 0.$$

Case 3. A cold reservation of the system components

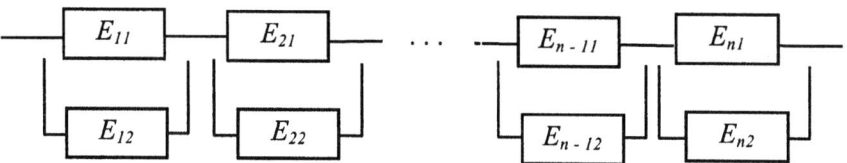

Fig. 8.15. The scheme of a series system with components having cold reservation

$$T = \min_{1 \le i \le n} \{ \sum_{j=1}^{2} T_{ij} \},$$

$$R_n^{(3)}(t) = [1 - [F(t)] * [F(t)]]^n = [[1 + \lambda t] \exp[-\lambda t]]^n$$

$$= [1 + \lambda t]^n \exp[-n\lambda t], \quad t \ge 0.$$

Case 4. A mixed reservation of the system components

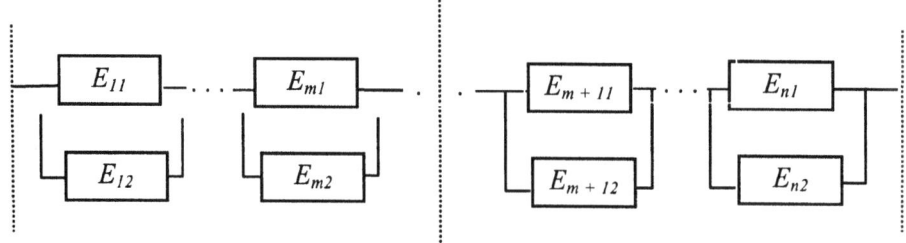

Fig. 8.16. The scheme of a series system with components having mixed reservation

$$T = \min\{\min_{1\leq i\leq m}\{\sum_{j=1}^{2}T_{ij}\},\ \min_{m+1\leq i\leq n}\{\max_{1\leq j\leq 2}\{T_{ij}\}\}\},$$

$$R_n^{(4)}(t) = [1-[F(t)]*[F(t)]]^m\ [1-R^2(t)]^{n-m}$$

$$= [[1+\lambda t]\exp[-\lambda t]]^m\ [1-[1-\exp[-\lambda t]]^2\ ]^{n-m}$$

$$= [1+\lambda t]^m\ \exp[-\lambda mt][2\exp[-\lambda t]-\exp[-2\lambda t]]^{n-m}$$

$$= [1+\lambda t]^m\ \exp[-\lambda nt][2-\exp[-\lambda t]]^{n-m},\ \ t\geq 0.$$

Case 5. A hot system reservation

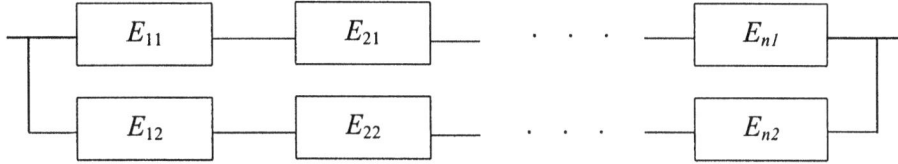

Fig. 8.17. The scheme of a series system with hot reservation

$$T = \max_{1\leq j\leq 2}\{\min_{1\leq j\leq n}\{T_{ij}\}\},$$

$$R_n^{(5)}(t) = 1-[1-[R(t)]^n]^2\ = 1-[1-\exp[-n\lambda t]]^2,\ \ t\geq 0.$$

Case 6. A cold system reservation

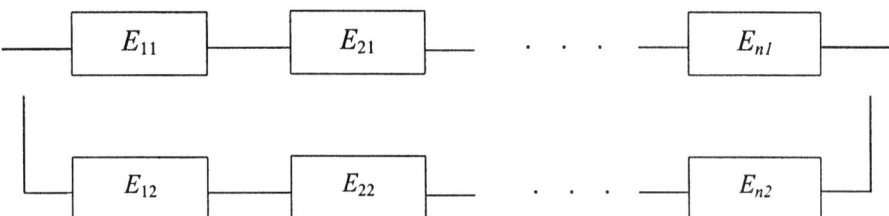

Fig. 8.18. The scheme of a series system with cold reservation

$$T = \sum_{j=1}^{2}\min_{1\leq i\leq n}\{T_{ij}\},$$

$$R_n^{(6)}(t) = 1-[1-[R(t)]^n]*[1-[R(t)]^n]\ = [1+n\lambda t]\exp[-n\lambda t],\ \ t\geq 0.$$

The difficulty arises when selecting the right method of improvement of reliability for a large system. This problem may be simplified and approximately solved by the application of the asymptotic approach. Comparisons of the limit reliability functions of the systems with different types of reserve and such systems with improved components allow us to find the value of the components' decreasing failure rate factor ρ, which gives rise to an equivalent effect on the system reliability improvement. Similar results are obtained under comparison of the system lifetime mean values. As an example we will present the asymptotic approach to the above methods of improving reliability for homogeneous two-state series systems.

Proposition 8.9
Case 1. If

$$a_n = 1/\lambda \rho n, \ b_n = 0,$$

then

$$\mathcal{R}^{(1)}(t) = 1 \ \text{for} \ t < 0 \ \text{and} \ \mathcal{R}^{(1)}(t) = \exp[-t] \ \text{for} \ t \geq 0,$$

is the limit reliability function of the homogeneous exponential series system with reduced failure rates of its components, i.e.

$$R_n^{(1)}(t) = \mathcal{R}^{(1)}(\lambda \rho n t) = \exp[-\lambda \rho n t] \ \text{for} \ t \geq 0$$

and

$$T^{(1)} = E[T] = \frac{1}{\lambda \rho n}.$$

Case 2. If

$$a_n = 1/\lambda \sqrt{n}, \ b_n = 0,$$

then

$$\mathcal{R}^{(2)}(t) = 1 \ \text{for} \ t < 0 \ \text{and} \ \mathcal{R}^{(2)}(t) = \exp[-t^2] \ \text{for} \ t \geq 0,$$

is the limit reliability function of the homogeneous exponential series system with hot reservation of its components, i.e.

$$R_n^{(2)}(t) \cong \mathcal{R}^{(2)}(\lambda \sqrt{n} t) = \exp[-\lambda^2 n t^2] \ \text{for} \ t \geq 0$$

and

$$T^{(2)} = E[T] \cong \Gamma(\frac{3}{2})[\lambda^2 n]^{-1/2} = \Gamma(\frac{3}{2}) \frac{1}{\lambda \sqrt{n}}.$$

Case 3. If

$$a_n = \sqrt{2} / \lambda \sqrt{n}, \ b_n = 0,$$

then

$$\mathscr{R}^{(3)}(t) = 1 \text{ for } t < 0 \text{ and } \mathscr{R}^{(3)}(t) = \exp[-t^2] \text{ for } t \geq 0,$$

is the limit reliability function of the homogeneous exponential series system with cold reservation of its components, i.e.

$$R_n^{(3)}(t) \cong \mathscr{R}^{(3)}(\lambda \sqrt{\frac{nt}{2}}) = \exp[-\frac{1}{2}\lambda^2 nt^2] \text{ for } t \geq 0$$

and

$$T^{(3)} = E[T] \cong \Gamma(\frac{3}{2})[\frac{1}{2}\lambda^2 n]^{-1/2} = \Gamma(\frac{3}{2})\frac{1}{\lambda}\sqrt{\frac{2}{n}}.$$

Case 4. If

$$a_n = \frac{1}{\lambda}\sqrt{\frac{2}{2n-m}}, \ b_n = 0,$$

then

$$\mathscr{R}^{(4)}(t) = 1 \text{ for } t < 0 \text{ and } \mathscr{R}^{(4)}(t) = \exp[-t^2] \text{ for } t \geq 0,$$

is the limit reliability function of the homogeneous exponential series system with mixed reservation of its components, i.e.

$$R_n^{(4)}(t) \cong \mathscr{R}^{(4)}(\lambda \sqrt{\frac{2n-m}{2}}) = \exp[-\frac{2n-m}{2}\lambda^2 t^2] \text{ for } t \geq 0$$

and

$$T^{(4)} = E[T] \cong \Gamma(\frac{3}{2})[\frac{2n-m}{2}\lambda^2]^{-1/2} = \Gamma(\frac{3}{2})\frac{1}{\lambda}\sqrt{\frac{2}{2n-m}}.$$

Case 5. If

$$a_n = \frac{1}{\lambda n}, \ b_n = 0,$$

then

$$\mathcal{R}^{(5)}(t) = 1 \text{ for } t < 0 \text{ and } \mathcal{R}^{(5)}(t) = 1 - [1 - \exp[-t]]^2 \text{ for } t \geq 0,$$

is the limit reliability function of the homogeneous exponential series system with hot reservation, i.e.

$$R_n^{(5)}(t) = \mathcal{R}^{(5)}(\lambda n t) = 1 - [1 - \exp[-\lambda n t]]^2 \text{ for } t \geq 0$$

and

$$T^{(5)} = E[T] = \frac{3}{2\lambda n}.$$

Case 6. If

$$a_n = \frac{1}{\lambda n}, \ b_n = 0,$$

then

$$\mathcal{R}^{(6)}(t) = 1 \text{ for } t < 0 \text{ and } \mathcal{R}^{(6)}(t) = [1 + t]\exp[-t] \text{ for } t \geq 0,$$

is the limit reliability function of the homogeneous exponential series system with cold reservation, i.e.

$$R_n^{(6)}(t) = \mathcal{R}^{(6)}(\lambda n t) = [1 + \lambda n t]\exp[-\lambda n t] \text{ for } t \geq 0$$

and

$$T^{(6)} = E[T] = \frac{2}{\lambda n}.$$

Motivation: We will prove the proposition parts concerned with the considered system limit reliability functions only. The approximate formulae for the system reliability functions follow directly from (1.1) and the expressions for the system lifetimes are easy to obtain by direct integration of the system approximate reliability functions.

Case 1. Since

$$a_n t + b_n = \frac{t}{\lambda \rho n} < 0 \text{ for } t < 0 \text{ and } a_n t + b_n = \frac{t}{\lambda \rho n} \geq 0 \text{ for } t \geq 0,$$

then

$$F(a_n t + b_n) = 0 \text{ for } t < 0 \text{ and } F(a_n t + b_n) = 1 - \exp[-\frac{t}{n}] \text{ for } t \geq 0.$$

Hence

$$\bar{V}(t) = \lim_{n \to \infty} nF(a_n t + b_n) = 0 \text{ for } t < 0$$

and

$$\bar{V}(t) = \lim_{n \to \infty} nF(a_n t + b_n) = \lim_{n \to \infty} n[1 - \exp[-\frac{t}{n}]]$$

$$= \lim_{n \to \infty} n[\frac{t}{n} + o(\frac{1}{n})] = t \text{ for } t \geq 0,$$

which from Lemma 4.1 completes the proof in this case.
Case 2. Since

$$a_n t + b_n = \frac{t}{\lambda \sqrt{n}} < 0 \text{ for } t < 0 \text{ and } a_n t + b_n = \frac{t}{\lambda \sqrt{n}} \geq 0 \text{ for } t \geq 0,$$

then

$$F(a_n t + b_n) = 0 \text{ for } t < 0 \text{ and } F(a_n t + b_n) = 1 - \exp[-\frac{t}{\sqrt{n}}] \text{ for } t \geq 0.$$

Hence

$$V(t) = \lim_{n \to \infty} n[F(a_n t + b_n)]^2 = 0 \text{ for } t < 0$$

and

$$V(t) = \lim_{n \to \infty} n[F(a_n t + b_n)]^2 = \lim_{n \to \infty} n[1 - \exp[-\frac{t}{\sqrt{n}}]]^2$$

$$= \lim_{n\to\infty} n[\frac{t^2}{n} + o(\frac{1}{n})] = t^2 \text{ for } t \geq 0,$$

which from Lemma 4.18 completes the proof in this case.

Case 3. Since

$$R(t) = 1 \text{ for } t < 0 \text{ and } R(t) = [1 + \lambda t]\exp[-\lambda t] \text{ for } t \geq 0$$

and

$$a_n t + b_n = \frac{\sqrt{2}t}{\lambda\sqrt{n}} < 0 \text{ for } t < 0 \text{ and } a_n t + b_n = \frac{\sqrt{2}t}{\lambda\sqrt{n}} \geq 0 \text{ for } t \geq 0,$$

then

$$F(a_n t + b_n) = 0 \text{ for } t < 0 \text{ and } F(a_n t + b_n) = 1 - [1 + \sqrt{\frac{2}{n}}t]\exp[-\sqrt{\frac{2}{n}}t] \text{ for } t \geq 0.$$

Hence

$$\overline{V}(t) = \lim_{n\to\infty} nF(a_n t + b_n) = 0 \text{ for } t < 0$$

and

$$\overline{V}(t) = \lim_{n\to\infty} nF(a_n t + b_n) = \lim_{n\to\infty} n[1 - [1 + \sqrt{\frac{2}{n}}t]\exp[-\sqrt{\frac{2}{n}}t]]$$

$$= \lim_{n\to\infty} n[\frac{2}{n}t^2 - \frac{1}{2}\cdot\frac{2}{n}t^2 + o(\frac{1}{n})] = t^2 \text{ for } t \geq 0,$$

which from Lemma 4.1 completes the proof in this case.

Case 4. We will motivate this case using the expressions for the system reliability sequence. Since

$$R_n^{(4)}(t) = 1 \text{ for } t < 0 \text{ and } R_n^{(4)}(t) = [1 + \lambda t]^m \exp[-\lambda nt][2 - \exp[-\lambda t]]^{n-m} \text{ for } t \geq 0$$

and

$$a_n t + b_n = \frac{t}{\lambda}\sqrt{\frac{2}{2n - m}} < 0 \text{ for } t < 0 \text{ and } a_n t + b_n = \frac{t}{\lambda}\sqrt{\frac{2}{2n - m}} \geq 0 \text{ for } t \geq 0,$$

then

$$\lim_{n \to \infty} R_n^{(4)}(a_n t + b_n) = 1 \quad \text{for } t < 0$$

and

$$\lim_{n \to \infty} R_n^{(4)}(a_n t + b_n)$$

$$= \lim_{n \to \infty} [1 + \sqrt{\frac{2}{2n-m}} t]^m \exp[-n\sqrt{\frac{2}{2n-m}} t] [2 - \exp[-\sqrt{\frac{2}{2n-m}} t]]^{n-m}$$

$$= \lim_{n \to \infty} \exp[m \log[1 + \sqrt{\frac{2}{2n-m}} t] - n\sqrt{\frac{2}{2n-m}} t$$

$$+ (n-m) \log[1 + [1 - \exp[-\sqrt{\frac{2}{2n-m}} t]]]$$

$$= \lim_{n \to \infty} \exp[m\sqrt{\frac{2}{2n-m}} t - \frac{m}{2n-m} t^2 + o(1) - n\sqrt{\frac{2}{2n-m}} t$$

$$+ (n-m)[1 - \exp[-\sqrt{\frac{2}{2n-m}} t]] - (n-m)\frac{1}{2}[1 - \exp[-\sqrt{\frac{2}{2n-m}} t]]^2 + o(1)]$$

$$= \lim_{n \to \infty} \exp[m\sqrt{\frac{2}{2n-m}} t - \frac{m}{2n-m} t^2 - n\sqrt{\frac{2}{2n-m}} t$$

$$+ (n-m)\sqrt{\frac{2}{2n-m}} t - \frac{n-m}{2n-m} t^2 - \frac{n-m}{2n-m} t^2 + o(1)]$$

$$= \lim_{n \to \infty} \exp[-t^2 + o(1)] = \exp[-t^2] \quad \text{for } t \geq 0,$$

which completes the proof in this case.
Case 5. Since

$$a_n t + b_n = \frac{t}{\lambda n} < 0 \quad \text{for } t < 0 \text{ and } a_n t + b_n = \frac{t}{\lambda n} \geq 0 \quad \text{for } t \geq 0,$$

then

$$R(a_n t + b_n) = 1 \quad \text{for } t < 0 \quad \text{and} \quad R(a_n t + b_n) = \exp[-\frac{t}{n}] \quad \text{for } t \geq 0.$$

Hence

$$V(t) = \lim_{n \to \infty} [R(a_n t + b_n)]^n = 1 \text{ for } t < 0$$

and

$$V(t) = \lim_{n \to \infty} [R(a_n t + b_n)]^n = \lim_{n \to \infty} [\exp[-\frac{t}{n}]]^n = \exp[-t] \text{ for } t \geq 0,$$

which from Lemma 4.13 completes the proof in this case.

Case 6. We will motivate this case using the expressions for the system reliability sequence. Since

$$R_n^{(6)}(t) = 1 \text{ for } t < 0 \text{ and } R_n^{(6)}(t) = [1 + n\lambda t] \exp[-n\lambda t] \text{ for } t \geq 0.$$

and

$$a_n t + b_n = \frac{t}{\lambda n} < 0 \text{ for } t < 0 \text{ and } a_n t + b_n = \frac{t}{\lambda n} \geq 0 \text{ for } t \geq 0,$$

then

$$\lim_{n \to \infty} R_n^{(6)}(a_n t + b_n) = 1 \text{ for } t < 0$$

and

$$\lim_{n \to \infty} R_n^{(6)}(a_n t + b_n) = \lim_{n \to \infty} [1 + t] \exp[-t] = [1 + t] \exp[-t] \text{ for } t \geq 0,$$

which completes the proof in this case. □

Corollary 8.5

Comparison of the system reliability functions

$$\mathscr{R}^{(i)}(t) = \mathscr{R}^{(1)}(t), \, i = 2,3,...,6$$

results respectively in the following values of the factor ρ:

$$\rho = \rho(t) = \lambda t \text{ for } i = 2,$$

$$\rho = \rho(t) = \frac{1}{2} \lambda t \text{ for } i = 3,$$

$$\rho = \rho(t) = \frac{2n - m}{2n} \lambda t \quad \text{for } i = 4,$$

$$\rho = \rho(t) = 1 - \log[2 - \exp[-\lambda nt]] \quad \text{for } i = 5,$$

$$\rho = \rho(t) = 1 - \frac{1}{\lambda nt} \log[1 + \lambda nt] \quad \text{for } i = 6,$$

while comparison of the system lifetimes

$$T^{(i)}(t) = T^{(1)}(t), \quad i = 2,3,\dots,6$$

results respectively in the following values of the factor ρ:

$$\rho = \frac{1}{\Gamma(\frac{3}{2})\sqrt{n}} \quad \text{for } i = 2,$$

$$\rho = \frac{1}{\Gamma(\frac{3}{2})\sqrt{2n}} \quad \text{for } i = 3,$$

$$\rho = \frac{1}{\Gamma(\frac{3}{2})n\sqrt{\frac{2}{2n-m}}} \quad \text{for } i = 4,$$

$$\rho = \frac{2}{3} \quad \text{for } i = 5,$$

$$\rho = \frac{1}{2} \quad \text{for } i = 6.$$

Example 8.7
We consider a simplified bus service company, which is similar to a municipal transportation system used in Gdynia, composed of 81 communication lines. We suppose that there is one bus operating on each communication line and that all buses are of the same type with the exponential reliability function

$$R(t) = 1 \text{ for } t < 0, \quad R(t) = \exp[-\lambda t] \text{ for } t \geq 0, \quad \lambda > 0.$$

Additionally we assume that this communication system is working when all its buses arc not failed, i.e. it is failed when any of the buses are failed. The failure rate of the buses evaluated on statistical data coming from the operational process of Gdynia municipal transportation system is assumed to be equal to $0.0049 \ h^{-1}$.

Under these assumptions the considered transportation system is a homogeneous series system made up of components with a reliability function

$$R(t) = 1 \text{ for } t < 0, \quad R(t) = \exp[-0.0049t] \text{ for } t \geq 0.$$

Here we will use four sensible methods from those considered for system reliability improvement. Namely, we apply the four previously considered cases.
Case 1. Improving the reliability of the system components by reducing their failure rates by a factor ρ.
Applying Proposition 8.8 with normalising constants

$$a_{81} = \frac{1}{0.0049 \cdot 81\rho} = \frac{1}{0.397\rho}, \quad b_{81} = 0,$$

we conclude that

$$\mathscr{R}^{(1)}(t) = \exp[-t] \text{ for } t \geq 0,$$

is the limit reliability function of the system, i.e.

$$R_n^{(1)}(t) = \mathscr{R}^{(1)}(0.397\rho t) = \exp[-0.397\rho t] \text{ for } t \geq 0$$

and

$$T^{(1)} = E[T] = \frac{1}{0.397\rho} \text{ h.}$$

Case 2. Improving the reliability of the system by a single hot reservation of its components.
This means that each of 81 communication lines has at its disposal two identical buses it can use and its task is performed if at least one of the buses is not failed.
Applying Proposition 8.8 with normalising constants

$$a_{81} = \frac{1}{0.0049 \cdot \sqrt{81}} = \frac{1}{0.0441}, \quad b_{81} = 0,$$

we conclude that

$$\mathscr{R}^{(2)}(t) = \exp[-t^2], \, t \geq 0.$$

is the limit reliability function of the system, i.e.

$$R_{81}^{(2)}(t) \cong \mathscr{R}^{(2)}(0.0441t) \cong \exp[-0.0019t^2], \quad t \geq 0,$$

and

$$T^{(2)} = E[T] \cong \Gamma(\frac{3}{2}) \frac{1}{0.0049\sqrt{81}} \cong 20.10 \ h.$$

Case 4. Improving the reliability of the system by a single mixed reservation of its components.

This means that each of 81 communication lines has at its disposal two identical buses. There are $m = 50$ communication lines with small traffic which are using one bus permanently and after its failure it is replaced by the second bus (a cold reservation) and $n - m = 81 - 50 = 31$ communication lines with large traffic which are using two buses permanently (a hot reservation).

Applying Proposition 8.8 with normalising constants

$$a_n = \frac{1}{0.0049}\sqrt{\frac{2}{112}} = \frac{1}{0.0367}, \quad b_n = 0,$$

we conclude that

$$\mathfrak{R}^{(4)}(t) = \exp[-t^2] \ \text{ for } t \ge 0,$$

is the limit reliability function of the system, i.e.

$$R_n^{(4)}(t) \cong \mathfrak{R}^{(4)}(0.0367t) = \exp[-0.00135t^2] \ \text{ for } \ t \ge 0$$

and

$$T^{(4)} \cong \Gamma(\frac{3}{2}) \frac{1}{0.0049}\sqrt{\frac{2}{112}} \cong 24.15 \ h.$$

Case 5. Improving the reliability of the system by a single hot reservation.

This means that the transportation system is composed of two independent companies, each of them operating on the same 81 communication lines and having at their disposal one identical bus for use on each line.

Applying Proposition 8.8 with normalising constants

$$a_{81} = \frac{1}{0.0049 \cdot 81} = \frac{1}{0.397}, \quad b_{81} = 0,$$

we conclude that

$$\mathfrak{R}^{(5)}(t) = 1 - [1 - \exp[-t]]^2 \ \text{ for } t \ge 0,$$

is the limit reliability function of the system, i.e.

$$R_n^{(5)}(t) = \mathcal{H}^{(5)}(0.397t) = 1 - [1 - \exp[-0.397t]]^2 \quad \text{for } t \geq 0$$

and

$$T^{(5)} = E[T] = \frac{3}{2 \cdot 0.0049 \cdot 81} \cong 3.78 \text{ h.}$$

Comparing the system reliability functions for considered cases of improvement, from Corollary 8.5, results in the following values of the factor ρ:

$$\rho = \rho(t) = 0.0049t \quad \text{for } i = 2,$$

$$\rho = 0.0340t \quad \text{for } i = 4,$$

$$\rho = \rho(t) = 1 - \log[2 - \exp[-0.397t]] \quad \text{for } i = 5,$$

while comparison of the system lifetimes results respectively in:

$$\rho = 0.1254 \quad \text{for } i = 2,$$

$$\rho = 0.1043 \quad \text{for } i = 4,$$

$$\rho = 0.6667 \quad \text{for } i = 5.$$

Methods of system reliability improvement presented here supply practitioners with simple mathematical tools, which can be used in everyday practice ([83]). The methods may be useful not only in the operation processes of real technical objects but also in designing new operation processes and especially in optimising these processes. Only the case of series systems made up of components having exponential reliability functions with a single reservations of their components and subsystems is considered. It seems to be possible to extend these results to systems that have more complicated reliability structures, and made up of components with different from the exponential reliability functions.

8.7. Reliability of large systems in their operation processes

This section proposes an approach to the solution of the practically very important problem of linking systems' reliability and their operation processes. To connect the interactions between the systems' operation processes and their reliability structures that are changing in time a semi-markov model ([37], [92]) of the system operation processes is applied. This approach gives a tool that is practically important and not difficult for everyday use for evaluating reliability of systems with changing reliability structures during their operation processes. Application of the proposed methods is illustrated here in the reliability evaluation of the port grain transportation system.

We assume that the system during its operation process is performing a repertory of tasks. Namely, the system at each moment t, $t \in <0, \theta>$, where θ is its operation time, is performing at most w tasks. We denote the process of change of the system task repertory by

$$Z(t) = [Z_1(t), Z_2(t), \ldots, Z_w(t)],$$

where

$$Z_j(t) = \begin{cases} 1, \text{ if the system is executing the jth task at the moment } t \\ 0, \text{ if the system is not executing the jth task at the moment } t, \ j = 1, 2, \ldots, w. \end{cases}$$

Thus $Z(t)$ is the process with continuous time t, $t \in <0, \theta>$, and discrete states from the set of states $\{0, 1\}^w$. We number the states of the process $Z(t)$ assuming that it has v different states from the set

$$Z = \{z^1, z^2, \ldots, z^v\}$$

and they are of the form

$$z^k = [z_1^k, z_2^k, \ldots z_w^k], \ k = 1, 2, \ldots, v,$$

where

$$z_j^k \in \{0, 1\}, \ j = 1, 2, \ldots, w.$$

If the process of change of the system task repertory $Z(t)$ is semi-markov ([37], [92]) with its conditional sojourn time θ^{kl} at the state z^k when its next state is z^l, $k, l = 1, 2, \ldots, v$, $k \neq l$, then it may be described by:

– the vector of probabilities of the initial states

$$[p^k(0)]_{1 \times v} = [p^1(0), p^2(0), \ldots, p^v(0)],$$

where

$$p^k(0) = P(Z(0) = z^k) \text{ for } k = 1, 2, \ldots, v,$$

– the matrix of probabilities of its transitions between states

$$[p^{kl}]_{v \times v} = \begin{bmatrix} p^{11} & p^{12} & \dots & p^{1v} \\ p^{21} & p^{22} & \dots & p^{2v} \\ \cdot & \cdot & \cdot & \cdot & \cdot & \cdot \\ p^{v1} & p^{v2} & \dots & p^{vv} \end{bmatrix},$$

where

$$p^{kk} = 0 \text{ for } k = 1,2,...,v,$$

– the matrix of conditional distribution functions of the sojourn times θ^{kl}

$$[H^{kl}(t)]_{v \times v} = \begin{bmatrix} H^{11}(t) & H^{12}(t) & \dots & H^{1v}(t) \\ H^{21}(t) & H^{22}(t) & \dots & H^{2v}(t) \\ \cdot & \cdot & \cdot & \cdot & \cdot & \cdot & \cdot & \cdot & \cdot \\ H^{v1}(t) & H^{v2}(t) & \dots & H^{vv}(t) \end{bmatrix},$$

where

$$H^{kl}(t) = P(\theta^{kl} < t) \text{ for } k,l = 1,2,...,v, \ k \neq l,$$

and

$$H^{kk}(t) = 0 \text{ for } k = 1,2,...,v.$$

Then, the sojourn time θ^{kl} mean values are given by

$$E[\theta^{kl}] = \int_0^\infty t dH^{kl}(t), \ k,l = 1,2,...,v, \ k \neq l. \tag{8.11}$$

The unconditional distribution functions of the sojourn times θ^k of the process $Z(t)$ at the states z^k, $k = 1,2,...,v$, are given by

$$H^k(t) = \sum_{l=1}^{v} p^{kl} H^{kl}(t), \ k = 1,2,...,v. \tag{8.12}$$

The mean values $E[\theta^k]$ of the unconditional sojourn times θ^k are given by

$$E[\theta^k] = \sum_{l=1}^{v} p^{kl} E[\theta^{kl}], \quad k = 1,2,...,v, \tag{8.13}$$

where $E[\theta^{kl}]$ are defined by (8.11).

Limit values of the transient probabilities at the states $p^k(t) = P(Z(t) = z^k)$ are given by

$$p^k = \lim_{t \to \infty} p^k(t) = \frac{\pi^k E[\theta^k]}{\sum_{l=1}^{v} \pi^l E[\theta^l]}, \quad k = 1,2,...,v, \tag{8.14}$$

where the probabilities π^k of the vector $[\pi^k]_{1 \times v}$ satisfy the system of equations

$$\begin{cases} [\pi^k] = [\pi^k][p^{kl}] \\ \sum_{l=1}^{v} \pi^l = 1. \end{cases}$$

In the case when the sojourn times θ^{kl}, $k, l = 1,2,...,v$, $k \neq l$, have exponential distributions with the transition rates between the states γ^{kl}, i.e. if

$$H^{kl}(t) = P(\theta^{kl} < t) = 1 - \exp[-\gamma^{kl}t], \quad t > 0, \tag{8.15}$$

for $k, l = 1,2,...,v$, $k \neq l$, then their mean values are determined by

$$E[\theta^{kl}] = \frac{1}{\gamma^{kl}}, \quad k, l = 1,2,...,v, \quad k \neq l, \tag{8.16}$$

and the probabilities of transitions between the states are given by

$$p^{kl} = \frac{\gamma^{kl}}{\sum_{j \neq k} \gamma^{kj}}, \quad k, l = 1,2,...,v, \quad k \neq l. \tag{8.17}$$

The unconditional distribution functions of the process $Z(t)$ sojourn times θ^k at the states z^k, $k = 1,2,...,v$, according to (8.12) and (8.15) are given by

$$H^k(t) = 1 - \sum_{l=1}^{v} p^{kl} \exp[-\gamma^{kl}t], \quad t > 0, \quad k = 1,2,...,v, \tag{8.18}$$

and their mean values, from (8.13) and (8.16), are

$$M^k = E[\theta^k] = \sum_{l=1}^{v} p^{kl} \frac{1}{\gamma^{kl}}, \quad k = 1, 2, ..., v. \tag{8.19}$$

Limit values of the transient probabilities $p^k(t)$ at the states z^k, according to (8.14), are given by

$$p^k = \lim_{t \to \infty} p^k(t) = \frac{\pi^k \cdot M^k}{\sum_{l=1}^{v} [\pi^l \cdot M^l]}, \quad k = 1, 2, ..., v, \tag{8.20}$$

where the probabilities π^k of the vector $[\pi^k]_{1 \times v}$ satisfy the system of equations

$$\begin{cases} [\pi^k] = [\pi^k][p^{kl}] \\ \sum_{l=1}^{v} \pi^l = 1, \end{cases}$$

with $[p^{kl}]$ and M^k given by (8.17) and (8.19) respectively.

As an example we will analyse here the reliability of the port grain transportation system in its operation process. This system, described in Chapter 6, is composed of four non-homogeneous series-parallel subsystems. Therefore we will need the following modifications of Proposition 7.4 (Case 2) reduced to two-state systems.

Proposition 8.9
If components of the non-homogeneous regular two-state series-parallel system have exponential reliability functions

$$R^{(i,j)}(t) = 1 \text{ for } t < 0, \ R^{(i,j)}(t) = \exp[-\lambda_{ij}t] \text{ for } t \geq 0, \ i = 1, 2, ..., a, j = 1, 2, ..., e_i,$$

and

$$k_n \to k, \ l_n \to \infty,$$

$$a_n = \frac{1}{\lambda l_n}, \ b_n = 0,$$

where

$$\lambda_i = \sum_{j=1}^{e_i} p_{ij} \lambda_{ij}, \quad \lambda = \min_{1 \leq i \leq a} \{\lambda_i\},$$

then

$$\mathcal{R}'_9(t) = 1 \text{ for } t < 0, \ \mathcal{R}'_9(t) = 1 - \prod_{i=1}^{a}[1 - \exp[-\frac{\lambda_i}{\lambda}t]]^{q_ik} \text{ for } t \geq 0, \ u = 1,2,...,z,$$

is its limit reliability function.

To consider jointly the port transportation system reliability and its operation process we assume that lifetimes of the components E_{ij} of the non-homogeneous series-parallel system defined in Chapter 2 depend on the states of its process of change of tasks $Z(t)$. The changes of the process $Z(t)$ states have influence on the system components' E_{ij} reliability and on the system reliability structure as well. Thus, we denote the conditional reliability function of the system component E_{ij}, and the conditional system reliability function while the system is performing the task z^k, $k = 1,2,...,v$, respectively by

$$[R^{(i,j)}(t)]^{(k)} = P(T_{ij}^{(k)} \geq t / Z(t) = z^k), \ t \in <0,\infty), \ i = 1,2,...,a, \ j = 1,2,...,e_i,$$

and

$$R^{(k)}(t) = P(T^{(k)} \geq t / Z(t) = z^k), \ t \in <0,\infty).$$

The reliability function $[R^{(i,j)}(t)]^{(k)}$ is the conditional probability that the component E_{ij} lifetime $T_{ij}^{(k)}$ is not less than t, while the process $Z(t)$ is at the operation state z^k. Similarly, the reliability function $R^{(k)}(t)$ is the conditional probability that the system lifetime $T^{(k)}$ is not less than t, while the process $Z(t)$ is at the operation state z^k.

Then, the unconditional reliability function of the system

$$R(t) = P(T > t), \ t \in <0,\infty),$$

where T is the unconditional lifetime of the system, is given by

$$R(t) = \sum_{k=1}^{v} p^k(t) R^{(k)}(t), \ t \in <0,\infty). \tag{8.21}$$

In the case when the system operation time θ is large enough, the transient probability $p^k(t)$ can be replaced by p^k given by (8.14) or (8.20) respectively and the last formula takes the form

$$R(t) = \sum_{k=1}^{v} p^k R^{(k)}(t), \quad t \in <0, \infty).$$

(8.22)

The formula (8.22) for the regular non-homogeneous series-parallel system takes the form

$$R_{k_n, l_n}(t) = \sum_{k=1}^{v} p^k R^{(k)}_{k_n, l_n}(t), \quad t \in <0, \infty),$$

(8.23)

where $R^{(k)}_{k_n, l_n}(t)$ is this system's conditional reliability function while it is performing the task z^k, $k = 1, 2, ..., v$.

In the particular exponential case, i.e. if the component conditional reliability functions while the system is performing the task z^k, $k = 1, 2, ..., v$, are given by

$$[R^{(i,j)}(t)]^{(k)} = \exp[-\lambda_{ij}^k t] \text{ for } t \ge 0, \ \lambda_{ij}^k > 0, \ i = 1, 2, ..., a, j = 1, 2, ..., e_i,$$

(8.24)

according to (2.14) and (2.15), we get

$$R^{(k)}_{k_n, l_n}(t) = 1 - \prod_{i=1}^{a} [1 - \exp[-l_n p_{ij} \lambda_{ij}^k t]]^{q_i k_n}, \quad t \ge 0.$$

(8.25)

In this case the mean value and the variance of the regular non-homogeneous series-parallel system lifetime are

$$m = \sum_{k=1}^{v} p^k E[T^k],$$

(8.26)

and

$$\sigma^2 = \sum_{k=1}^{v} p^k \sigma^2[T^k],$$

(8.27)

where for $k = 1, 2, ..., v$,

$$E[T^k] = \int_0^{\infty} R^{(k)}_{k_n, l_n}(t) dt, \quad t \ge 0,$$

(8.28)

and

$$\sigma^2[T^k] = 2 \int_0^{\infty} t R^{(k)}_{k_n, l_n}(t) dt - [E[T^k]]^2, \quad t \ge 0,$$

(8.29)

and p^k are given either by (8.14) or by (8.20).

The grain elevator described in Chapter 6 is composed of the following transportation subsystems:

S_1 – horizontal conveyors of the first type,

S_2 – vertical bucket elevators,

S_3 – horizontal conveyors of the second type,

S_4 – worm conveyors.

To illustrate the problem we simplify our considerations by assuming that all elevator components have exponential distributions of their lifetimes.

Subsystem S_1 is composed of two identical belt conveyors. In each conveyor there is one belt with a reliability function

$$R^{(1,1)}(t) = \exp[-0.0125t] \text{ for } t \geq 0,$$

two drums with reliability functions

$$R^{(1,2)}(t,1) = \exp[-0.0015t] \text{ for } t \geq 0,$$

117 channelled rollers with reliability functions

$$R^{(1,3)}(t,1) = \exp[-0.005t] \text{ for } t \geq 0,$$

and nine supporting rollers with reliability functions

$$R^{(1,4)}(t,1) = \exp[-0.004t] \text{ for } t \geq 0.$$

Subsystem S_2 is composed of three identical bucket elevators. In each elevator there is one belt with a reliability function

$$R^{(1,1)}(t) = \exp[-0.025t] \text{ for } t \geq 0,$$

two drums with reliability functions

$$R^{(1,2)}(t) = \exp[-0.0015t] \text{ for } t \geq 0,$$

740 buckets with reliability functions

$$R^{(1,3)}(t) = \exp[-0.03t] \text{ for } t \geq 0.$$

Subsystem S_3 is composed of two identical belt conveyors. In each conveyor there is one belt with a reliability function

$$R^{(1,1)}(t) = \exp[-0.0125t] \text{ for } t \geq 0,$$

two drums with reliability functions

$R^{(1,2)}(t) = \exp[-0.0015t]$, for $t \geq 0$,

117 channelled rollers with reliability functions

$R^{(1,3)}(t) = \exp[-0.005t]$ for $t \geq 0$,

and 19 supporting rollers with reliability functions

$R^{(1,4)}(t) = \exp[-0.004t]$ for $t \geq 0$.

Subsystem S_4 is composed of three chain conveyors, each composed of a wheel driving the belt, a reversible wheel and 160, 160 and 240 links respectively. The subsystem consists of three conveyors. Two of them have 162 components and the remaining one has 242 components. Thus it is a non-homogeneous non-regular multi-state series-parallel system. In order to make it a regular system we conventionally complete two first conveyors that have 162 components with 80 components that do not fail. After this supplement the subsystem consists of $k_n = 3$ conveyors, each composed of $l_n = 242$ components. In two of them there are two driving wheels with reliability functions

$R^{(1,1)}(t) = \exp[-0.005t]$ for $t \geq 0$,

160 links with reliability functions

$R^{(1,2)}(t) = \exp[-0.012t]$ for $t \geq 0$,

and 80 components with "reliability functions"

$R^{(1,3)}(t) = \exp[-\lambda t]$ for $t \geq 0$, where $\lambda = 0$.

The third conveyor is composed of two driving wheels with reliability functions

$R^{(2,1)}(t) = \exp[-0.022t]$ for $t \geq 0$,

and 240 links with reliability functions

$R^{(2,2)}(t) = \exp[-0.034t]$ for $t \geq 0$.

Taking into account the operation process of the considered transportation system we distinguish the following as its three tasks:

    task 1 – the system operation with the largest efficiency when all components of the subsystems S_1, S_2, S_3 and S_4 are used, task 2 – the system operation with less efficiency system when the first conveyor of subsystem S_1, the first and second elevators of subsystem S_2, the first conveyor of subsystem S_3 and the first and second conveyors of subsystem S_4 are used,

task 3 – the system operation with least efficiency when only the first conveyor of subsystem S_1, the first elevator of subsystem S_2, the first conveyor of subsystem S_3 and the first conveyor of subsystem S_4 are used.

Since the system tasks are disjoint then its operation states belong to the set

$$Z = \{z^1, z^2, z^3\},$$

where

$$z^1 = [1,0,0], \quad z^2 = [0,1,0], \quad z^3 = [0,0,1].$$

Moreover, we arbitrarily assume the following matrix of the conditional distribution functions of the system sojourn times θ^{kl}, $k,l = 1,2,3$,

$$[H^{kl}(t)] = \begin{bmatrix} 0 & 1-e^{-5t} & 1-e^{-10t} \\ 1-e^{-40t} & 0 & 1-e^{-50t} \\ 1-e^{-10t} & 1-e^{-20t} & 0 \end{bmatrix}.$$

Hence, from (8.17), the probabilities of transitions between the states are given by

$$[p^{kl}] = \begin{bmatrix} 0 & \dfrac{1}{3} & \dfrac{2}{3} \\ \dfrac{4}{9} & 0 & \dfrac{5}{9} \\ \dfrac{1}{3} & \dfrac{2}{3} & 0 \end{bmatrix}$$

and further, according to (8.18), the unconditional distribution functions of the process $Z(t)$ sojourn times θ^k in the states z^k, $k = 1,2,3$, are given by

$$\begin{cases} H^1(t) = 1 - \dfrac{1}{3}\exp[-5t] - \dfrac{2}{3}\exp[-10t] \\ H^2(t) = 1 - \dfrac{4}{9}\exp[-40t] - \dfrac{5}{9}\exp[-50t] \\ H^3(t) = 1 - \dfrac{1}{3}\exp[-10t] - \dfrac{2}{3}\exp[-20t] \end{cases}$$

and their mean values, from (8.19), are

$$\begin{cases} M^1 = E[\theta^1] = \dfrac{1}{3}\dfrac{1}{5} + \dfrac{2}{3}\dfrac{1}{10} = \dfrac{6}{45} \\[2mm] M^2 = E[\theta^2] = \dfrac{4}{9}\dfrac{1}{40} + \dfrac{5}{9}\dfrac{1}{50} = \dfrac{1}{45} \\[2mm] M^3 = E[\theta^3] = \dfrac{1}{3}\dfrac{1}{10} + \dfrac{2}{3}\dfrac{1}{20} = \dfrac{3}{45} \end{cases}$$

Since from the system of equations

$$\begin{cases} [\pi^1, \pi^2, \pi^3] = [\pi^1, \pi^2, \pi^3] \begin{bmatrix} 0 & \dfrac{1}{3} & \dfrac{2}{3} \\[2mm] \dfrac{4}{9} & 0 & \dfrac{5}{9} \\[2mm] \dfrac{1}{3} & \dfrac{2}{3} & 0 \end{bmatrix} \\[6mm] \pi^1 + \pi^2 + \pi^3 = 1 \end{cases}$$

we get

$$\pi^1 = \frac{17}{61}, \quad \pi^2 = \frac{21}{61}, \quad \pi^3 = \frac{23}{61},$$

then the limit values of the transient probabilities $p^k(t)$ at the operational states z^k, according to (8.20), are given by

$$p^1 = \frac{34}{64}, \quad p^2 = \frac{7}{64}, \quad p^3 = \frac{23}{64}. \tag{8.30}$$

The structure of the port grain transportation system at operation state 1 is given in Figure 8.19.

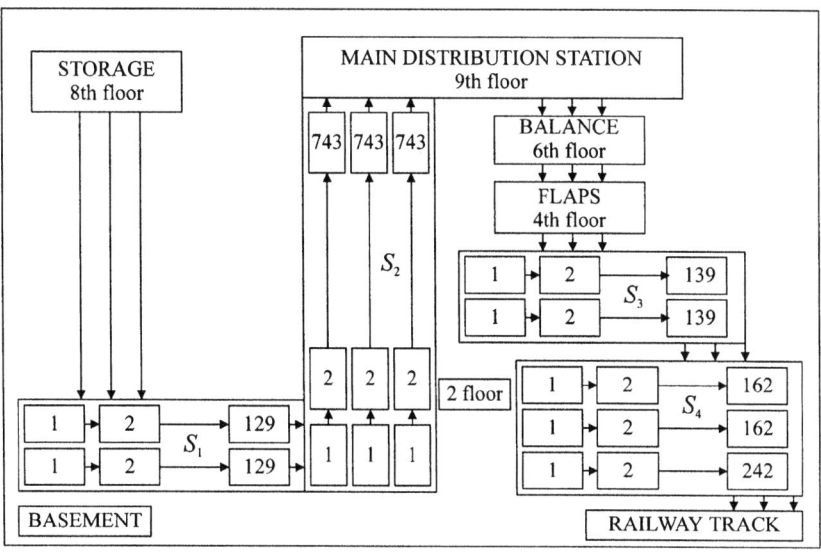

Fig. 8.19. The scheme of the grain transportation system structure at operation state 1

At system operational state 1, subsystem S_1 becomes a non-homogeneous regular series-parallel system with parameters

$k_n = k = 2,\ l_n = 129,\ a = 1,\ q_1 = 1,\ e_1 = 4,$

$p_{11} = 1/129,\ p_{12} = 2/129,\ p_{13} = 117/129,\ p_{14} = 9/129,$

$\lambda_{11} = 0.0125,\ \lambda_{12} = 0.0015,\ \lambda_{13} = 0.005,\ \lambda_{14} = 0.004.$

Since

$$\lambda_1 = \sum_{j=1}^{4} p_{1j}\lambda_{1j} = \frac{1}{129}0.0125 + \frac{2}{129}0.0015 + \frac{117}{129}0.005 + \frac{9}{129}0.004 = 0.0049,$$

$$\lambda = \min\{0.0049\} = 0.0049,$$

then applying Proposition 8.9 with normalising constants

$$a_n = \frac{1}{0.0049 \cdot 129} \cong \frac{1}{0.6365},\ b_n = 0,$$

we conclude that

$$\mathscr{R}_9^{(1)}(t) = 1 - [1 - \exp[-t]]^2 \text{ for } t \geq 0$$

is the subsystem S_1 limit reliability function and from (1.1), we get

$$R_{2,129}^{(1)}(t) = \mathscr{R}_9^{(1)}(0.6365t) = 1 - [1 - \exp[-0.6365t]]^2 \text{ for } t \geq 0.$$

At system operational state 1, subsystem S_2 becomes a non-homogeneous regular series-parallel system with parameters

$k_n = k = 3, l_n = 743, a = 1, q_1 = 1, e_1 = 3,$

$p_{11} = 1/743, p_{12} = 2/743, p_{13} = 740/743,$

$\lambda_{11} = 0.025, \lambda_{12} = 0.0015, \lambda_{13} = 0.03.$

Since

$$\lambda_1 = \sum_{j=1}^{3} p_{1j}\lambda_{1j} = \frac{1}{743}0.025 + \frac{2}{743}0.0015 + \frac{740}{743}0.03 = 0.0299,$$

$$\lambda = \min\{0.0299\} = 0.0299,$$

then applying Proposition 8.9 with normalising constants

$$a_n = \frac{1}{0.0299 \cdot 743} \cong \frac{1}{22.228}, \quad b_n = 0,$$

we conclude that

$$\mathscr{R}_9^{(1)}(t) = 1 - [1 - \exp[-t]]^3 \text{ for } t \geq 0$$

is the subsystem S_2 limit reliability function and from (1.1), we get

$$R_{3,743}^{(1)}(t) = \mathscr{R}_9^{(1)}(22.228t) = 1 - [1 - \exp[-22.228t]]^3 \text{ for } t \geq 0.$$

At system operational state 1, subsystem S_3 is a non-homogeneous regular series-parallel system with parameters

$k_n = k = 2, l_n = 139, a = 1, q_1 = 1, e_1 = 4,$

$p_{11} = 1/139, p_{12} = 2/139, p_{13} = 117/139, p_{14} = 19/139,$

$\lambda_{11} = 0.0125,\ \lambda_{12} = 0.0015,\ \lambda_{13} = 0.005,\ \lambda_{14} = 0.004.$

Since

$$\lambda_1 = \sum_{j=1}^{4} p_{1j}\lambda_{1j} = \frac{1}{139}0.0125 + \frac{2}{139}0.0015 + \frac{117}{139}0.005 + \frac{19}{139}0.004 = 0.00487,$$

$$\lambda = \min\{0.00487\} = 0.00487,$$

then applying Proposition 8.9 with normalising constants

$$a_n = \frac{1}{0.00487 \cdot 139} \cong \frac{1}{0.6765},\ b_n = 0,$$

we conclude that

$$\mathfrak{R}_9^{(1)}(t) = 1 - [1 - \exp[-t]]^2 \text{ for } t \geq 0$$

is the subsystem S_3 limit reliability function and from (1.1), we get

$$R_{2,139}^{(1)}(t) = \mathfrak{R}_9^{(1)}(0.6765t) = 1 - [1 - \exp[-0.6765t]]^2 \text{ for } t \geq 0.$$

At system operational state 1, subsystem S_4 becomes a non-homogeneous regular series-parallel system with parameters

$k_n = k = 3,\ l_n = 242,\ a = 2,\ q_1 = 2/3,\ q_2 = 1/3,$

$e_1 = 3,\ p_{11} = 2/242,\ p_{12} = 160/242,\ p_{13} = 80/242,$

$\lambda_{11} = 0.005,\ \lambda_{12} = 0.012,\ \lambda_{13} = 0,$

$e_2 = 2,\ p_{21} = 2/242,\ p_{22} = 240/242,$

$\lambda_{21} = 0.022,\ \lambda_{22} = 0.034.$

Since

$$\lambda_1 = \sum_{j=1}^{3} p_{1j}\lambda_{1j} = \frac{2}{242}0.005 + \frac{160}{242}0.012 + \frac{80}{242}0 = 0.007975,$$

$$\lambda_2 = \sum_{j=1}^{2} p_{2j}\lambda_{2j} = \frac{2}{242}0.022 + \frac{240}{242}0.034 = 0.0339,$$

$\lambda = \min\{0.007975, 0.0339\} = 0.007975$,

then applying Proposition 8.9 with normalising constants

$$a_n = \frac{1}{0.007975 \cdot 242} \cong \frac{1}{1.93}, \quad b_n = 0,$$

we conclude that

$$\mathscr{R}_9^{(1)}(t) = 1 - [1 - \exp[-t]]^2 \text{ for } t \geq 0$$

is the subsystem S_4 limit reliability function and from (1.1), we get

$$R_{3,242}^{(1)}(t) \cong \mathscr{R}_9^{(1)}(1.93t) = 1 - [1 - \exp[-1.93t]]^2 \text{ for } t \geq 0.$$

Since the considered subsystems create a series structure in a reliability sense, then the reliability function of the whole transportation system, for $t \geq 0$, is given by

$$\overline{R}^{(1)}(t) \cong R_{2,129}^{(1)}(t) R_{3,743}^{(1)}(t) R_{2,139}^{(1)}(t) R_{3,242}^{(1)}(t)$$

$$= 24 \exp[-25.471t] - 24 \exp[-47.699t] - 12 \exp[-26.1075t]$$

$$- 12 \exp[-27.401t] + 12 \exp[-48.3355t] + 12 \exp[-49.629]$$

$$- 12 \exp[-26.1475t] + 12 \exp[-48.3755t] + 8 \exp[-69.927t]$$

$$+ 6 \exp[-28.0375t] - 6 \exp[-50.2655t] - 6 \exp[-50.3055t]$$

$$+ 6 \exp[-26.784t] + 6 \exp[-28.0775t] - 6 \exp[-49.012t]$$

$$- 4 \exp[-70.6035t] - 4 \exp[-70.5635t] - 4 \exp[-71.857t]$$

$$+ 3 \exp[-50.942t] - 3 \exp[-28.714t] + 2 \exp[-72.4935t]$$

$$+ 2 \exp[-71.24t] + 2 \exp[-72.5335t] - \exp[-73.17t] \qquad (8.31)$$

and according to (8.28) and (8.29) the system lifetime mean value and its standard deviation are

$$E[T^{(1)}] \cong 0.0807, \quad \sigma(T^{(1)}) \cong 0.057. \qquad (8.32)$$

The structure of the port grain transportation system at operation state 2 is given in Figure 8.20.

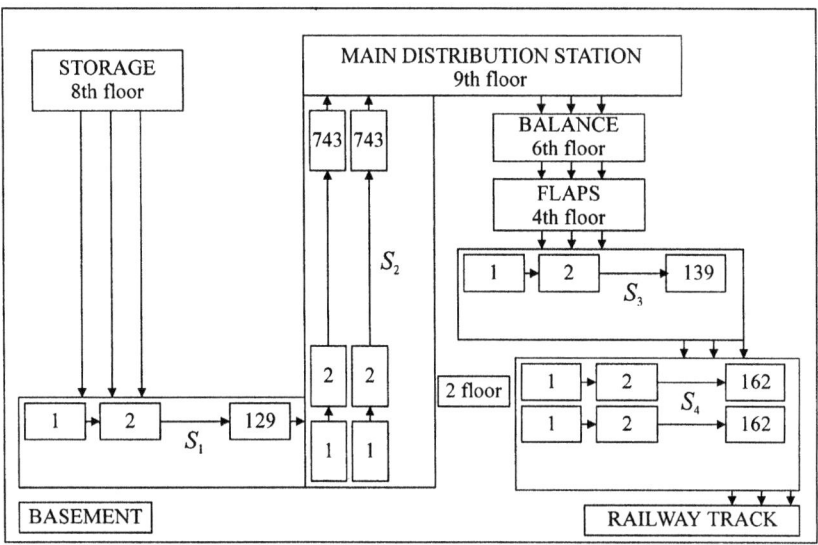

Fig. 8.20. The scheme of the grain transportation system structure at operation state 2

At system operational state 2, subsystem S_1 becomes a non-homogeneous regular series-parallel system with parameters

$$k_n = k = 1,\ l_n = 129,\ a = 1,\ q_1 = 1,\ e_1 = 4,$$

$$p_{11} = 1/129,\ p_{12} = 2/129,\ p_{13} = 117/129,\ p_{14} = 9/129,$$

$$\lambda_{11} = 0.0125,\ \lambda_{12} = 0.0015,\ \lambda_{13} = 0.005,\ \lambda_{14} = 0.004.$$

Since

$$\lambda_1 = \sum_{j=1}^{4} p_{1j}\lambda_{1j} = \frac{1}{129}0.0125 + \frac{2}{129}0.0015 + \frac{117}{129}0.005 + \frac{9}{129}0.004 = 0.0049,$$

$$\lambda = \min\{0.0049\} = 0.0049,$$

then applying Proposition 8.9 with normalising constants

$$a_n = \frac{1}{0.0049 \cdot 129} \cong \frac{1}{0.6365},\ b_n = 0,$$

we conclude that

$$\mathscr{R}_9^{(2)}(t) = 1 - [1 - \exp[-t]] = \exp[-t] \text{ for } t \geq 0$$

is the subsystem S_1 limit reliability function and from (1.1), we get

$$R_{1,129}^{(2)}(t) = \mathscr{R}_9^{(2)}(0.6365t) = \exp[-0.6365t] \text{ for } t \geq 0.$$

At system operational state 2, subsystem S_2 becomes a non-homogeneous regular series-parallel system with parameters

$$k_n = k = 2, l_n = 743, a = 1, q_1 = 1, e_1 = 3,$$

$$p_{11} = 1/743, p_{12} = 2/743, p_{13} = 740/743,$$

$$\lambda_{11} = 0.025, \lambda_{12} = 0.0015, \lambda_{13} = 0.03.$$

Since

$$\lambda_1 = \sum_{j=1}^{3} p_{1j}\lambda_{1j} = \frac{1}{743}0.025 + \frac{2}{743}0.0015 + \frac{740}{743}0.03 = 0.0299,$$

$$\lambda = \min\{0.0299\} = 0.0299,$$

then applying Proposition 8.9 with normalising constants

$$a_n = \frac{1}{0.0299 \cdot 743} \cong \frac{1}{22.228}, \quad b_n = 0,$$

we conclude that

$$\mathscr{R}_9^{(2)}(t) = 1 - [1 - \exp[-t]]^2 \text{ for } t \geq 0$$

is the subsystem S_2 limit reliability function and from (1.1), we get

$$R_{2,743}^{(2)}(t) = \mathscr{R}_9^{(2)}(22.228t) = 1 - [1 - \exp[-22.228t]]^2 \text{ for } t \geq 0.$$

At system operational state 2, subsystem S_3 is a non-homogeneous regular series-parallel system with parameters

$$k_n = k = 1, l_n = 139, a = 1, q_1 = 1, e_1 = 4,$$

$p_{11} = 1/139$, $p_{12} = 2/139$, $p_{13} = 117/139$, $p_{14} = 19/139$,

$\lambda_{11} = 0.0125$, $\lambda_{12} = 0.0015$, $\lambda_{13} = 0.005$, $\lambda_{14} = 0.004$.

Since

$$\lambda_1 = \sum_{j=1}^{4} p_{1j}\lambda_{1j} = \frac{1}{139}0.0125 + \frac{2}{139}0.0015 + \frac{117}{139}0.005 + \frac{19}{139}0.004 = 0.00487,$$

$\lambda = \min\{0.00487\} = 0.00487$,

then applying Proposition 8.9 with normalising constants

$$a_n = \frac{1}{0.00487 \cdot 139} \cong \frac{1}{0.6765}, \quad b_n = 0,$$

we conclude that

$$\mathfrak{R}_9^{(2)}(t) = 1 - [1 - \exp[-t]] = \exp[-t] \text{ for } t \geq 0$$

is the subsystem S_3 limit reliability function and from (1.1), we get

$$R_{1,139}^{(2)}(t) = \mathfrak{R}_9^{(2)}(0.6765t) = \exp[-0.6765t] \text{ for } t \geq 0.$$

At system operational state 2, subsystem S_4 becomes a non-homogeneous regular series-parallel system with parameters

$k_n = k = 2$, $l_n = 242$, $a = 1$, $q_1 = 1$,

$e_1 = 3$, $p_{11} = 2/242$, $p_{12} = 160/242$, $p_{13} = 80/242$,

$\lambda_{11} = 0.005$, $\lambda_{12} = 0.012$, $\lambda_{13} = 0$.

Since

$$\lambda_1 = \sum_{j=1}^{3} p_{1j}\lambda_{1j} = \frac{2}{242}0.005 + \frac{160}{242}0.012 + \frac{80}{242}0 = 0.007975,$$

$\lambda = \min\{0.007975\} = 0.007975$,

then applying Proposition 8.9 with normalising constants

$$a_n = \frac{1}{0.007975 \cdot 242} \cong \frac{1}{1.93}, \quad b_n = 0,$$

we conclude that

$$\mathfrak{R}_9^{(2)}(t) = 1 - [1 - \exp[-t]]^2 \text{ for } t \geq 0$$

is the subsystem S_4 limit reliability function and from (1.1), we get

$$R_{2,242}^{(2)}(t) = \mathfrak{R}_9^{(2)}(1.93t) = 1 - [1 - \exp[-1.93t]]^2 \text{ for } t \geq 0.$$

Since the considered subsystems create a series structure in a reliability sense, then the reliability function of the whole transportation system is given by

$$\overline{R}^{(2)}(t) = R_{1,129}^{(2)}(t) R_{2,743}^{(2)}(t) R_{1,139}^{(2)}(t) R_{2,242}^{(2)}(t)$$

$$= 4\exp[-25.471t] - 2\exp[-27.401t] - 2\exp[-47.699t]$$

$$+ \exp[-49.629t] \text{ for } t \geq 0, \tag{8.33}$$

and according to (8.28) and (8.29) the system lifetime mean value and its standard deviation are

$$E[T^{(2)}] = 0.0623, \quad \sigma(T^{(2)}) = 0.0466. \tag{8.34}$$

The structure of the port grain transportation system at operation state 3 is given in Figure 8.21. At system operational state 3, subsystem S_1 becomes a non-homogeneous regular series-parallel system with parameters

$$k_n = k = 1, l_n = 129, a = 1, q_1 = 1, e_1 = 4,$$

$$p_{11} = 1/129, p_{12} = 2/129, p_{13} = 117/129, p_{14} = 9/129,$$

$$\lambda_{11} = 0.0125, \lambda_{12} = 0.0015, \lambda_{13} = 0.005, \lambda_{14} = 0.004.$$

Since

$$\lambda_1 = \sum_{j=1}^{4} p_{1j}\lambda_{1j} = \frac{1}{129}0.0125 + \frac{2}{129}0.0015 + \frac{117}{129}0.005 + \frac{9}{129}0.004 = 0.0049,$$

$$\lambda = \min\{0.0049\} = 0.0049,$$

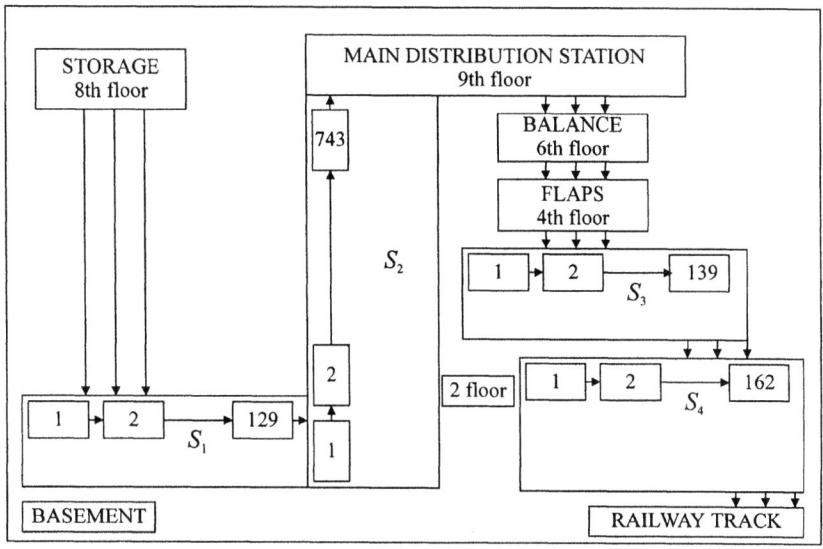

Fig. 8.21. The scheme of the grain transportation system structure at operation state 3

then applying Proposition 8.9 with normalising constants

$$a_n = \frac{1}{0.0049 \cdot 129} \cong \frac{1}{0.6365}, \quad b_n = 0,$$

we conclude that

$$\mathcal{R}_9^{(3)}(t) = 1 - [1 - \exp[-t]] = \exp[-t] \text{ for } t \geq 0$$

is the subsystem S_1 limit reliability function and from (1.1), we get

$$R_{1,129}^{(3)}(t) = \mathcal{R}_9^{(3)}(0.6365t) = \exp[-0.6365t] \text{ for } t \geq 0.$$

At system operational state 3, subsystem S_2 becomes a non-homogeneous regular series-parallel system with parameters

$$k_n = k = 1, l_n = 743, a = 1, q_1 = 1, e_1 = 3,$$

$$p_{11} = 1/743, p_{12} = 2/743, p_{13} = 740/743,$$

$$\lambda_{11} - 0.025, \lambda_{12} - 0.0015, \lambda_{13} = 0.03.$$

Since

$$\lambda_1 = \sum_{j=1}^{3} p_{1j} \lambda_{1j} = \frac{1}{743} 0.025 + \frac{2}{743} 0.0015 + \frac{740}{743} 0.03 = 0.0299,$$

$$\lambda = \min\{0.0299\} = 0.0299,$$

then applying Proposition 8.9 with normalising constants

$$a_n = \frac{1}{0.0299 \cdot 743} \cong \frac{1}{22.228}, \quad b_n = 0,$$

we conclude that

$$\mathfrak{R}_9^{(3)}(t) = 1 - [1 - \exp[-t]] = \exp[-t] \text{ for } t \geq 0$$

is the subsystem S_2 limit reliability function and from (1.1), we get

$$R_{1,743}^{(3)}(t) = \mathfrak{R}_9^{(3)}(22.228t) = \exp[-22.228t] \text{ for } t \geq 0.$$

At system operational state 3, subsystem S_3 is a non-homogeneous regular series-parallel system with parameters

$$k_n = k = 1, l_n = 139, a = 1, q_1 = 1, e_1 = 4,$$

$$p_{11} = 1/139, p_{12} = 2/139, p_{13} = 117/139, p_{14} = 19/139,$$

$$\lambda_{11} = 0.0125, \lambda_{12} = 0.0015, \lambda_{13} = 0.005, \lambda_{14} = 0.004.$$

Since

$$\lambda_1 = \sum_{j=1}^{4} p_{1j} \lambda_{1j} = \frac{1}{139} 0.0125 + \frac{2}{139} 0.0015 + \frac{117}{139} 0.005 + \frac{19}{139} 0.004 = 0.00487,$$

$$\lambda = \min\{0.00487\} = 0.00487,$$

then applying Proposition 8.9 with normalising constants

$$a_n = \frac{1}{0.00487 \cdot 139} \cong \frac{1}{0.6765}, \quad b_n = 0,$$

we conclude that

$$\mathcal{R}_9^{(3)}(t) = 1 - [1 - \exp[-t]] = \exp[-t] \text{ for } t \geq 0$$

is the subsystem S_3 limit reliability function and from (1.1), we get

$$R_{1,139}^{(3)}(t) = \mathcal{R}_9^{(3)}(0.6765t) = \exp[-0.6765t] \text{ for } t \geq 0.$$

At system operational state 3, subsystem S_4 becomes a non-homogeneous regular series-parallel system with parameters

$$k_n = k = 1, \, l_n = 242, \, a = 1, \, q_1 = 1,$$

$$e_1 = 3, \, p_{11} = 2/242, \, p_{12} = 160/242, \, p_{13} = 80/242,$$

$$\lambda_{11} = 0.005, \, \lambda_{12} = 0.012, \, \lambda_{13} = 0.$$

Since

$$\lambda_1 = \sum_{j=1}^{3} p_{1j}\lambda_{1j} = \frac{2}{242}0.005 + \frac{160}{242}0.012 + \frac{80}{242}0 = 0.007975,$$

$$\lambda = \min\{0.007975\} = 0.007975,$$

then applying Proposition 8.9 with normalising constants

$$a_n = \frac{1}{0.007975 \cdot 242} \cong \frac{1}{1.93}, \quad b_n = 0,$$

we conclude that

$$\mathcal{R}_9^{(3)}(t) = 1 - [1 - \exp[-t]] = \exp[-t] \text{ for } t \geq 0$$

is the subsystem S_4 limit reliability function and from (1.1), we get

$$R_{1,242}^{(3)}(t) = \mathcal{R}_9^{(3)}(1.93t) = \exp[-1.93t] \text{ for } t \geq 0.$$

Since the considered subsystems create a series structure in a reliability sense, then the reliability function of the whole transportation system is given by

$$\overline{R}^{(3)}(t) = R_{1,129}^{(3)}(t) \, R_{1,743}^{(3)}(t) \, R_{1,139}^{(3)}(t) \, R_{1,242}^{(3)}(t)$$

$$= \exp[-25.471t] \text{ for } t \geq 0 \tag{8.35}$$

and according to (8.28) and (8.29) the system lifetime mean value and its standard deviation are

$$E[T^{(3)}] = 0.0393, \quad \sigma(T^{(3)}) = 0.0393. \tag{8.36}$$

Finally, considering (8.22) and (8.30), the system unconditional reliability is given by

$$R(t) \cong \frac{34}{64} \overline{R}^{(1)}(t) + \frac{7}{64} \overline{R}^{(2)}(t) + \frac{23}{64} \overline{R}^{(3)}(t),$$

where $\overline{R}^{(1)}(t)$, $\overline{R}^{(2)}(t)$ and $\overline{R}^{(3)}(t)$ respectively are given by (8.31), (8.33) and (8.35). Hence, applying (8.26)–(8.27) and (8.32), (8.34) and (8.36), we get the mean value and the standard deviation of the system unconditional lifetime respectively given by

$$m \cong \frac{34}{64} \cdot 0.0807 + \frac{7}{64} \cdot 0.0623 + \frac{23}{64} \cdot 0.0393 \cong 0.0638 \text{ years,}$$

$$\sigma \cong \sqrt{\frac{34}{64}(0.057)^2 + \frac{7}{64}(0.0466)^2 + \frac{23}{64}(0.0393)^2} \cong 0.0502 \text{ years.}$$

The reliability data concerned with the operation process and component reliability functions of the port grain transportation system are not precise. They come from experts and are concerned with the mean lifetimes of the system components and with the conditional sojourn times of the system in the operation states under the arbitrary assumption that their distributions are exponential. By further development of the proposed method it seems to be possible to obtain more general results useful in the complex technical systems and their operation processes reliability and availability evaluation, improvement and optimisation.

SUMMARY

In this book the asymptotic approach to the reliability evaluation of homogeneous and non-homogeneous series and parallel systems, homogeneous "m out of n" systems and homogeneous and non-homogeneous regular series-parallel and parallel-series systems has been completely analysed. For these systems, in the case where their components are two-state as well in the case where they are multi-state, the classes of limit reliability functions have been fixed. Moreover, the auxiliary theorems useful for finding limit reliability functions of real technical systems composed of components having any reliability functions have been formulated and motivated. The series-parallel and parallel-series systems have been considered in the case where their reliability structures are regular. However, this fact does not restrict the completeness of the performed analysis, since by conventional joining of a suitable number of failed components in parallel subsystems of the non-regular parallel-series systems we get the regular non-homogeneous parallel-series systems considered in the book. Similarly, conventional joining of a suitable number of components which do not fail, in series sub-systems of the non-regular series-parallel systems, leads us to the regular non-homogeneous series-parallel systems considered in the book. Thus the problem has been analysed exhaustively.

In addition to the general solutions, a practically important case when the components of the considered systems have exponential reliability functions has been considered separately. In this case the class of limit reliability functions for the considered multi-state systems has been fixed and other practically useful theorems have been proposed. The results obtained in this case may play the role of an easy-to-use guide necessary in quick reliability evaluations of real large technical systems, as well as during their operations and when they are designed. There are proposed algorithms presented in the form of tables giving simple sequential steps in systems reliability evaluation. To make these algorithms an easy and useful tool for reliability practitioners their usage is illustrated by various examples of their application to the evaluation of the real system reliability characteristics. Thinking about the practitioners using computers, on the basis of the algorithms the computer program has been elaborated and the results of its practical use have been illustrated in the book.

Theoretical results have been illustrated by many practical examples of their application in reliability evaluation of an extensive range of large systems. The evaluations of the real technical systems presented here have been obtained on the basis of non-precise component reliability data and therefore first of all they should be treated as an illustration of a wide possibility of applications of the proposed asymptotic approach in system reliability analysis. Reliability data come from experts and from the literature and are concerned with component's mean lifetimes and their expected reliability function types. These evaluations, despite not being precise may be a very useful, simple and quick tool in approximate reliability evaluation, especially during the design of large systems, and when planning and improving their safety and effectiveness operation processes. Optimisation of the reliability structures of large systems with respect to their safety and costs is complicated and often not possible to perform by practitioners because of the mathematical complexity of the exact methods.

The proposed method offers enough simplified formulae to allow significant simplifying of reliability optimising calculations. This is testified by the recent partial and preliminary results presented in the last chapter. Especially important are results concerned with exponential systems. Their joining with non-precise data coming from experts concerned with the component mean lifetimes and expected reliability function only allow the constructors to evaluate and to optimise the system structures and their operation processes during designing and before including them in everyday practice ([44]).

The results presented in the book have become the basis of investigations on domains of attraction of system limit reliability functions and initiated the problem of the speed at which system reliability function sequences reach their limit reliability functions. The problem of the domains of attraction for fixed limit reliability functions, presented partly in the book, has been completely solved for two-state systems in [81] and generalised to multi-state systems in [71]. The solutions deliver the necessary and sufficient conditions that the reliability functions of the system particular components have to satisfy in order that the system limit reliability function is one of the reliability functions from the fixed class of possible limit reliability functions for this system. This way, on the basis of data about the types of system component reliability functions and the system shape it is possible to expect which of the reliability functions is its limit reliability function.

Another significant problem in applying the proposed method to the reliability evaluation of real technical system is its accuracy. This problem has been completely solved for the considered two-state systems in [71]. Practical examples presented in the book testify that the mistakes in the approximation of the exact system reliability functions by their limiting forms are not significant in practice, often for not very large systems. Moreover, in the asymptotic approach to system reliability evaluation it is possible to get lower and upper bounds of the exact system reliability characteristics, which is illustrated in the investigation of the ship-rope elevator. The complete solution of the proposed method, i.e. the evaluation of the speed of convergence of the system exact reliability sequences to their limiting forms presented partly in the book for two-state systems, is easy to transfer to multi-state systems. The way of proceeding in this transfer is commented on in the book and solved in [71]. The asymptotic method accuracy evaluations are complicated, so it is probably not possible to use them in everyday practice by reliability practitioners. However, it seems to be practically possible to make the accuracy evaluation supported by computer calculations.

Additionally, the main results of the book have initiated and become the basis for further investigations on limit reliability functions. Especially the investigations on limit reliability functions of practically important large series-"*m* out of *n*" and "*m* out of *n*"-series systems and hierarchical systems have been recently significantly developed ([95], [25]).

Another problem concerned with the methods of the improving of large systems reliability, which are briefly presented in the last chapter, has been recently completely solved for series systems in [83].

The results presented in the book suggest that it seems reasonable to continue the investigations focusing on:

– finding the classes of limit reliability functions for series-"*m* out of *n*" and "*m* out of *n*"-series systems, series-parallel and parallel-series hierarchical systems and systems with cold reserve,
– methods of improving reliability for large two-state and multi-state systems,
– methods of reliability optimisation for large two-state and multi-state systems related to costs and safety of the system operation processes,
– availability and maintenance of large systems,
– elaboration of universal practical tools in the form of computer program package addressed to large systems operators, allowing them to evaluate and optimise automatically large systems reliability, availability and safety,
– elaboration of the offer of training courses in the scope of safety, reliability and availability of large system operation processes based on the computer program package.

BIBLIOGRAPHY

[1] Abouammoh A., Al-Kadi M.: Component relevancy in multi-state reliability models. *IEEE Transactions on Reliability* 40, 1991, 370–375.

[2] Amari S. V., Misra R. B.: Comment on: Dynamic reliability analysis of coherent multi-state systems. *IEEE Transactions on Reliability* 46, 1997, 460–461.

[3] Anderson C. W.: Extreme value theory for a class of discrete distributions with applications to some stochastic processes. *J. Appl. Prob.* 7, 1970, 99–113.

[4] Aven T.: Reliability evaluation of multi-state systems with multi-state components. *IEEE Transactions on Reliability* 34, 1985, 473–479.

[5] Aven T, Jensen U.: *Stochastic Models in Reliability*. Springer-Verlag, New York, 1999.

[6] Aven T.: On performance measures for multi-state monotone systems. *Reliability Engineering and System Safety* 41, 1993, 259–266.

[7] Barlow R. E., Proschan F.: *Statistical Theory of Reliability and Life Testing. Probability Models*. Holt Rinehart and Winston, Inc., New York, 1975.

[8] Barlow R. E., Wu A. S.: Coherent systems with multi-state components. *Mathematics of Operations Research* 4, 1978, 275–281.

[9] Barndorff-Nielsen O.: On the limit behaviour of extreme order statistics. *Ann. Math. Statist.* 34, 1963, 992–1002.

[10] Berman S. M.: Limiting distribution of the maximum term in sequences of dependent random variables. *Ann. Math. Statist.* 33, 1962, 894–908.

[11] Berman S. M.: Limit theorems for the maximum term in stationary sequences. *Ann. Math. Statist.* 35, 1964, 502–516.

[12] Block H. W., Savitis T. H.: A decomposition for multi-state monotone systems. *J. Applied Probability* 19, 1982, 391–402.

[13] Bobrowski D.: *Mathematical Models in Reliability Theory* (in Polish). WNT, Warsaw, 1985.

[14] Boedigheimer R., Kapur K.: Customer-driven reliability models for multi-state coherent systems. *IEEE Transactions on Reliability* 43, 1994, 45–50.

[15] Bausch A.: Calculation of critical importance for multi-state components. *IEEE Transactions on Reliability* 36, 1987, 247–249.

[16] Brunelle R. D., Kapur K. C.: Review and classification of reliability measures for multi-state and continuum models. *IEEE Transactions* 31, 1999, 1117–1180.

[17] Butler D.: A complete importance ranking for components of binary coherent systems with extension to multi-state systems. *Naval Research Logistics* 26, 1979, 556–578.

[18] Butler D.: Bounding the reliability of multi-state systems. *Operations Research* 30, 1982, 530–544.

[19] Cardalora L.: Coherent systems with multi-state components. *Nucl. Eng. Design* 58, 1980, 127–139.

[20] E. Castillo. *Extreme Value Theory in Engineering*. Boston Academic Press, Boston, 1988.

[21] Chernoff H., Teicher H.: Limit distributions of the minimax of independent identically distributed random variables. *Proc. Americ. Math. Soc.* 116, 1965, 474–491.

[22] Certificate for steel wire rope. Casar Drahtseilwerk Saar BMBH, Kirkel, 1996.

[23] Cichocki A., Kurowicka D., Milczek B.: On limit reliability functions of large systems. Chapter 12, *Statistical and Probabilistic Models in Reliability*. Ionescu D. C. and Limnios N. Eds., Birkhauser, Boston, 1998, 184–193.

[24] Cichocki A.: Limit reliability functions of some homogeneous regular series-parallel and parallel-series systems of higher order. *Applied Mathematics and Computation* 120, 2001, 55–72.

[25] Cichocki A.: *Determining of Limit Reliability Functions of Hierarchical Systems under Power Standardisation* (in Polish). PhD Thesis. Gdynia Maritime University-System Research Institute Warsaw, 2003.

[26] Collet J.: Some remarks on rare-event approximation. *IEEE Transactions on Reliability* 45, 1996, 106–108.

[27] Daniels H. E.: The statistical theory of the strength of bundles of threads. *J. Proc. Roy. Soc.* 183, 1945, 404–435.

[28] De Haan L.: *On Regular Variation and Its Application to the Weak Convergence of Sample Extremes*. Math. Centr. Tracts 32, Mathematics Centre, Amsterdam, 1970.

[29] Dziubdziela W.: Limit distributions of order statistics (in Polish). *Applied Mathematics* 9, 1977, 45–71.

[30] Ebrahimi N.: Multistate reliability models. *Naval Res. Logistics* 31, 1984, 671–680.

[31] El-Neweihi E., Proschan F., Setchuraman J.: Multi-state coherent systems. *J. Applied Probability* 15, 1978, 675–688.

[32] Fardis M. N., Cornell C. A.: Analysis of coherent multistate systems. *IEEE Transactions on Reliability* 30, 1981, 117–122.

[33] Fisher R. A., Tippett L. H. C.: Limiting forms of the frequency distribution of the largest and smallest member of a sample. *Proc. Cambr. Phil. Soc.* 24, 1928, 180–190.

[34] Frechet M.: Sur la loi de probabilite de l'ecart maximum. *Ann. de la Soc. Polonaise de Math.* 6, 1927, 93–116.

[35] Galambos J.: Limit laws for mixtures with applications to asymptotic theory of extremes. *Z. Wahrscheinlichkeitstheorie Verw. Gebiete* 32, 1975, 197–207.

[36] Gniedenko B. W.: Sur la distribution limite du terme maximum d'une serie aleatoire. *Ann. of Math.* 44, 1943, 432–453.

[37] Grabski F.: *Semi-Markov Models of Systems Reliability and Operations. Monograph. Analysis* (in Polish). Monograph. System Research Institute, Polish Academy of Science, Warsaw, 2002.

[38] Griffith W. S.: Multi-state reliability models. *J. Applied Probability* 17, 1980, 735–744.

[39] Gumbel E. J.: Les valeurs extremes des distributions statistiques. *Ann. Inst. H. Poincare* 4, 1935, 115.

[40] Gumbel E. J.: *Statistics of Extremes*. New York, 1962.

[41] Harlow D. G., Phoenix S. L.: Approximations for the strength distribution and size effects in an idealised lattice model of material breakdown. *Journal of Mechanics and Physics of Solids* 39, 1991, 173–200.

[42] Harlow D. G.: Statistical properties of hybrid composites: asymptotic distributions for strain. *Reliability Engineering and System Safety* 56, 1997, 197–208.

[43] Harris R.: An application of extreme value theory to reliability theory. *Ann. Math. Statist.* 41, 1970, 1456–1465.

[44] Hryniewicz O.: Lifetime tests for imprecise data and fuzzy reliability requirements. *Reliability and Safety Analyses under Fuzziness*. Onisawa T. and Kacprzyk J., Eds., Physica Verlag, Heidelberg, 1995, 169–182.

[45] Huang J., Zuo M. J., Wu Y.: Generalized multi-state k-out-of-n:G systems. *IEEE Transactions on Reliability* 49, 2000, 105–111.

[46] Hudson J.C., Kapur K. C.: Reliability theory for multistate systems with multistate components. *Microelectronics and Reliability* 22, 1982, 1–7.

[47] Hudson J. C., Kapur K. C.: Reliability analysis of multistate systems with multistate components. *Transactions of Institute of Industrial Engineers* 15, 1983, 127–135.

[48] Hudson J., Kapur K. C.: Modules in coherent multistate systems. *IEEE Transactions on Reliability* 32, 1983, 183–185.

[49] Hudson J., Kapur K.: Reliability bounds for multistate systems with multistate components. *Operations Research* 33, 1985, 735–744.

[50] Jaźwinski J., Fiok-Ważyńska K.: *Reliability of Technical Systems* (in Polish). PWN, Warsaw, 1990.

[51] Karpiński J., Korczak E.: *Methods of Reliability Evaluation of Two-state Technical Systems* (in Polish). System Research Institute, Polish Academy of Science, Warsaw, 1990.

[52] Kaufman L. M., Dugan J. B., Johnson B. W.: Using statistics of the extremes for software reliability analysis. *IEEE Transactions on Reliability* 48, 3, 1999, 292–299.

[53] Kołowrocki K.: On a class of limit reliability functions of some regular homogeneous series-parallel systems. *Reliability Engineering and System Safety* 39, 1993, 11–23.

[54] Kołowrocki K.: On asymptotic reliability functions of series-parallel and parallel-series systems with identical components. *Reliability Engineering and System Safety* 41, 1993, 251–257.

[55] Kołowrocki K.: On a class of limit reliability functions of some regular homogeneous series-parallel systems. *Applied Mathematics* 36, 1993, 55–69.

[56] Kołowrocki K.: *On a Class of Limit Reliability Functions for Series-parallel and Parallel-series Systems*. Monograph. Maritime University Press, Gdynia, 1993.

[57] Kołowrocki K.: A remark on the class of limit reliability functions of series-parallel systems. *Exploitation Problems of Machines* 2, 98, 1994, 279–296.

[58] Kołowrocki K.: The classes of asymptotic reliability functions for series-parallel and parallel-series systems. *Reliability Engineering and System Safety* 46, 1994, 179–188.

[59] Kołowrocki K.: Limit reliability functions of some series-parallel and parallel-series systems. *Applied Mathematics and Computation* 62, 1994, 129–151.

[60] Kołowrocki K.: Limit reliability functions of some non-homogeneous series-parallel and parallel-series systems. *Reliability Engineering and System Safety* 46, 1994, 171–177.

[61] Kołowrocki K.: On limiting forms of the reliability functions sequence of the series-parallel and parallel-series systems. *Applied Mathematics and Computer Science* 4, 1994, 575–590.

[62] Kołowrocki K.: On a class of limit reliability functions for series-parallel and parallel-series systems. *International Journal of Pressure Vessels and Piping* 61, 1995, 541–569.

[63] Kołowrocki K.: Asymptotic reliability functions of some non-homogeneous series-parallel and parallel-series systems. *Applied Mathematics and Computation* 73, 1995, 133–151.

[64] Kołowrocki K.: On applications of asymptotic reliability functions to the reliability and risk evaluation of pipelines. *International Journal of Pressure Vessels and Piping* 75, 1998, 545–558.

[65] Kołowrocki K.: On limit reliability functions of large systems. Chapter 11. *Statistical and Probabilistic Models in Reliability*. Ionescu D. C. and Limnios N. Eds., Birkhauser, Boston, 1999, 153–183.

[66] Kołowrocki K.: On reliability and risk of large multi-state systems with degrading components. *Exploitation Problems of Machines*, 1999, 189–210.

[67] Kołowrocki K.: On asymptotic approach to multi-state systems reliability evaluation. Chapter 11. *Recent Advances in Reliability Theory: Methodology, Practice and Inference*. Limnios N. and Nikulin M. Eds., Birkhauser, Boston, 2000, 163–180.

[68] Kołowrocki K.: Weibull distribution applications to reliability evaluation of transportation systems. *Archives of Transport* 12, 2, 2000, 17–31.

[69] Kołowrocki K.: Asymptotic approach to reliability evaluation of piping and rope transportation systems. *Exploitation Problems of Machines* 2, 122, 2000, 111–133.

[70] Kołowrocki K.: Reliability evaluation of port transportation systems (in Polish). Consultants: Baranowski Z., Cichosz J., Jewasiński D. Report. Maritime Academy, Gdynia, 2000.

[71] Kołowrocki K. et al.: Limit reliability functions of large multistate systems and their applications to transportation and durability problems. Parts 1, 2, 3 (in Polish). Report. Project founded by the Polish Committee for Scientific Research. Maritime University, Gdynia, 2000, p. 567.

[72] Kołowrocki K., Kurowicka D.: Limit reliability functions for non-homogeneous systems. Chapter 3. *Techniques in Representing High Dimensional Distributions*. Delft University, 2001, 63–86.

[73] Kołowrocki K.: Asymptotic approach to reliability evaluation of a rope transportation system. *Reliability Engineering and System Safety* 71, 1, 2001, 57–64.

[74] Kołowrocki K.: *Asymptotic Approach to System Reliability Analysis* (in Polish). Monograph. System Research Institute, Polish Academy of Science, Warsaw, 2001.

[75] Kołowrocki K.: On limit reliability functions of large multi-state systems with ageing components. *Applied Mathematics and Computation* 121, 2001, 313–361.

[76] Kołowrocki K.: An asymptotic approach to reliability evaluation of large multi-state systems with applications to piping transportation systems. *International Journal of Pressure Vessels and Piping* 80, 2003, 59–73.

[77] Kołowrocki K.: Asymptotic approach to reliability analysis of large systems with degrading components. *International Journal of Reliability, Quality and Safety Engineering* 10, 3, 2003, 249–288.

[78] Kossow A., Preuss W.: Reliability of linear consecutively-connected systems with multistate components. *IEEE Transactions on Reliability* 44, 1995, 518–522.

[79] Krajewski B., Pawluk C.: An opinion on reliability and exploitation of ship rope elevator (in Polish). Naval Shipyard, Gdynia, 1999.

[80] Kurowicka D.: Domains of attraction of asymptotic reliability functions of some homogeneous series-parallel systems. *Applied Mathematics and Computation* 98, 1998, 61–74.

[81] Kurowicka D.: *Techniques in Representing High Dimensional Distributions*. PhD Thesis. Gdynia Maritime University-Delft University, 2001.

[82] Kwiatuszewska-Sarnecka B.: On a class of limit reliability functions of large series-parallel systems with assisting components. *Applied Mathematics and Computation* 123, 2001, 155–177.

[83] Kwiatuszewska-Sarnecka B.: *Reliability Analysis of Reservation Effectiveness in Series Systems* (in Polish). PhD Thesis. Gdynia Maritime University-System Research Institute Warsaw, 2003.

[84] Leadbetter M. R.: On extreme values in stationary sequences. *Z. Wahrscheinlichkeitstheorie Verw. Gebiete* 28, 1974, 289–303.

[85] Levitin G., Lisnianski A., Ben Haim H., Elmakis D.: Redundancy optimisation for multistate series-parallel systems. *IEEE Transactions on Reliability* 47, 1998, 165–172.

[86] Levitin G., Lisnianski A.: Joint redundancy and maintenance optimisation for series-parallel multistate systems. *Reliability Engineering and System Safety* 64, 1998, 33–42.

[87] Levitin G., Lisnianski A.: Importance and sensitivity analysis of multi-state systems using universal generating functions method. *Reliability Engineering and System Safety* 65, 1999, 271–282.

[88] Levitin G., Lisnianski A.: Optimisation of imperfect preventive maintenance for multistate systems. *Reliability Engineering and System Safety* 67, 2000, 193–203.

[89] Levitin G., Lisnianski A.: Optimal replacement scheduling in multi-state series-parallel systems. *Quality and Reliability Engineering International* 16, 2000, 157–162.

[90] Levitin G., Lisnianski A.: Structure optimisation of multi-state system with two failure modes. *Reliability Engineering and System Safety* 72, 2001, 75–89.

[91] Lisnianski A., Levitin G.: *Multi-state System Reliability. Assessment, Optimisation and Applications*. World Scientific Publishing Co., New Jersey, London, Singapore, Hong Kong, 2003.

[92] Limnios N., Oprisan G.: *Semi-Markov Processes and Reliability*. Birkhauser, Boston, 2001.

[93] Meng F.: Component-relevancy and characterisation in multi-state systems. *IEEE Transactions on Reliability* 42, 1993, 478–483.

[94] Milczek B.: On the class of limit reliability functions of homogeneous series-"k out of n" systems. *Applied Mathematics and Computation* 137, 2001, 161–174.

[95] Milczek B.: *Reliability of Large Series-"k out of n" Systems* (in Polish). PhD Thesis. Gdynia Maritime University-System Research Institute Warsaw, 2004.

[96] Natvig B.: Two suggestions of how to define a multi-state coherent system. *Adv. Applied Probability* 14, 1982, 434–455.

[97] Natvig B., Streller A.: The steady-state behaviour of multi-state monotone systems. *J. Applied Probability* 21, 1984, 826–835.

[98] Natvig B.: Multi-state coherent systems. *Encyclopaedia of Statistical Sciences*, Wiley and Sons, New York, 1984.

[99] Ohio F., Nishida T.: On multi-state coherent systems. *IEEE Transactions on Reliability* 33, 1984, 284–287.

[100] Pantcheva E.: *Limit Theorems for Extreme Order Statistics Under Non-linear Normalisation.* Lecture Notes in Mathematics, 1155, 1984, 284–309.

[101] Piasecki S.: *Elements of Reliability Theory and Multi-state Objects Exploitation* (in Polish). System Research Institute, Polish Academy of Science, Warsaw, 1995.

[102] Polish Norm PN-68/M-80-200. Steel Ropes. Classification and Construction (in Polish).

[103] Polish Norm PN-81/M-46-650. Bucket Conveyors. Classification and Construction (in Polish).

[104] Pourret O., Collet J., Bon J-L.: Evaluation of the unavailability of a multistate-component system using a binary model. *IEEE Transactions on Reliability* 64, 1999, 13–17.

[105] Prasad V., Kuo W., Kim K.: Optimal allocation of s-identical multi-functional spares in a series systems. *IEEE Transactions on Reliability* 48, 2, 1999, 118–126.

[106] Resnick S. I.: Limit laws for recorded values. *Stochastic Processes and their Applications* 1, 1973, 67–82.

[107] Ross S.: Multi-valued state component systems. *Annals of Probability* 7, 1979, 379–383.

[108] Smirnow N. W.: *Predielnyje Zakony Raspredielenija dla Czlienow Wariacjonnogo Riada.* Trudy Matem. Inst. im. W. A. Stieklowa, 1949.

[109] Smith R. L.: The asymptotic distribution of the strength of a series-parallel system with equal load-sharing. *Ann. of Prob.* 10, 1982, 137–171.

[110] Smith R. L.: Limit theorems and approximations for the reliability of load sharing systems. *Adv. Appl. Prob.* 15, 1983, 304–330.

[111] Sutherland L. S., Soares C. G.: Review of probabilistic models of the strength of composite materials. *Reliability Engineering and System Safety* 56, 1997, 183–196.

[112] Tata M. N.: On outstanding values in a sequence of random variables. *Z.Wahrscheinlichkeists-theorie Verw. Gebiete* 12, 1969, 9–20.

[113] Trade Norm BN-75/2118-01. Cranes. Instructions for Expenditure Evaluation of Steel Ropes (in Polish).

[114] Von Mises R.: La distribution de la plus grande de n valeurs. *Revue Mathematique de l'Union Interbalkanique* 1, 1936, 141–160.

[115] Watherhold R. C.: Probabilistic aspects of the strength of short fibre composites with planar distributions. *Journal of Materials Science* 22, 1987, 663–669.

[116] Xue J.: On multi-state system analysis. *IEEE Transactions on Reliability* 34, 1985, 329–337.

[117] Xue J., Yang K.: Dynamic reliability analysis of coherent multi-state systems. *IEEE Transactions on Reliability* 4, 44, 1995, 683–688.

[118] Xue J., Yang K.: Symmetric relations in multi-state systems. *IEEE Transactions on Reliability* 4, 44, 1995, 689–693.

[119] Yu K., Koren I., Guo Y.: Generalised multistate monotone coherent systems. *IEEE Transactions on Reliability* 43, 1994, 242–250.

ACKNOWLEDGEMENTS
The author would like to thank all his friends and colleagues involved in the research activity of ESREL, KONBiN and MMR conferences and the Winter School of Reliability for their inspiration to write this book. Especially, the author would like to thank Prof. Jerzy Jaźwiński, the outstanding Polish leader in the reliability field, for his amicable assistance and support in preparing the book.

INDEX